T0181910

Undergraduate Lecture Notes in Physics

Series Editors

Neil Ashby, University of Colorado, Boulder, CO, USA

William Brantley, Department of Physics, Furman University, Greenville, SC, USA

Michael Fowler, Department of Physics, University of Virginia, Charlottesville, VA, USA

Morten Hjorth-Jensen, Department of Physics, University of Oslo, Oslo, Norway

Michael Inglis, Department of Physical Sciences, SUNY Suffolk County Community College, Selden, NY, USA

Barry Luokkala ⓘ, Department of Physics, Carnegie Mellon University, Pittsburgh, PA, USA

Undergraduate Lecture Notes in Physics (ULNP) publishes authoritative texts covering topics throughout pure and applied physics. Each title in the series is suitable as a basis for undergraduate instruction, typically containing practice problems, worked examples, chapter summaries, and suggestions for further reading.

ULNP titles must provide at least one of the following:

- An exceptionally clear and concise treatment of a standard undergraduate subject.
- A solid undergraduate-level introduction to a graduate, advanced, or non-standard subject.
- A novel perspective or an unusual approach to teaching a subject.

ULNP especially encourages new, original, and idiosyncratic approaches to physics teaching at the undergraduate level.

The purpose of ULNP is to provide intriguing, absorbing books that will continue to be the reader's preferred reference throughout their academic career.

More information about this series at https://link.springer.com/bookseries/8917

Gabor Kunstatter · Saurya Das

A First Course on Symmetry, Special Relativity and Quantum Mechanics

The Foundations of Physics

Second Edition

 Springer

Gabor Kunstatter
University of Winnipeg
Winnipeg, MB, Canada

Saurya Das
University of Lethbridge
Lethbridge, AB, Canada

ISSN 2192-4791   ISSN 2192-4805 (electronic)
Undergraduate Lecture Notes in Physics
ISBN 978-3-030-92345-7   ISBN 978-3-030-92346-4 (eBook)
https://doi.org/10.1007/978-3-030-92346-4

This Springer imprint is published by the registered company Springer Nature Switzerland AG
The registered company address is: Gewerbestrasse 11, 6330 Cham, Switzerland

Preface

Special relativity and quantum mechanics provide the beautiful but sometimes counterintuitive underpinnings of twenty-first century physics. A lesser known crucial ingredient is the notion of *symmetry*. Just as symmetry is intimately linked to humans' perception of beauty, it also plays a key role in determining the beauty and desirability of physical theories. More important than the aesthetic significance of symmetry is its remarkable utility. Through the powerful theorem proven by Emmy Noether in 1915, symmetry leads to conservation laws that give us confidence that some things do remain constant despite the rapid and often dizzying changes occurring in the world. More practically, symmetry provides invaluable tools for decoding complicated structures and simplifying problems that would otherwise be intractable.

The purpose of this text is to provide undergraduate students with a comprehensive and mathematically rigorous introduction to special relativity and quantum mechanics. A novel aspect of the presentation is that symmetry is given its rightful prominence as an integral part of the foundation of physics. Students are given a conceptual understanding of symmetry and the important role it plays in physics. They are also provided with mathematical tools that allow for quantitative applications.

The primary target audience consists of second-year physics majors who have completed an introductory calculus course and a first-year physics course that includes Newtonian mechanics and electrostatics. Some knowledge of linear algebra is useful but not essential.

The book is intended to provide material for a self-contained two-semester course on the foundations of physics. It contains clearly marked supplementary sections that introduce more advanced topics such as variational mechanics and a proof of Noether's theorem, four-vectors and tensors and relativistic quantum mechanics.

We have included many pedagogical descriptions of relevant topics of general interest. These include the role of symmetry in the discovery of special and general relativity, the connection between symmetry and conservation laws, and the geometrical nature of Einstein's theory of general relativity. Other specific subjects of current interest are gravitational waves, cosmology, quantum computers, Bell's theorem, entanglement and the associated "spooky" action at a distance in quantum mechanics.

There are many worked Examples and Exercises for the student interspersed throughout the text. The Examples are solved within the text, while solutions to the Exercises are provided in a separate solutions manual. A first reading of the text can be accomplished with only a cursory examination of the Examples and Exercises, but a complete understanding of the material requires the student to work through them carefully.

Winnipeg, Canada   Saurya Das
Lethbridge, Canada   Gabor Kunstatter

Acknowledgements

The authors thank Pasquale Bosso, Kevin Brown, Ramin Daghigh, Esmat Elhami, Richard Epp, Valerio Faraoni, Mary Kunstatter, Vesna Milosevic-Zdjelar, Elias C. Vagenas, Dwight Vincent, Mark Walton and Jonathan Ziprick for invaluable discussions and feedback.

SD thanks Paritosh Kumar Das, Jayasri Das, Sangeeta Barua, Aalok Banerjee and Brishti Das for their unwavering support and encouragement.

GK is indebted to the many undergraduates at the University of Winnipeg, who had the dubious honour of being test subjects for much of the material in this textbook.

In addition, GK is deeply grateful to Mary Kunstatter for blessing him with her love, support and encouragement during the preparation of this text and also during the previous 45 years spent learning to appreciate the symmetry and beauty of Nature. Last but by no means least, GK thanks David and Shauna Kunstatter for their love and support, and for being a continual source of inspiration.

Contents

Conventions and Notations

- We denote geometrical vectors (i.e. without indices) by bold face lower and upper case letters: $\{\mathbf{A}, \mathbf{a}, \mathbf{B}, \mathbf{b}, \mathbf{L}, \mathbf{P}, ...\}$.
- Cartesian basis vectors are denoted by $\mathbf{i}, \mathbf{j}, \mathbf{k}$.
- The magnitude of a vector is denoted by the corresponding upper or lower case letter, not bold face $\{A := |\mathbf{A}|, b = |\mathbf{b}|, B := |\mathbf{B}|, ...\}$.
- When referring to components, the letter name will be the same as for the geometrical object but not bold face: $\mathbf{L} = L_x\mathbf{i} + L_y\mathbf{j} + L_z\mathbf{k}$.
- Matrices and transformations without indices are generally denoted by upper case letters in the latter half of the alphabet: $\{R, S, T, U\}$. The identity matrix is denoted by I, and is an exception. For example, a rotation by $30°$ in the $x - y$ plane is given by the matrix:

$$R := \begin{bmatrix} \cos(30) & \sin(30) \\ -\sin(30) & \cos(30) \end{bmatrix} \tag{1}$$

- Hats are placed on operators in quantum mechanics: $\hat{p}, \hat{H}$.
- The notation $A := B$ signifies that the expression on the right defines the expression on the left. The notation emphasizes that definition is directional, in contrast to equality or equivalence.
- Terms that correspond to important concepts are highlighted in *italics*.

Some Useful Constants and Units

- We primarily use MKS units (Meters, Kilograms, Seconds).
- Speed of Light $c = 3 \times 10^8$ m/s
- Gravitational acceleration: $g = 10$ m/s^2
- Newton's Constant: $G = 6.67 \times 10^{-11}$ Nm^2/kg^2
- Boltzmann's Constant: $k_B = 1.38 \times 10^{-23}$ J/K
- Stefan–Boltzmann Constant: $\sigma = 5.67 \times 10^{-8} \frac{\text{W}}{\text{m}^2\text{K}^4}$

- Planck Constant:

  - Planck Constant: $h = 6.63 \times 10^{-34}$ J $\cdot$ s
  - Reduced Planck Constant: $\hbar = 1.05 \times 10^{-34}$ J $\cdot$ s
  - A useful combination: $hc = 1242$ e V $\cdot$ nm

- Mass of electron: 9.09×10^{-31} kg $= 511$ keV$/c^2$
- Compton wavelength of electron: $\lambda_c = 2.4 \times 10^{-12}$ m
- Mass of proton: 1.673×10^{-27} kg $= 938.3$ MeV$/c^2$
- Mass of neutron: 1.675×10^{-27} kg $= 938.6$ MeV$/c^2$
- Bohr radius: $r_0 = 5.29 \times 10^{-11}$ m
- Nucleon radius: $r_N = 1.2 \times 10^{-15}$ m
- Units:

  - 1 Joule $= 1.60 \times 10^{-19}$ eV
  - 1 atomic unit: 1 u $= 1.661 \times 10^{-27}$ kg
  - 1 light year (lt-yr) $= 9.5 \times 10^{12}$ km
  - 1 Parsec (psc) $= 3.26$ lt-yr $= 3.1 \times 10^{13}$ km.

List of Figures

List of Tables

Chapter 1
Introduction

1.1 The Goal of Physics

In very broad terms, the ultimate goal of physics is to understand the contents of our Universe and the rules governing its evolution starting from just a few underlying principles and assumptions. An important step towards this goal involves *synthesis*, by which we mean taking diverse physical phenomena and describing them within the context of a single model or theory. For the resulting theory to be successful, it must provide predictions that can be compared to experiment so as to confirm or negate the validity of the theory. In other words, the theory must be falsifiable. No theory can ever be proved completely correct. The best one can do is show that it is consistent with observations within a certain realm of application. In order to prove that a theory is not valid within a given range of parameters, one merely needs to show that one of its predictions fails to match observations.

Examples of successful synthesis in the past include the following:

- **Newtonian mechanics** accurately predicted the trajectory of any object subjected to one or more forces.
- **Newton's Law of Universal Gravitation** successfully described the motion of projectiles on Earth and orbits of celestial bodies (the moon, planets, etc.) within a single theory.
- **Maxwell's Theory of Electromagnetism** unified electricity, magnetism and optics.
- **Einstein's Theory of Special Relativity** modified Newtonian mechanics to reconcile it with Maxwell's theory of electromagnetism.
- **Electroweak Theory**: The so-called "weak" interactions play an important role in the processes leading to energy production in the Sun. In 1979, the Nobel Prize in Physics was awarded to Sheldon Lee Glashow, Abdus Salam and Steven Weinberg for their contributions to the formulation of a theory that unified the weak interactions with Maxwell's theory of electromagnetism. The cornerstone

© The Author(s), under exclusive license to Springer Nature Switzerland AG 2022
G. Kunstatter and S. Das, *A First Course on Symmetry, Special Relativity and Quantum Mechanics*, Undergraduate Lecture Notes in Physics,
https://doi.org/10.1007/978-3-030-92346-4_1

of the resulting "electro-weak theory" was the existence of a particle that filled
the vacuum of empty space thereby providing mass for fundamental constituents
of the theory that would otherwise have been massless. The Higgs particle was
detected forty-three years later, in 2012, by experimental physicists using the Large
Hadron Collider[1] at CERN (the European Center for Nuclear Research).[2] A year
later, the Nobel prize in physics was awarded to François Englert and Peter Higgs
for their theoretical formulation of the mechanism by which the Higgs particle
bestows mass.

Often a theory starts making incorrect predictions beyond a certain range of parameters such as energy or size. It is then time to find a better theory that incorporates the
old one where it was proven valid and makes correct predictions where the old one
started failing.[3] Examples include predictions of Newton's second law when particle
speeds approach the speed of light and that of Maxwell's equations for the orbiting
electron in a Hydrogen atom. While the first problem was addressed by Special Relativity and eventually General Relativity, both formulated by Einstein, the second
was resolved by the discovery of quantum mechanics, whose foundations were laid
at the turn of the last century.[4]

1.2   The Connection Between Physics and Mathematics

The fundamental laws of physics are given in terms of mathematical equations. This
is not just a matter of convenience. Physics requires, among other things, rigorous,
logical arguments that start with a few basic assumptions and definitions and ultimately lead to unambiguous conclusions. In this sense, mathematics and physics are
two sides of the same coin. Mathematics provides the logical structures that underly
physics, which in turn provides motivation for new mathematics that are inspired by
structures implemented by nature. It is therefore impossible to completely disentan-
gle physics from mathematics. Thus, mathematics is in many ways more than simply
the "language of physics" as often claimed.

One example is that the laws of physics are expressed in terms of differential
equations of mathematics (ordinary or partial, usually first or second order in the
derivatives). The reason for this is that these laws must be able to determine the 'rate
of change' in space or with time of physical quantities so that they can predict the
values of these quantities in other regions of space or at later times. Rates of change
are (partial) derivatives with respect to space or time, as appropriate. For instance,
Newton's second law of motion relates the rate of change with time of the velocity of

---

[1] For a good description of the Higgs particle, see https://home.cern/topics/higgs-boson.

[2] https://home.cern/topics/higgs-boson.

[3] This will be discussed further in Sect. 1.4.

[4] For an excellent account of the formulation of general relativity, the reader may consult *Subtle is
the Lord*, by A. Pais and for a thought provoking history of quantum mechanics *Quantum Leaps* by
Jeremy Bernstein, Cambridge, Massachusetts: Belknap Press of Harvard University Press, 2009.

a particle (its acceleration) to the forces acting on it. Solving the resulting differential equation allows one to predict the trajectory of the particle given knowledge of its initial position and velocity at some particular instant in time. Newton (and Leibnitz almost simultaneously) were forced to invent calculus in order to describe the non-uniform motion of particles under various forces, including gravity.

There are two crucial differences between physics and mathematics. The first of these differences is that the construction of a physical theory must start with an initial "guess" of an underlying structure that is relevant to the problem at hand.[5]

The second is that any serious physical theory of nature must yield falsifiable predictions. One example of an ultimately incorrect theory that was not falsifiable based on observations of the day was the description of the cosmos by the ancient Greek astronomers. It was based on two principles.[6] The first, due to Plato, was that the motions of heavenly bodies could only be based on the most perfect of geometrical shapes, namely circles and spheres.[7] The second, advocated by Plato's student Aristotle, was that the Earth lies at the center of the Universe. These two principles led to the natural conclusion that the planets must move in circles around the Earth. Stars were considered as fixed on the Heavenly sphere surrounding the Earth.

Observations of the day implied that most planetary orbits were not circles centered on the Earth. The Greek astronomer Ptolemy was nonetheless able to produce a very accurate model using circular orbits with the Earth at the center. Ptolemy figured out that the orbits could be accurately described in terms of circles whose centers move on other circles which in turn moved on yet more circles ("epicycles"). The last and most fundamental circle in this series was centered on the Earth. Despite the accuracy of the model, it turned out to be incorrect. Although the underlying shape from which orbits were described (i.e. the circle) was simple and beautiful, the descriptions of the orbits themselves became very complicated by requiring many epicycles, in some cases close to a hundred. These epicycles are illustrated in Fig. 1.1. In the fifteenth century, Nicolaus Copernicus realized that putting the Sun at the centre of the solar system led to planetary orbits around the Sun that were astoundingly simple and, ironically, in many cases very close to circular.

To recap a very important point: physical laws, as opposed to mathematical theorems, cannot be proven by pure reason. They are only accepted after a wealth of experimental evidence. Logical consistency is necessary, but not sufficient.

There are many examples of physical theories that are inexorably linked to a particular branch of mathematics. *Partial differential equations*, as mentioned above, are ubiquitous in physics. They are a crucial part of not only Newtonian mechanics but also electromagnetism, quantum mechanics, thermodynamics, fluid mechanics and general relativity. *Fourier transform theory* (see Appendix 15.3) allows virtually

---

[5] By *structure*, we mean mainly a new action containing a set of dynamical variables that leads to equations describing the relevant physical phenomena. This process will be described in more detail in the following. See for example Sect. 3.6.

[6] See the textbook *ASTRO, Canadian Edition* by D.E. Backman, M.A. Seeds, S. Ghose, V. Milosevic-Zdjelar, and L. A. Reed, Nelson (2013), p. 50 for a nice pedagogical description of Greek astronomy.

[7] In Sect. 3.2 we will explain in what sense these shapes in particular are considered "perfect".

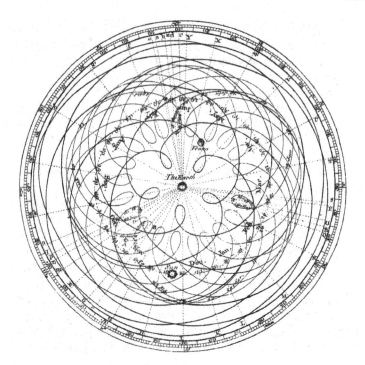

Fig. 1.1 Ptolemy's epicycle model became very complicated once observations of the orbits became more accurate. *Credit* Encyclopaedia Britannica (1st Edition, 1771; facsimile reprint 1971), Volume 1, Fig. 2 of Plate XL facing page 449

any function of space and time to be expressed as a linear combination of pure waves. It is also at the heart of complementarity (i.e. the Heisenberg uncertainty principle) in quantum mechanics. General relativity is based on *geometry*, which is mathematically described in terms of *differential geometry* and *tensor calculus*. In Chap. 3 it is shown that the set of symmetries of a given system provides a physical manifestation of mathematical objects called *groups*.

Finally, we mention that physics often makes use of mathematical techniques and results even before they are rigorously developed by mathematicians. Newton's calculus is perhaps the most obvious example. Another more obscure but fascinating example concerns a particular type of approximation used in theoretical particle physics. It consists of systematically summing up the terms of a series that is similar in many ways to a Taylor series. The odd thing is that while the first few terms in the series produce reliable approximations, if one calculates too many terms the series gets more inaccurate rather than getting closer to the right answer. Such approximations seemed arbitrary and opportunistic at first, but were successfully used in physics before mathematicians ultimately made sense of them as *asymptotic series*.

1.3 Paradigm Shifts

Every now and then our understanding of how the Universe works gets turned on its head. Such *paradigm shifts* (a phrase coined by Thomas Kuhn in 1962) have a deep and long term impact on science and society. The following lists some of the most important paradigm shifts in physics:

1. As previously mentioned, the ancient Greeks thought that the Earth was the centre of the Universe and that the heavenly bodies, stars and planets moved around it on trajectories based on the most perfect geometrical shape, the circle. Copernicus showed in the sixteenth century that the motions of the planets could most easily be understood by assuming that all the planets, including the Earth, move around the Sun. This was a radical change of perspective since it implied that humans no longer occupied a special place in the Universe.
2. In the seventeenth century, Newton showed that mechanistic laws governed the behaviour of both terrestrial and heavenly bodies. The resulting worldview of a predictive "clockwork Universe" paved the way for a scientific, cultural and educational revolution that re-defined western society.[8]
3. Quantum mechanics, developed in the early 1900s and subsequently verified experimentally to great accuracy, implied that the outcomes of microscopic experiments are determined by probabilities and cannot in most cases be uniquely predicted, even in principle. Such a probabilistic Universe was counter to Einstein's view that there must exist an objective reality whose attributes can in principle at least be determined with certainty. This belief led him to explain during a heated discussion with the famous Danish physicist, Niels Bohr, that surely "God does not play dice". Bohr is said to have told Einstein to "stop telling God what to do." In addition to its inherently probabilistic nature, the quantum world was later shown to contain an even more bizarre feature, namely a spooky action at a distance called *quantum entanglement* that connects spatially separated parts of a system. See Chap. 13.

   Quantum mechanics provided a paradigm shift that to some extent negated the cherished predictability of Newtonian mechanics and the "clockwork Universe" and replaced it by probabilities and uncertainty. It is important to stress that quantum mechanics is nonetheless a predictive theory at a deeper level.
4. The formulation of special relativity in 1905 by Einstein was based on the assumption that all non-accelerating (i.e. inertial) observers must observe the same laws of physics. This led to the conclusion that all inertial observers must measure the same speed of light, irrespective of their motion or that of the source. More remarkably, it implied that time is not absolute as assumed by Newton, but relative in the sense that the time between two events as measured by two different observers depends on their state of relative motion. Finally, Einstein's theory

---

[8] See "The Clockwork Universe: Isaac Newton, the Royal Society, and the Birth of the Modern World" by Edward Dolnick (Harper 2011).

united time with space to form a single "spacetime continuum" that corresponds to the stage on which all events play out.

5. The theory of general relativity, first published by Einstein[9] in 1915, showed that the spacetime continuum is not merely a rigid stage but in fact one of the players. The structure of spacetime itself is determined by equations of motion analogous to Maxwell's equations, but much more complicated.

6. Paradigm shifts of the future?

   Despite the tremendous success of quantum mechanics and general relativity in describing physics at very short and long distances, respectively, the two theories seem incompatible. At the moment there exists no complete, falsifiable theory that unifies quantum mechanics and general relativity into a single framework.[10] Superstring theory, loop quantum gravity and holography are bold modern attempts in this direction that have made important theoretical progress and are based on as yet unverified world-views. For example superstring theory states that the most fundamental entities in our Universe are tiny strings, while loop quantum gravity requires altering the structure of space itself at tiny scales. Holography postulates that all information pertaining to the observed three dimensional world can be encoded in a two dimensional "hologram". These theories will remain speculative until they provide concrete new predictions that are verified by experiment. If and when a successful theory of quantum gravity emerges, we will undoubtedly witness a huge paradigm shift that will once again turn our view of the world on its head.

1.4   The Correspondence Principle

It is sometimes said that once a new paradigm or theory is developed the old theory must be "thrown out". This is not the way physics works, however. Any previously accepted theory must have been verified by experiment in at least some circumstances, perhaps for a particular range of energy scales, or distance scales. This was and continues to be true of Newtonian gravity which provides an excellent description of the motions of projectiles, satellites and planets and is adequate for describing a great deal of astrophysics. It is only at very high speeds, large gravitational forces or when great accuracy is required (such as in tracking GPS satellite systems), that Newtonian gravity breaks down.

Einstein's theory of gravity has a much larger realm of validity with respect to distances, speeds and strength of gravitational fields. It explains the orbit of Mercury, something Newtonian gravity was unable to do, and predicted new phenomena such as black holes that are outside the realm of Newtonian gravity. Nonetheless, the first

---

[9] It is a remarkable fact that a single person, namely Einstein, was instrumental in the development of three major paradigm shifts in the early 20th Century.

[10] Theories that attempt to reconcile quantum mechanics and general relativity are called theories of *quantum gravity*.

important test of any new theory, including Einstein's gravity, is to verify that it makes approximately the same predictions as the previous theory (i.e. Newtonian gravity) where the latter has been successfully tested. In other words, the new theory must reduce to the old theory for some appropriate range of parameters and data. If one adopts the popular metaphor of a physical theory as a vase that contains within it an accurate description of a wide range of physical phenomena, then any new theory must be an even bigger vessel than the old one, so that it contains all the successful predictions of the old vase, and still has room for new phenomena that the old theory did not contain. One then should not break, or throw away, the old vase, since it remains valid and useful, albeit in a restricted, smaller regime than the new one.

Chapter 2
Symmetry and Physics

2.1 Learning Outcomes

Conceptual

1. What is symmetry?
2. Why is it important?
3. When are symmetries broken and why is it potentially a good thing?

2.2 What Is Symmetry?

Physicists are guided in their quest for synthesis by somewhat subjective notions of simplicity and elegance. A crucial concept in this context is symmetry. Everyone has an intuitive understanding of symmetry. For example, a floor made of a regular alternating pattern of black and white tiles appears to have a high degree of symmetry, but the same tiles arranged randomly do not appear to be symmetric. Symmetry plays an important role in art and aesthetics: shapes, paintings and buildings tend to be more pleasing to the eye if they have some symmetry, but not too much.

In physics, too, theories and objects that have a high degree of symmetry are preferred, both for their simplicity and their elegance. For symmetry to be a useful concept in physics there must be a way to define it that allows for rigorous, quantitative analysis. As discussed in more detail in Sect. 3.2, there does exist a definition of symmetry that goes beyond the subjective, aesthetic notions with which we are familiar. Very roughly, this definition states that a symmetry of an object, system of objects or even a set of equations of motion, corresponds to some sort of operation that you can perform that leaves the system looking exactly the same as before. This definition matches our intuition. For example, we think of a human face as being symmetrical if the left and right sides of the face look more or less the same, which means they are roughly mirror images of each other. In the context of the above

© The Author(s), under exclusive license to Springer Nature Switzerland AG 2022
G. Kunstatter and S. Das, *A First Course on Symmetry, Special Relativity and Quantum Mechanics*, Undergraduate Lecture Notes in Physics,
https://doi.org/10.1007/978-3-030-92346-4_2

definition, this implies that the mirror reflection of the face looks almost identical to the unreflected face.

It is important to distinguish at the outset between two types of symmetries: the first and most common is the symmetry of objects such as geometrical shapes, faces, paintings, buildings or even the Universe as a whole. The second is less well known but has deep and important consequences for physics. It is the symmetry of the theories or equations that describe a given system. In this case one is looking for an operation which leaves the form of the equations unchanged. This operation could be a rotation of the physical system about some axis, or in the case of a set of identical particles, interchanging the particles amongst themselves.

Both types of symmetry play several vital roles in 21st century physics, three of which are discussed briefly in the remainder of this section.[1]

2.3 Role of Symmetry in Physics

2.3.1 Symmetry as a Guiding Principle

Symmetry is often used as a guiding principle to decide on the "attractiveness" of a theory and to provide guidelines for constructing new theories. It is generally held that the more symmetry a theory has, the simpler and more elegant it is. As will be seen in Chap. 5, symmetry played a pivotal role in the formulation by Einstein of both special relativity and general relativity as well as in the standard model describing the fundamental particles of high energy physics. Most recently string theory, whose conjectured starting point is that the world is made up of identical, microscopic vibrating strings instead of different types of elementary particles, is often purported to be the most symmetric, and hence beautiful, theory possible.

Certain symmetries are thought by some to be "sacred" in the sense that they must be respected by any viable theory. For example, the laws of physics should be the same at all places and all times, and should not change if an isolated laboratory is rotated by 30°, for example. There are sometimes big surprises, however. As will be discussed in Sect. 2.4.2, it was shown experimentally in the 1950s that, much as Alice found in Wonderland,[2] the world can look quite different through the looking glass. Although no Mad Hatters appear, the microscopic laws of physics turn out to change quite dramatically when they are reflected through a mirror, something that many physicists of the day found shocking.

---

[1] In many undergraduate textbooks the term "modern physics" is used to refer collectively to special relativity and quantum mechanics which were developed in the early twentieth century. We avoid the term "modern physics" in this context because we feel that it is somewhat misleading to refer to discoveries dating back to the early 1900s as "modern". It is nonetheless the case that these theories provide the underlying foundation for much of the truly modern developments in physics.

[2] This is a reference to the 1865 children's book "Alice in Wonderland" written by Charles Lutwidge Dodgson under the pseudonym Lewis Carroll.

2.3.2 Symmetry and Conserved Quantities

One of the most important features of symmetry is its connection to conserved quantities. Physical quantities are said to be conserved if they do not change with time as a system evolves. Noether's theorem (Emmy Noether 1915) proves that there exists a conserved quantity for every *continuous global symmetry* of the equations describing a given system. A continuous global symmetry is one that is specified by a single arbitrary real-valued parameter that is independent of space and time. Examples of continuous symmetries include rotations, with angle as the parameter, and translations in space, with distance moved as the parameter.

According to Noether's theorem, the conservation of angular momentum is a direct consequence of the fact that the laws of physics do not care what direction the experimental lab happens to face. They are unchanged by arbitrary rotations of the entire lab. Similarly, linear momentum conservation derives from symmetry under spatial translations, and energy is conserved if the laws of physics do not change their form with the passage of time.

Noether's theorem is of great significance at a fundamental level and also provides a practical tool for solving complicated partial differential equations. A simple proof of Noether's theorem as it applies to time translation invariance and energy conservation is presented in Sect. 3.6.3.

2.3.3 Symmetry as a Tool for Simplifying Problems

Symmetry is an indispensable practical tool for simplifying calculations. If the system under consideration is known to have symmetries, then the equations can be greatly simplified by choosing coordinates adapted to these symmetries. Consider for example a particle moving in a central potential, such as the electron in the hydrogen atom. The quantum mechanical equations describing the system are very complicated in Cartesian coordinates, but become quite tractable in spherical coordinates. Another example is related to the life cycle of black holes. The relevant equations are very complicated in general and virtually intractable. On the other hand, if one assumes that the black hole is perfectly round and smooth (has spherical symmetry) and chooses suitable coordinates, then the complexity of the equations is immensely reduced. While this symmetry is almost certainly absent in real astrophysical black holes, it is a good approximation in certain theoretical contexts.

2.4 Symmetries Were Made to Be Broken

2.4.1 Spacetime Symmetries

The laws of physics are thought to be the same everywhere in space and at all times. Also, if a laboratory is well insulated from outside effects, the experimental results should not be affected by the rotation of the laboratory through some arbitrary angle. In this case, we say that these operations, namely translations in space, translations in time and rotations, are symmetries of the laws of physics.

We know by just looking around us that the Universe does not look the same everywhere or in every direction. The Earth is populated by an incredible variety of life forms. On larger scales, we see forests, mountains and rivers. As you move away from the Earth you see the planets and asteroids inhabiting the solar system. Further out there are the stars that make up the galaxy, and then clusters of galaxies.

Given the high degree of symmetry obeyed by the equations describing the Universe, the question that comes to mind is: What is the source of the observed structure and lack of symmetry?

Before attempting an answer we note that symmetry of equations does not require the physical system itself (i.e. the solutions) to possess the same degree of symmetry. What it does guarantee, however, is that if a physical system initially possesses the same symmetry as the equations, then those symmetries will be preserved as the system evolves. Given this observation, the following partial answer to the above question presents itself: The structure we see came from the initial conditions that existed at the start of the Universe. This is of course not a complete answer because it is legitimate to ask: Where did the initial conditions come from? This is a very difficult question, but not in principle unanswerable within the context of science. It may be that the consistency of the ultimate theory, should one exist, requires the Universe to have started in only one, or at least one of a finite set, of initial states.

Different symmetries manifest themselves at different scales. For example, despite the complicated and beautiful structure we observe on Earth, it is nonetheless true that when observed from 10,000 m above its surface, it looks almost perfectly round and smooth (albeit very colourful). Similarly, we will see in Sect. 7.8, that when looking at the Universe over very large distances of the order of hundreds of millions of light years[3] it looks not only the same in all directions (*isotropic*), but the same at all locations (*homogeneous*). The approximate spherical shape of the Earth is understood to be caused by the attractive force of gravity, which tends to smooth out bumps and wiggles as large objects form via gravitational collapse. The origin of the high degree of symmetry of the observed Universe at large scales is not known for certain, although there is one conjectured explanation that goes by the name of *inflation*.

If one goes in the other direction in terms of length scales and examines matter at microscopic scales, one finds again new symmetries that are not observed when

---

[3] A light year is the distance light travels in a year, namely 10^{16} m.

looking at objects on the surface of the Earth. Many solids occur in crystalline form, with molecules that are arranged in lattice structures that have a high degree of symmetry. Delving even further into the atomic scale, one observes, a hidden symmetry of hydrogen atoms that plays a major role in determining the properties of the periodic table (see Chap. 11). At subatomic scales, even more symmetry appears. This symmetry is associated not with spatial transformations such as rotations, but with the internal structure of the fundamental constituents of Nature as we currently know them.

Such symmetries appear to be fundamental to our description of the world around us. On the other hand, just as the spherical symmetry of the Earth turns out to be only approximate on closer inspection, most, if not all, symmetries in nature are only approximate. This is a good thing. If the Universe were perfectly symmetric, it would not only be boring, but also impenetrable to our scrutiny.

2.4.2 Parity Violation

Given that the laws of physics are the same in all directions (possess *spherical symmetry*), it seems natural to expect the laws of physics to respect *reflection symmetry* as well. Reflection symmetry means that the laws should not look different after a coordinate transformation $x \to -x$ corresponding to reflection through the $y - z$ plane, or $y \to -y$ (reflection through the $x - z$ plane) or $z \to -z$ (reflection through the $x - y$ plane). Invariance of physical laws under a simultaneous application of all of the above transformations, namely $(x, y, z) \to (-x, -y, -z)$, is called invariance under *parity*. Note that such reflections cannot be achieved by a succession of rotations. You should convince yourself of this for reflection through the $x - z$ plane by looking at Fig. 2.1.

One physical implication of reflection symmetry is that if a physical phenomenon, such as the decay of one particle into two others, is observed in nature, then the mirror image of this process should also be observed with equal probability. While we know that macroscopic objects such as trees or human beings (our internal organs are not symmetrically placed) do not look the same under reflection, one might nonetheless expect the microscopic world and the fundamental laws governing it to respect this fundamental symmetry.

In fact, three of the four fundamental forces of nature, namely electromagnetism, gravity and the strong nuclear force, do respect parity. One can see this for example by noting that the Coulomb force in electromagnetism and the gravitational force in Newtonian gravity depend on the square of the distance, r, between them. In terms of fields, the electric (gravitational) field of a charge (mass) situated at the origin of the coordinate system goes as $1/r^2 = 1/(x^2 + y^2 + z^2)$, which is invariant under reflection through any plane that contains the origin and hence under parity. All phenomena following from these laws, however complex they may seem, are expected to be fundamentally reflection invariant.

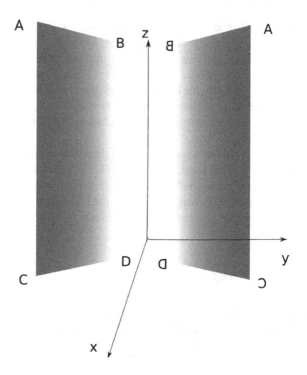

Fig. 2.1 A piece of paper with corners marked A, B, C, D is reflected through the $x - z$ plane. No series of rotations starting from the figure on the left can produce the figure on the right

The situation changes when one considers the *weak nuclear force*, which is responsible for some forms of radioactive decay, such as beta decay, that induce nuclear transmutation. This force changes a proton into a neutron, positron and neutrino, or a neutron into a proton, electron and anti-neutrino. These particles all possess *intrinsic angular momentum*, or *spin* (see Sect. 13.4 for details), and are called fermions. This spin contributes to the total angular momentum of a system, but is a quantum property that, unlike regular angular momentum, is not directly associated with orbital motion of the particle. The magnitude of the spin of an electron is $h/4\pi$ where $h = 6.626 \times 10^{-34}$ Js is the fundamental universal constant associated with quantum mechanics, just as Newton's constant, G, is the fundamental constant associated with gravity. The ridiculously small numerical value of h in MKS units[4] is why we do not experience quantum effects in our everyday lives. Planck's constant, h, will play an important role in subsequent chapters.

Given the existence of spin, the notion of rotational symmetry is modified so that the quantity conserved via Noether's Theorem is the **total** angular momentum of a particle, consisting of the sum of both the orbital angular moment and the intrinsic spin. This is confirmed by experiments.

Fermions have another interesting property. *Spinors*, which are the mathematical objects describing particles with spin, can be decomposed into two components, *left*

---

[4] MKS stands for Meter, Kilogram and Second. This system of units has gradually been adopted by the scientific community as the universal standard.

handed and *right handed*. These components switch under the parity operation, just as our left and right hands do under reflection. Roughly speaking, the handedness tells us whether the particle is spinning in a clockwise or counterclockwise sense along its direction of motion. In general, a spinor may be in a linear combination of a left and a right handed state. Again one might be justified in expecting that an equal number of right handed and left handed fermions should be observed in nature. However, in the 1950s, physicists T. D. Lee and C. N. Yang suspected that this is not the case, and that nature is not ambidextrous, as it were. They based their skepticism on the observation that certain weak reactions appeared to violate parity symmetry.

Lee and Yang's proposal was considered very radical at that time, but almost immediately after they made it, physicist C. S. Wu and her collaborators[5] demonstrated convincingly that this is indeed the case. In this experiment, cobalt nuclei were carefully aligned in an external magnetic field, so that their spin was known to lie along the field. These nuclei decayed into nickel nuclei by changing a neutron into a proton with the emission of an electron and an anti-neutrino. It turned out that there were an overwhelming number of electrons emitted in the direction opposite to the initial nuclear spin. Momentum conservation required anti-neutrinos to be emitted in the opposite direction, namely along the external magnetic spin. From angular momentum conservation it followed that only right handed anti-neutrinos participated in the reaction. Left-handed anti-neutrinos, which in principle should exist in equal numbers, were not observed, nor have they been directly observed since.

Figure 2.2 illustrates schematically the experiment. The magnetic field was provided by a current loop, I. In the mirror image experiment, shown on the right hand side of the figure, the current runs in the opposite direction. The preferential upward emission of electrons and corresponding downward emission of anti-neutrinos violates reflection symmetry. The results on the right of the mirror (solid line) in Fig. 2.2 are not the mirror image of the results on the left, even though the experimental setups are mirror images!

We note that the results shown in Fig. 2.2 are nonetheless consistent with rotational symmetry. Rotating the experiment by 180° around an axis perpendicular to the plane of Fig. 2.2 flips the current loop over, which is equivalent to the effect on the experimental setup of the reflections through the mirror. The difference between the rotation and reflection is that the rotation changes the direction of electron emission from upward to downward, while the reflection leaves it unchanged. Rotating the experiment therefore gives rotated results, confirming that the laws of physics describing this process are invariant under rotations. Reflection of the experiment, however, does not give reflected results, so that parity is violated by the process. The only way for this experiment to respect both parity and rotational invariance is for the emitted particles to be distributed equally in all directions, which in turn would require the existence of both left-handed and right-handed neutrinos, as expected by physicists of the day. The experiment therefore caused great consternation.

The parity violating results have been subsequently verified with increasing precision in a variety of experiments, including in the capture of polarized neutrons by cadmium emitting gamma rays in a preferential direction and the preferential

---

[5] Wu et al. [1].

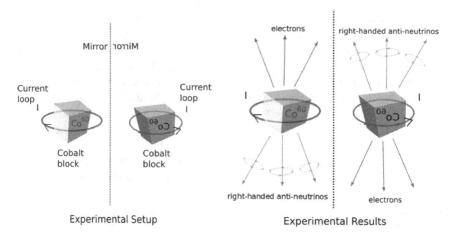

Experimental Setup   Experimental Results

Fig. 2.2 Schematic of experiment by Wu and collaborators verifying that parity is not conserved. The mirror image experiment shown in the first figure on the right should produce mirror image results. The second figure shows that this is not the case, so that parity (reflection through a mirror) is not a symmetry respected by nature

scattering of polarized protons from randomly oriented protons in a target. Nature indeed violates parity via weak interactions, and in fact does so maximally in the sense that we do not directly detect either left-handed anti-neutrinos or right-handed neutrinos. While we do not know why this is the case, there nonetheless exists a theory that describes it mathematically and predicts other verifiable results. It is called the Standard Model of particle physics. Lee and Yang were awarded the Nobel prize in physics in 1957 for their discovery. While C.S. Wu and her collaborators did not receive a Nobel Prize for their experimental confirmation of parity violation, Dr. Wu's contribution to this very important discovery was publicly acknowledged in 1978 when she was awarded the prestigious inaugural Wolf Prize.

2.4.3 Spontaneously Broken Symmetries

If one poses a question about a situation that has a high degree of symmetry and answers it using equations that also possess the same symmetry, one expects the solution to be symmetric. It turns out that this very intuitive expectation is sometimes not correct. Although there always will be a symmetric solution, there may be other solutions with less symmetry that are more relevant physically under certain circumstances. This phenomenon is called "spontaneous symmetric breaking".

A simple example of symmetry breaking is the following. Consider four towns located on the four corners of a square with sides a km in length, as in Fig. 2.3. We call the towns NW, NE, SE and SW, corresponding to the geographic corner (i.e. north-west, north-east, etc.) of the square that they occupy. An engineer is called

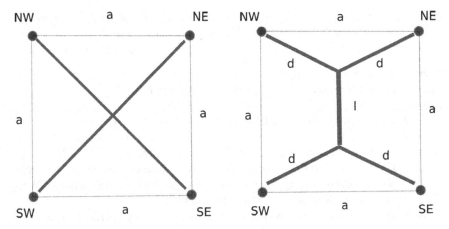

Fig. 2.3 Four towns joined by roads

in to design a road connecting all four towns at minimal cost, which really means with minimum concrete and hence minimum total road length. What configuration of roads would meet this criterion?

Given that the towns are indistinguishable for the purpose of the construction, this problem is invariant under 90° rotations, and multiples thereof (i.e. it has the symmetries of a square). One would expect that the minimum length roadway would also have these symmetries. The simplest possibility is to join the towns directly along the perimeter of the square. The total length would be $4a$. A bit of reflection (in the sense of "contemplation", pun intended) reveals that a more economical option would be to join the towns along the diagonals as shown in the left hand side of Fig. 2.3. The driving distance between neighbouring towns on the perimeter would be greater, but the total length of roadway required in this case is $2\sqrt{a^2 + a^2} = 2\sqrt{2}a \sim 2.83a$. Thus, the engineer wisely chooses the diagonals over the perimeter to save money on pavement.

A physics student working during the summer for the engineer comes up with a radical third possibility, shown on the right hand side of Fig. 2.3. This configuration has less symmetry than the previous two. It looks the same under 180° rotations but not 90°. For example, the driving distance from NW to SW is greater than that from NW to NE. The student points out that before comparing this lay-out to the diagonals, one needs to determine the length, l, of the piece of road running north-south that minimizes the total length of road. Although it is not at all obvious without doing the calculation, the answer is that the optimal length, l, is:

$$l = \left(1 - \frac{1}{\sqrt{3}}\right) a \qquad (2.1)$$

The surprising fact is that with this choice of l, the total length of road required is $(\sqrt{3}+1)a \sim 2.73a$. This is about 3% less than the diagonal scheme. If the towns are $a = 20$ km apart, this represents 2 km of pavement, a significant financial saving.

Exercise 1 Calculate the length D of roadway in Fig. 2.3 as a function of l and a, and verify that it is minimized by the value of l given in Eq. (2.1) above.

So where is the symmetry in this case? It is clear that if you rotate the road configuration by 90° so that the north-south strip of road runs east-west, the total length of roadway required does not change. So in fact there are two different but equally valid solutions to the problem related by a symmetry operation. It would require a political decision as to whether one wants the driving distance from NW to NE and SW to SE to be shorter, or NW to SW and NE to SE.

To recap, in the absence of symmetry breaking, symmetry operations that leave the equations invariant also leave the physically relevant solutions invariant. This would be the case for the present problem if the configuration on the left hand of Fig. 2.3 were the physically relevant (least expensive) configuration. In the present case, however, the symmetry is broken, in that the configuration requiring the least pavement does not have all the symmetries of the original problem. The symmetry in this case manifests itself by the presence of more than one independent solution such that the action of the broken symmetry operations change one solution into another. The configuration of roads on the right hand side of Fig. 2.3 is invariant under rotations by 180°, but not by 90° or 270°. If one rotates this solution (just the pavement, not the towns) by either 90° or 270°, one obtains a different solution in which NW and SW are connected directly, but not NW and NE.

Exercise 2 Consider three towns, called N (for North), SW and SE, respectively, located a distance a apart at the vertices of an equilateral triangle, as shown in Fig. 2.4. We wish to build a network of roads connecting all three towns in such a way that the roads (shown in blue in e-version) meet at an arbitrary point P along the line joining the center of the triangle to the northern town at some distance l from N. Show that the minimum total length for such a configuration of roads occurs when $l = a/\sqrt{3}$, so that the least expensive way to join the towns is to have the three segments of road meet at the center, C. Is there symmetry breaking in this case? Explain.

Hint: Use the law of cosines to figure out the distance from P to the other two towns SW and SE. This should give you an expression for the total length of the three roads that join P to N, SW and SE. Finally, minimize the total length of pavement with respect to the parameter l and show that the minimum occurs at $l = a/\sqrt{3}$.

Spontaneous symmetry breaking plays an important role in particle physics. The main role of the famous Higgs particle, whose detection led to the 2013 Nobel Prize in physics, is to spontaneously break the (internal) symmetry associated with the equations describing the unification of electromagnetism and the weak interaction. It is this symmetry breaking process that imparts other particles in the theory with mass. See https://www.nobelprize.org/nobel_prizes/physics/laureates/2013/press.html.

Fig. 2.4 Three towns to be
joined by shortest road
(Colour figure online)

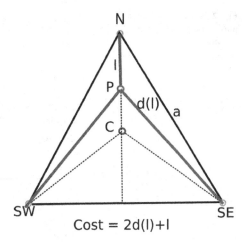

$$\text{Cost} = 2d(l) + l$$

2.4.4   Variational Calculations: Lifeguards and Light Rays

The above determination of the value of l is an example of a *variational calculation* that finds the minimum of a "cost function" in order to determine a physically relevant configuration. Such calculations are vital to physics, since all the known laws of nature can be obtained by minimizing an appropriate cost function. One of the best and intuitively most clear examples of this is *Fermat's principle*, which says that when light travels between two points it chooses the path that minimizes the transit time required. Knowing that light travels at different speeds in two different mediums, say air and water, one can use Fermat's principle to derive the law of reflection (angle of incidence equals the angle of reflection) as well as the law of refraction, otherwise known as *Snell's law*.

A simple way to understand Fermat's principle is to consider the dilemma facing a lifeguard on a beach who needs to rescue a swimmer, S, apparently drowning in the water some distance from the shore. See Fig. 2.5.

The lifeguard swims at a speed V_S that is slower than the speed V_G that she can run, and must therefore choose the quickest, not shortest path to the swimmer. Of the three paths shown in Fig. 2.5, Path 1 covers less distance on the land, where the lifeguard can move fastest, than in the water where she is slowest. The total distance she needs to travel is, however, the least of the three paths. Path 3, on the other hand, covers more distance on the land than in the water, but is the longest of the three paths. It is reasonable to suppose that the optimal path is somewhere in between. That is, for any given values of the speeds V_S and V_G, there is a path, such as Path 2, somewhere in between 1 and 3, that gets the lifeguard to the swimmer in the least possible time. Note that this will not be the shortest distance, namely a straight line between the lifeguard and swimmer, unless the speeds on land and in water are equal. It is relatively straightforward (see Exercise 3) to show that the angles θ_S and θ_G (see Fig. 2.5) that define the optimal path are determined by the simple relationship:

Fig. 2.5 The lifeguard, G must choose the quickest, not shortest, path to the swimmer, S

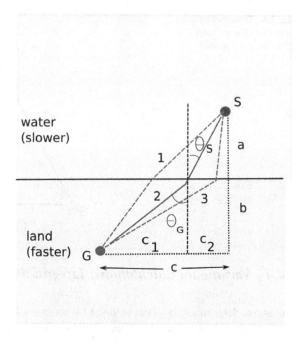

$$V_S \sin(\theta_G) = V_G \sin(\theta_S) \tag{2.2}$$

Exercise 3 Derive Eq. (2.2) using the diagram in Fig. 2.5. The distances a, b and c are given. You can use the fact that the total time T_{tot} taken along Path 2 is:

$$T_{tot} = \frac{\sqrt{c_1^2 + b^2}}{V_G} + \frac{\sqrt{c_2^2 + a^2}}{V_S} \tag{2.3}$$

and find the equation for c_1 that minimizes T_{tot}. *Hint*: in order to solve the resulting equation, you may need:

$$\frac{c_1}{b} = \tan(\theta_G)$$

$$\frac{c_2}{a} = \tan(\theta_S) \tag{2.4}$$

Those who have taken geometric optics may recognize Eq. (2.2) as *Snell's Law*, which determines the path taken by a light ray between two points, one in a less dense and hence quicker medium, say air, and the other in a more dense and hence slower medium, say water. The above analogy shows that the light ray bends as it passes from one medium to another because it takes the path that is the quickest given the properties of the two mediums. This condition of taking the quickest path is known as Fermat's principle and it applies equally to the light ray and to the life guard.

It is not surprising that an experienced lifeguard can determine the quickest path when trying to reach a drowning swimmer. An interesting philosophical question concerns how the light ray chooses the quickest path. Of course light rays do not choose paths. We are using this suggestive wording to highlight an interesting feature of Fermat's principle and its generalization, the Principle of Least Action (see Sect. 3.6.1). The choice of path must be made from the beginning and light rays cannot sample all paths before hand,[6] or make adjustments along the way. It is Nature that provides the rules (equations) that light rays must obey. Snell's law determines how much the path is deflected at the interface between two materials. All light rays obey this relationship. Using the language of the lifeguard example, one would require many light rays running off in different directions. All would refract at the interface according to Snell's law, so only the one going in the right direction initially would get to the swimmer.

In the supplementary Sect. 3.6, we will show how to use a variational calculation to derive Newton's second law for the motion of a particle subject to a conservative force, and also to prove Noether's theorem in the context of energy conservation.

Reference

1. C.S. Wu, E. Ambler, R.W. Hayward, D.D. Hoppes, R.P. Hudson, Experimental test of parity conservation in beta decay. Phys. Rev. **105**(4), 1413–1415 (1957)

---

[6] In fact, in the quantum description a light ray does in some sense get to sample all possible paths. The mechanism is the Feynman path integral of the probability for a particular trajectory. All paths are possible, but the path that is quickest has the highest probability.

Chapter 3
Formal Aspects of Symmetry

3.1 Learning Outcomes

Conceptual:

- Weyl's definition of symmetry operations.
- Examples of symmetry operations.
- Noether's theorem: statement and significance.
- Importance of symmetry in physics.

Acquired skills:

- Construction and use of the multiplication table for simple groups of symmetry operations.

3.2 Symmetries as Operations

3.2.1 Definition of a Symmetry Operation

Symmetry plays a vital role in 21st century physics. In order to be useful, it must be quantifiable. In particular, what does it mean to say that one object, or one spacetime, or one theory, is more symmetric than another? The basis for such a quantification is the following definition due originally to Hermann Weyl, who was one of the most brilliant mathematicians of the twentieth century[1]:

---

[1] A brief description of Weyl's life and achievements can be found at https://plato.stanford.edu/entries/weyl/#LifAch.

© The Author(s), under exclusive license to Springer Nature Switzerland AG 2022

G. Kunstatter and S. Das, *A First Course on Symmetry, Special Relativity and Quantum Mechanics*, Undergraduate Lecture Notes in Physics, https://doi.org/10.1007/978-3-030-92346-4_3

A symmetry operation is an action on an object, set of objects or a set of equations (laws of physics) that leaves the system unchanged or *invariant*. Symmetry operations that act on physical objects or systems of equations are often referred to as transformations. We will use the two terms interchangeably.

This definition is consistent with and quantifies our intuitive notion of symmetry. For example an equilateral triangle looks the same after any rotation by 120°, or integer multiple thereof. It is also invariant under reflection through any axis that bisects one of the corner angles. As an example of invariance of physical laws, Maxwell's equations are invariant under translations and rotations of the spatial coordinates. More importantly for the considerations in this book, once special relativity is taken into account, Maxwell's equations look the same to all observers moving at any constant velocity. This last symmetry is called invariance under "boosts" from one inertial frame to another. Symmetry operations can also involve taking specific linear combinations of different fields or particles contained in a theory. The associated symmetries are called *internal symmetries* because they aren't directly related to spacetime coordinates. Such symmetries play an important role in particle physics.

Weyl's definition quantifies the notion of symmetry in the sense that it enables one to count the number of distinct symmetry operations that leave a given system unchanged. The object or system of equations that has a greater number of such operations is more symmetric. For example, consider a square and a rectangle. The square can be rotated by any multiple of ninety degrees and still look the same. If one includes rotating by 0°, i.e. doing nothing at all, which certainly leaves the square unchanged, then there are four symmetry operations in total. A rotation by 360° is the same as doing nothing so should not be counted separately. Similarly, a rotation by 90° + 360° is the same as a 90° rotation, etc. The rectangle, on the other hand, has only two rotational symmetry operations, namely rotations by 0° and 180°. According to Weyl's definition, then, a square is more symmetric than a rectangle, while a pentagon is more symmetric yet.

The above process can be continued by considering 4, 5, 6, . . . sided polygons, which are more and more symmetric according to the above definition, until one reaches the most symmetric geometrical shape, the circle. Circles can be rotated by any angle θ between 0° and 360° and therefore have an infinite number of symmetries.[2] It is no wonder that, as explained in Sect. 1.2, the ancient Greeks thought the circle to be the most perfect geometrical object, and therefore tried to construct a picture of the heavenly spheres based on circles.

---

[2] Strictly speaking the symmetries of an n-sided polygon in the limit that n goes to infinity does not quite have the same symmetries as the circle. The number of symmetries of the former are in one to one correspondence with the integers (infinite but "countable" in the jargon of mathematics). The symmetries of the circle are in one to one correspondence with the **real** numbers, which are both infinite and "uncountable". This will be discussed more thoroughly in Sect. 3.4.

3.2.2 Rules Obeyed by Symmetry Operations

We will use upper case latin letters, $\{A, B, C, \ldots\}$, to denote generic symmetry operations that act on some object or system of objects. For example this may be the set of four rotations that leave the square invariant. The states of the system or object on which they act will be denoted by $\{x, y, z, \ldots\}$. Thus

$$y = A \cdot x \tag{3.1}$$

denotes the new state or configuration of the system x after it has been acted on by the operation A.

One can now act on the transformed system by another operation B, say. The new configuration is then:

$$z = B \cdot y = B \cdot A \cdot x \tag{3.2}$$

The notation keeps track of the order in which the operations are performed. The operation closer to the original object x, and hence to the right, is performed first. If B and A are performed in the opposite order then the resulting configuration would be denoted $A \cdot B \cdot x$. In the following we will be concerned with the properties of the operations themselves, and won't care what they act on. We can therefore neglect the object x, and simply use the notation $A \cdot B$ as the operation that consists of operation B followed by operation A. As we will see, the order of operations can be important in some cases since $A \cdot B \neq B \cdot A$ in general.

Now consider the set of all possible symmetries of some object, and denote this set by $S := \{A, B, C, \ldots\}$. In the case of a square the set $S = \{R(0°), R(90°), R(180°), R(270°)\}$ where the notation $R(\theta)$ is shorthand for rotations counterclockwise by θ. No symmetries are missed in this list because any other rotation is equivalent to one already in S. For example $R(-90)$, which denotes a 90° rotation in the clockwise direction, is equivalent to $R(270°)$.

It is clear at an intuitive level that such a set of symmetry operations must obey the following properties simply by virtue of being symmetry operations:

1. **Completeness**: If one performs one symmetry operation, C, belonging to the set S,[3] and then follows it by another symmetry operation, B, then the combined operation must also leave the system invariant, i.e. be a symmetry operation. Thus, the operation $G := B \cdot C$ must also belong to S.

   Note: If the operations are done in the reverse order, then $F := C \cdot B$ must also belong to S but as mentioned above there is no reason for it to be the same operation as $G := B \cdot C$. If $C \cdot B = B \cdot C$, then B and C are said to commute. As shown in Fig. 3.1, a simple example of operations that do not commute is found in the case of the rotations of a cube about perpendicular axes.

---

[3] The mathematical notation for this is $C \in S$, which means "C belongs to the set S".

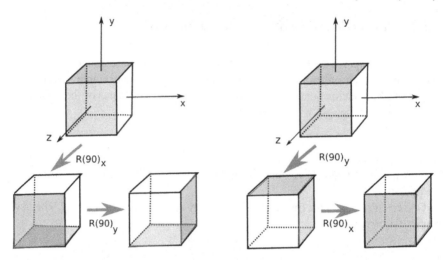

Fig. 3.1 Rotations of a cube about perpendicular axes. Two faces are coloured in order to distinguish the effects of the combined operations. Clearly order matters

If one first rotates the cube 90° according to the right-hand rule[4] about the $x-$axis, and then by 90°, also according to right-hand rule, about the $y-$axis as on the left side of Fig. 3.1 then what started as the front face ends up on the bottom. On the other hand if one does the rotation about the $y-$axis first, then as shown on the right side of Fig. 3.1, the front face ends up on the right-hand side. Clearly the order in which the rotations are done does make a difference in this case.

2. **Identity Operation**: The set S must contain an operation, normally referred to as I, that corresponds to doing nothing at all, which definitely leaves the object unchanged.

3. **Inverse**: Any given symmetry operation must be reversible. In other words for every operation, A, there exists another symmetry operation in S that, when performed directly after A, leads to a net operation equivalent to doing nothing at all. More technically for every A, there exists an H such that $H \cdot A = I$, where H is called the left inverse of A. There must also be an operation K, say, that is undone by A, so that $A \cdot K = I$. K is the right inverse of A. The definition does not require the left inverse and the right inverse to be the same operation. However, as we will prove directly below, consistency requires that $H = K$. In this case both left and right inverses are referred to simply as the inverse of A and denoted by A^{-1}, so that:

$$A^{-1} \cdot A = A \cdot A^{-1} = I \tag{3.3}$$

---

[4] The right hand rule says to point the thumb of one's right hand along the given axis (in this case the x-axis) and rotate in the direction that one's fingers are pointing. The left-hand rule is the same but with left-hand replacing right-hand.

4. **Associativity**: Suppose one performs three operations $A \cdot B \cdot C$ in succession. This must produce an unambiguous outcome that corresponds to some other symmetry operation belonging to S. We know from the completeness property that $(B \cdot C) = F$ must be a symmetry operation contained in S, as is $(A \cdot B) = G$. For the combination of three symmetry operations in a row to be unambiguous the result must be independent of how you choose to associate the three operations. That is $A \cdot F = A \cdot (B \cdot C)$ must be the same as $G \cdot C = (A \cdot B) \cdot C$. In other words, for any three operations A, B, C:

$$A \cdot (B \cdot C) \equiv (A \cdot B) \cdot C \tag{3.4}$$

We can now prove that for sets of symmetry operations that obey the above rules, the left and right inverses of all the operations must be the same. That is, every operation must have a unique inverse. This is required by associativity. Suppose $H \cdot A = I$ and $A \cdot K = I$. Then by associativity we must have that

$$(H \cdot A) \cdot K = H \cdot (A \cdot K)$$
$$\rightarrow I \cdot K = H \cdot I$$
$$\rightarrow K = H \tag{3.5}$$

3.2.3 Multiplication Tables

As the notation above suggests, the successive action of symmetry operations corresponds to a form of multiplication law, a generalization of the multiplication law for integers that is familiar to everyone. Just as with the integers, the properties of the multiplication law for symmetry operations can be summarized in a multiplication table. In general such a table is constructed as follows (see Table 3.3): First list all the operations across the top and down the left hand side. The elements across the top are those that act first, and those on the left side of the table are those that act second. For example the box located in the column below B and the row to the right of C must contain the operation that corresponds to B followed by C. Completeness requires that this joint action correspond to one of the symmetries listed along the top and along the left hand side. The other three properties must also be satisfied. Detailed examples will be given in Sect. 3.3

The following are some general properties of the multiplication table that are dictated by the four rules listed in Sect. 3.2.2:

1. The identity operation must appear exactly once in every row and column (not counting the headings along the top and on the left) in order to ensure that each operation has a unique inverse.

2. Every operation must appear exactly once in each row and column (not counting the headings).

   Proof The proof is by contradiction. Suppose $B \cdot C = F \cdot C = D$, for two different operations B and F so that D appears twice in the C column of the table. Given that C must have an inverse, say C^{-1}, we have that

$$B \cdot C = F \cdot C$$
$$\rightarrow B \cdot C \cdot C^{-1} = F \cdot C \cdot C^{-1}$$
$$\rightarrow B = F \tag{3.6}$$

   This contradicts the assumption that B and F are different, so the assumption is invalid.
3. If the table is not a symmetric matrix, it means that at least some operations do not commute.

3.2.4 Symmetry and Group Theory

The set of symmetry operations plus the multiplication table defined by their action define a mathematical object called a *group*. Our physically motivated study of symmetries has therefore naturally led us to consider the mathematical discipline called *group theory*. The rules for symmetry operations described in Sect. 3.2.2 come from rather intuitive notions about expected properties of successive symmetry operations. They also constitute precisely the four axioms underlying the definition of a mathematical group. This provides another example of the many ways that physics and mathematics are inexorably entwined.

3.3 Examples

3.3.1 The Identity Operation

For any system, doing nothing constitutes a rather trivial *set* consisting of symmetry operations: $S = \{I\}$. The multiplication table for a single operation that satisfies all of the properties listed in Sect. 3.2.2 is simply (Table 3.1):

Table 3.1 Multiplication table for trivial set of symmetry operations containing only the identity

	I
I	I

3.3.2 Permutations of Two Identical Objects

A permutation is an operation that interchanges the order, or location, of two objects. We consider the set of all possible permutations of two identical particles, labelled by a and b, respectively. Since the two particles are identical, the labels a and b are arbitrary, but they are needed so that we can keep track of the effects of the operations. There are in fact only two distinct operations that one can perform. The first corresponds to the operation of doing nothing at all, i.e. the identity operation, I. The second operation, which we will call P (for permutation), interchanges particle a with particle b. Thus the set of symmetry operations is $S = \{I, P\}$. You can verify that the operations obey the multiplications laws: $I \cdot I = I, I \cdot P = P \cdot I = P, P \cdot P = I$. The multiplication table is shown in Table 3.2. Note that P is its own inverse and that this group of operations is commutative: it does not matter what order you do any pair of operations.

Table 3.2 Multiplication table for permutation symmetries of two identical objects

	I	P
I	I	P
P	P	I

3.3.3 Permutations of Three Identical Objects

The set of all possible permutations of three identical objects (a, b, c) is given by:

$$
I : \begin{pmatrix} a \\ b \\ c \end{pmatrix} \to \begin{pmatrix} a \\ b \\ c \end{pmatrix} ; \qquad A : \begin{pmatrix} a \\ b \\ c \end{pmatrix} \to \begin{pmatrix} b \\ a \\ c \end{pmatrix}
$$

$$
B : \begin{pmatrix} a \\ b \\ c \end{pmatrix} \to \begin{pmatrix} a \\ c \\ b \end{pmatrix} ; \qquad C : \begin{pmatrix} a \\ b \\ c \end{pmatrix} \to \begin{pmatrix} c \\ b \\ a \end{pmatrix} \qquad (3.7)
$$

$$
F : \begin{pmatrix} a \\ b \\ c \end{pmatrix} \to \begin{pmatrix} c \\ a \\ b \end{pmatrix} ; \qquad G : \begin{pmatrix} a \\ b \\ c \end{pmatrix} \to \begin{pmatrix} b \\ c \\ a \end{pmatrix}
$$

Note that the operations are defined by their action on the elements in specific locations in the vector (i.e. A interchanges top and middle) and not by the action on specific letter names (interchange a with b).

This set of symmetry operations has the same number of elements as the symmetries of an equilateral triangle, if you include the three reflections as well as the three rotations. This is no coincidence. In fact there is a precise correspondence between the operations listed in Eq. (3.7) and the symmetry operations of the triangle. To verify this, simply label the vertices of the triangle by the letters a, b and c, and see what happens to them as you rotate and reflect.

The multiplication table for this group of symmetry operations is given in Table 3.3.

Table 3.3 Multiplication table for permutation symmetries of three identical objects

	I	A	B	C	G	F
I	I	A	B	C	G	F
A	A	I	F	G	C	B
B	B	G	I	F	A	C
C	C	F	G	I	B	A
G	G	B	C	A	F	I
F	F	C	C	B	I	G

Several interesting properties of the group are evident from the table:

- The operations A, B and C simply interchange, or permute, two of the three elements. For example A interchanges the top two. They are therefore their own inverses. Each pair $\{I, A\}$, $\{I, B\}$ and $\{I, C\}$ form a *sub-group* (the elements of a group contained totally within a larger group) that has a multiplication table identical in form to that of the permutations of two objects.
- F and G involve all three objects, and are called "cyclic" permutations. The set $\{I, F, G\}$ forms a subgroup as well in which F is the inverse of G and vice versa. This set is commutative.
- If one looks at the complete set of six operations, it becomes clear that the entire set is non-commutative, in that, for example: $C \cdot G = B$, while $G \cdot C = A$.

Exercise 1 Verify that the entries in the multiplication Table 3.3 follow from the actions of the symmetry operations listed in Eq. (3.7).

3.3.4 Rotations of Regular Polygons

1. The set of rotations that leave a square invariant:

$$S = \{\mathcal{R}(0°), \mathcal{R}(90°), \mathcal{R}(180°), \mathcal{R}(270°)\}. \tag{3.8}$$

Note that: $\mathcal{R}(360°) = \mathcal{R}(0°)$.

2. The set of rotations that leave a hexagon invariant:

$$S = \{\mathcal{R}(0°), \mathcal{R}(60°), \mathcal{R}(120°), \mathcal{R}(180°), \mathcal{R}(240°), \mathcal{R}(300°)\}. \tag{3.9}$$

3. The set of rotations that leave a circle invariant:

$$S = \{R(\theta), 0 \le \theta < 2\pi\} \tag{3.10}$$

Note: Rotations in a plane commute:

$$\mathcal{R}(\theta_2) \cdot \mathcal{R}(\theta_1) = \mathcal{R}(\theta_1 + \theta_2) = \mathcal{R}(\theta_1) \cdot \mathcal{R}(\theta_2) \tag{3.11}$$

Exercise 2 Construct the multiplication table for the set of rotations that leave a hexagon invariant.

3.4 Continuous Versus Discrete Symmetries

The sets of permutation symmetries described in the previous section contain a finite number of symmetry operations; two in the case of two objects and six in the case of three objects.[5] The number of distinct symmetry operations is also finite for the set of rotations that leave polygons looking the same. This type of symmetry is called *discrete* because you can count the number of operations. It is possible to have an infinite but nonetheless countable (in the mathematical sense) number of symmetry operations. Such operations are in one to one correspondence with the integers and are also referred to as discrete.

On the other hand, if one considers the complete set of symmetry operations (rotations) that leave a circle unchanged, one finds that there is one such operation for every angle between 0 and 2π. This in turn means that the operations are in one to one correspondence with the points on a circle. This set of symmetry operations therefore has an *uncountable infinity* of elements. It is clear that this set of symmetry operations has an infinite number of elements, but why "uncountable"? Just as in the case of the real line, between any two points you care to choose on the circle there is in principle a further infinity of points. No matter how small the interval between the two points you choose, there is still an infinite number of points between them. In the case of the integers, one can at least start counting them, although it is not possible to finish. In the case of the points on a circle or real line, it is impossible to even begin.

The symmetries that lead to conserved quantities must not only be uncountably infinite in number, but in fact they must be *continuous*, in the same way that the points on the real line are continuous. This condition states that one can choose points that are arbitrarily close together, and this in turn allows one to define differentiation and integration. Continuous groups of symmetries, including translations, rotations,

---

[5] The general formula for the number of independent permutations of n objects is in fact $n! := n \cdot (n-1)(n-2) \dots$. The reason for this is that when rearranging n identical objects you have n possible locations for the first, $(n-1)$ possible locations for the second and so on.

Lorentz transformations (see Sect. 6.2) and certain *internal symmetries* play a very important role in physics.

An interesting question relevant to theoretical physics is whether or not nature avails itself of the full continuum of real numbers that are mathematically available, or whether perhaps real numbers are mathematical abstractions that have no counterpart in the "real" (again, pun intended) world. Consider that solid objects are made of a large but finite number of molecules, and that the line your professor draws on the blackboard consists of a finite number of particles of chalk. As we will see in Chap. 11, the electron in a hydrogen atom can only have one of a discrete number of possible energies, and not a continuum of possible values as in Newton's theory. On a more abstract level, one can wonder whether it is in fact possible even in principle to move an arbitrarily small (infinitesimal) distance through space or time. Perhaps by applying the rules of quantum mechanics to gravity itself we may find that the spacetime through which we move also consists of a countable number of points.

3.5 Noether's Theorem

Noether's theorem is one of the most important, if not **the** most important, theorem in physics. Its discoverer Emmy Noether (1882–1935) made many other important contributions to abstract algebra as well as theoretical physics.[6]

The theorem states that a dynamical system possesses a conserved quantity or *charge* for every *continuous global symmetry* of the equations. We have defined *continuous symmetries* directly above. *A global symmetry* is one whose parameters do not depend on location. You must do precisely the same operation everywhere in space. This is the case for the translations, rotations and Lorentz transformations with which we will primarily be dealing, but there are other important symmetries, such as those of general relativity and particle physics, in which one can do a different transformation at each point in space and still leave the equations invariant. As mentioned in Sect. 2.3.2, conserved quantities are those that do not change with time as the system evolves. Applications of Noether's theorem are ubiquitous in physics:

- Conservation of linear momentum follows from the assumption that the laws of physics are the same everywhere or in other words, that spatial translations are a symmetry of the laws of physics.
- Energy conservation follows from the assumption that the laws of physics do not change with time and are therefore symmetric under time translations.
- Symmetry of the laws of physics under rotations yields conservation of angular momentum.
- As we will discuss in Chap. 9, the state of a particle is described in quantum mechanics by a complex function (*wave function*). Multiplying this wave function by a complex number of unit norm does not change the state. The conserved

---

[6] See for example *Noether: A Tribute to Her Life and Work*, edited by Emmy Noether, Martha K. Smith and James W. Brewer, Marcel Dekker, 1982.

quantity associated with this symmetry is the total probability that the particle is located somewhere. This total probability must be one at all times for quantum mechanics to make sense.

3.6 Supplementary: Variational Mechanics and the Proof of Noether's Theorem

Noether's theorem proves that for every continuous global symmetry of a system, there exists a conserved quantity. It is perhaps useful to provide a concrete example that is in some sense generic. If you understand how this works you will understand all conserved quantities.

Consider a single particle moving in one spatial dimension along a trajectory $x(t)$ under the influence of a conservative, position dependent force

$$F(x) = -\frac{dV(x)}{dx} \tag{3.12}$$

Newton's second law relates the acceleration a of the particle to the force $F(x)$ and mass m.

$$a := \frac{d^2x}{dt^2} = \frac{F(x)}{m} \tag{3.13}$$

We know that for such a particle, the total energy is conserved along the trajectory $x(t)$. That is for all times t, the quantity:

$$E := KE + V = \frac{1}{2}m[\dot{x}(t)]^2 + V(x(t)) = \text{constant} \tag{3.14}$$

In the above $V(x(t))$ refers to the value of the potential at the point $x(t)$.

Exercise 3 Show that $\frac{dE}{dt} = 0$ if $x(t)$ satisfies Eq. (3.13).

The conservation of energy in Eq. (3.14) is a direct consequence of the fact that the potential $V(x(t))$ is a function only of the position of the particle and does not depend explicitly on time.[7] This means that the force on the particle also does not depend explicitly on time, which in turn implies that Newton's equations do not change form under time translations, $t \to t + a$ for any constant a. Physically this symmetry guarantees that two identical experiments will yield precisely the same results irrespective of when they are performed.

---

[7] It depends on time implicitly, since its value at the particle location changes as the particle moves.

3.6.1 Variational Mechanics: Principle of Least Action

In order to understand how the proof works we first show that Newton's equations follow directly from the *Principle of Least Action*. The Principle of Least Action states that for any mechanical system, nature will always choose as the physical trajectory between any given initial and final state the unique trajectory that extremizes a suitable action, denoted $I[x(t)]$,[8] associated with the system.[9] The action, $I[x(t)]$, is in general defined to be the integral of the difference between the kinetic and potential energies of a system along the trajectory $x(t)$. For a single particle moving in a potential $V(x)$ the associated action, $I[x(t)]$ is:

$$I[(x(t)] = \int_{t_i}^{t_f} dt\, L(x(t), \dot{x}(t)) \tag{3.15}$$

where the integrand $L(x(t), \dot{x}(t))$ is called the *Lagrangian* for the system. It is a function of time, t, along the trajectory that is equal at each point to the difference between the kinetic and potential energies:

$$\begin{aligned} L(x(t), \dot{x}(t)) &:= KE(t) - V(x(t)) \\ &= \frac{1}{2} m\, (\dot{x}(t))^2 - V(x(t)) \end{aligned} \tag{3.16}$$

For example, for a simple harmonic oscillator consisting of a mass m attached to a spring with spring constant K oscillating about its equilibrium position (defined for convenience to be at $x = 0$) the potential energy at each point $x(t)$ along its trajectory is given by:

$$V(x(t)) = \frac{1}{2} K\, (x(t))^2 \tag{3.17}$$

Let's see how the action principle gives rise to Newton's equations in the case of the Lagrangian in Eq. (3.16). Consider an arbitrary trajectory for the particle $x(t)$ such that at some initial time t_i the particle starts at a particular fixed location, $x_i = x(t_i)$, and at a final time, t_f, it ends up at $x_f = x(t_f)$. The action principle states that nature will choose the unique path $x_p(t)$ between the initial and final points for which the value of the action is an extremum. The subscript p is used here to emphasize that

---

[8] The square brackets denote that $I[x(t)]$ is defined as an integral that depends on the values of some function $x(t)$ (the trajectory) at all points between a fixed initial time, t_i, and final time, t_f. It is therefore not a function of a single, or even a finite number of variables. It depends on an infinite number of variables, namely the values of $x(t)$ in the interval $[t_i, t_f]$. It is therefore called a "functional" to highlight this crucial difference.

[9] While the principle is called the principle of least action, in many physical situations the relevant trajectory is the maximum, not the minimum. In either case however, the trajectory is a stationary point so that it is a local extremum (either maximum or minimum). The word *local* has been added because one is only checking the slope and cannot always be certain that greater or smaller local extrema do not exist elsewhere in the solution space.

$x_p(t)$ refers to the physical path (i.e. the solution to the equations of motion). The action evaluated along this path is given by:

$$I[x_p(t)] = \int_{t_i}^{t_f} dt \left[\frac{1}{2} m(\dot{x}_p(t))^2 - V(x_p(t)) \right] \tag{3.18}$$

What does it mean that the action is an extremum when evaluated along the physical path $x_p(t)$? Recall that for functions of a single variable, such as $f(y)$, we say that y_p is an extremum if when making a small change dy to the argument

$$y_p \to y_p + dy \tag{3.19}$$

the change in the function vanishes as $dy \to 0$:

$$df = f(y_p + df) - f(y_p) \to 0 \tag{3.20}$$

Since by definition of the derivative

$$f(y_p + dy) - f(y_p) \sim \left. \frac{df(y)}{dy} \right|_p dy \tag{3.21}$$

one can restate this condition as:

$$\left. \frac{df(y)}{dy} \right|_p = 0 \tag{3.22}$$

Thus, the change in the value of the function under arbitrarily small displacements, dy, from an extremum, y_p, is arbitrarily small if and only if the first derivative of the function vanishes at that point. The vanishing of the first derivative is the usual definition of extremum. However, in order to deal with integrals of functions (called *functionals*) such as the action, we take the definition of *extremum* to be that: y_p is an extremum of a function $f(y)$ if and only if the change in the value of the function under arbitrarily small displacements, dy, away from y_p is arbitrarily small.

This definition is equivalent to Eq. (3.20) for ordinary functions.

Functionals can be thought of as functions of functions. It turns out that one can define a suitable *functional derivative* for integrals that re-enforces this equivalence. Functional derivatives look at the changes in a functional under small (infinitesimal) changes of the functions that are their arguments. This is the subject of a relatively new field of mathematics, inspired by physics, called *functional calculus*.

Let's now get back to the action principle. The analysis is more complicated because the action is a functional. Specifically it is defined as an integral of a function $x(t)$ of time evaluated along a given path $x(t)$ between specific end points. The numerical value of the action, and functionals in general, therefore depends on all values of its argument function between the end points. Consider a path that deviates slightly from the physical path $x_p(t)$:

$$x(t) = x_p(t) + \delta x(t) \tag{3.23}$$

where $\delta x(t)$ is an arbitrary function of t, except for two important restrictions. First, its magnitude must be small so that we can do the equivalent of a Taylor expansion in $\delta x(t)$ and neglect all but the leading order terms as done in ordinary calculus to obtain Eq. (3.21), for example. Second, $\delta x(t)$ must vanish at t_i and t_f so that the particle's start and end points do not change. The situation is illustrated in Fig. 3.2. We wish to evaluate the difference δI between the value of I evaluated along the path $x(t) = x_p(t) + \delta x(t)$ and the value of I evaluated along the path $x_p(t)$. That is:

$$\delta I := I[x_p(t) + \delta x(t)] - I[x_p(t)] \tag{3.24}$$

The value of I evaluated along the trajectory $x(t) = x_p(t) + \delta x(t)$ is:

$$I[x_p + \delta x] = \int_{x_i}^{x_f} dx \left[\frac{1}{2} m \left(\dot{x}_p(t) + \frac{d\delta x(t)}{dt} \right)^2 - V(x_p(t) + \delta x(t)) \right] \tag{3.25}$$

As suggested above, since $\delta x(t)$ is assumed to be a small change to the trajectory, we can calculate the approximate resulting small change to the action by only keeping terms in the action to first order in $\delta x(t)$. Thus we can neglect the quadratic term $(\frac{d\delta x(t)}{dt})^2$ in the kinetic energy and also do a Taylor expansion of the potential energy around $x_p(t)$ at each time:

$$V(x_p(t) + \delta x(t)) = V(x_p(t)) + \left. \frac{dV}{dx} \right|_{x_p(t)} \delta x(t) + \frac{1}{2} \left. \frac{d^2V}{dx^2} \right|_{x_p(t)} (\delta x(t))^2 \tag{3.26}$$

Neglecting terms of order in $(\delta x(t))^2$ and higher in the Taylor expansion, the expression for the difference between the value of the action evaluated along the different paths is:

$$\delta I := I[x_p + \delta x] - I[x_p]$$
$$= \int_{x_i}^{x_f} dx \left[\frac{1}{2} m \left(\dot{x}_p(t) + \frac{d\delta x(t)}{dt} \right)^2 - V(x_p(t) + \delta x(t)) - \left\{ \frac{1}{2} m(\dot{x}_p(t))^2 - V(x_p(t)) \right\} \right]$$
$$= \int_{x_i}^{x_f} dx \left[\frac{1}{2} m \left((\dot{x}_p(t))^2 + 2\dot{x}_p(t) \frac{d(\delta x)}{dt} \right) - V(x_p(t)) - \left. \frac{dV}{dx} \right|_{x_p(t)} \delta x(t) \right.$$
$$\left. - \left\{ \frac{1}{2} m(\dot{x}_p(t))^2 - V(x_p(t)) \right\} \right]$$
$$= \int_{x_i}^{x_f} dx \left[m\dot{x}_p(t) \frac{d(\delta x)}{dt} - \left. \frac{dV}{dx} \right|_{x_p(t)} \delta x(t) \right] \tag{3.27}$$

Fig. 3.2 Top figure illustrates a particular particle trajectory $x_p(t)$ between a fixed starting point $x(t_i)$ and end point $x(t_f)$. It also shows a different trajectory that has been slightly changed by the addition of the time dependent function $\delta x(t)$. The lower graph plots $\delta x(t)$, which vanishes at the beginning, t_i, and end, t_f, of the trajectory so as not to change the endpoints

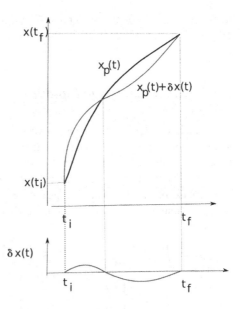

Integrating the first term by parts and recalling that $\delta x(t)$ must vanish at the endpoints t_f and t_i, one is left with:

$$\delta I = \int_{x_i}^{x_f} dx \left[-m\ddot{x}_p(t) - \frac{dV}{dx}\bigg|_{x_p(t)} \right] \delta x(t) \tag{3.28}$$

The Principle of Least Action requires nature to choose the path that extremizes $I[x(t)]$ which in turn implies that $\delta I[x(t)] = 0$ for arbitrary small variations in the path $\delta x(t)$. The only way that δI in Eq. (3.28) can vanish for arbitrary variations is if the sum of terms in the square brackets vanish for all values of t. Thus the true path $x_p(t)$ must satisfy the equation:

$$- m\ddot{x}_p(t) - \frac{dV}{dx}\bigg|_{x_p(t)} = 0$$

$$\text{or} \qquad m\ddot{x}_p(t) = -\frac{dV}{dx}\bigg|_{x_p(t)} \tag{3.29}$$

which is precisely Newton's equation for a particle moving in a potential.

The above is one specific application of the Principle of Least Action, which as claimed above applies to virtually all areas of physics, not only Newtonian mechanics. In the case of most, if not all, physical systems there exists a suitable action from which the equations of motion can be derived via the Principle of Least Action. The resulting equations are generally called the *Euler-Lagrange* equations, after the mathematicians Leonhard Euler and Joseph-Louis Lagrange who first developed the procedure in the middle of the eighteenth century.

3.6.2   Euler-Lagrange Equations

The Principle of Least Action applies very generally and is a powerful tool in mechanics. Consider a Lagrangian $L(x(t), \dot{x}(t))$ that is a function of the position $x(t)$ of a particle and its velocity $\dot{x}(t)$. In Newtonian mechanics the Lagrangian is often a quadratic function of the velocity, as in the example above, but this is not necessarily the case in general. The change in the action under arbitrary small changes in trajectory that vanish at the end points is then:

$$
\begin{aligned}
\delta I &= \int_{t_i}^{t_f} dt \left[\left. \frac{\partial L}{\partial x} \right|_{x_p(t)} \delta x(t) + \left(\left. \frac{\partial L}{\partial \dot{x}} \right|_{x_p(t)} \right) \delta \dot{x} \right] \\
&= \int_{t_i}^{t_f} dt \left[\left\{ \left. \frac{\partial L}{\partial x} \right|_{x_p(t)} - \frac{d}{dt} \left(\left. \frac{\partial L}{\partial \dot{x}} \right|_{x_p(t)} \right) \right\} \delta x + \frac{d}{dt} \left(\left. \frac{\partial L}{\partial \dot{x}} \right|_{x_p(t)} \delta x \right) \right]
\end{aligned}
$$

$$(3.30)$$

The last term again integrates to a boundary term that vanishes because we demand $\delta x(t)$ to vanish at the end points. Thus, in order for δI to vanish under such variations we require that the physical trajectory obey:

$$
\frac{\partial L}{\partial x_p} - \frac{d}{dt} \left(\frac{\partial L}{\partial \dot{x}_p} \right) = 0 \tag{3.31}
$$

Equation (3.31) is just the Euler-Lagrange equation for the more general particle action. In fact, it can be thought of as the general form of Newton's second law, with the first term as the force and the second term as the generalized mass times the acceleration.

Example 1   As an example consider a rigid object free to rotate about a single axis subject to an external constant torque τ. The kinetic energy is:

$$
\begin{aligned}
KE &= \frac{1}{2} I \omega^2(t) \\
&= \frac{1}{2} I \left(\frac{d\theta}{dt} \right)^2
\end{aligned}
\tag{3.32}
$$

where I is the moment of inertia about the rotation axis, $\theta(t)$ is the angular displacement at time t and $\omega(t) := d\theta(t)/dt$ is the corresponding angular velocity.

If the torque is constant with respect to the angular displacement, and positive in the $+\theta$ direction then the object's potential energy is:

$$V(\theta) = -\tau\theta \tag{3.33}$$

so that

$$
\begin{aligned}
L &= KE - V \\
&= \frac{1}{2}I\left(\frac{d\theta}{dt}\right)^2 + \tau\theta
\end{aligned} \tag{3.34}
$$

Then

$$
\begin{aligned}
\frac{\partial L}{\partial \theta} &= \tau \\
\frac{d}{dt}\left(\frac{\partial L}{\partial \dot{\theta}}\right) &= I\frac{d^2\theta}{dt^2}
\end{aligned} \tag{3.35}
$$

The Euler-Lagrange equations are therefore Newton's second law telling us that

$$I\frac{d^2\theta}{dt^2} = \tau \tag{3.36}$$

For complicated systems with many degrees of freedom, or possibly with constraints (a bead forced to move along a circular wire), all one needs to do is to write the Lagrangian $KE - V$ in whatever coordinates are most convenient, and then apply the principle of least action to obtain the Euler-Lagrange equations of motion. This is also where symmetry comes in: the most convenient coordinates are generally those that are compatible with the symmetry of the problem. For the bead on a frictionless circular wire, things are greatly simplified by using cylindrical coordinates (R, θ), where R is the radius of the wire, and θ is the angular coordinate locating the bead on the wire.

Exercise 4 Derive the equations of motion for a bead moving on a circular frictionless wire (see Fig. 3.3) using the principle of least action.

3.6.3 Proof of Noether's Theorem

In addition to allowing one to derive equations of motion in convenient coordinates, the principle of least action provides a method for proving Noether's theorem. To illustrate this we now prove that as long as the action, and hence equations of motion,

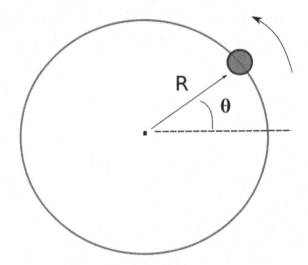

Fig. 3.3 Bead moving along frictionless circular wire

do not have explicit time dependence (i.e. all the time dependence is implicit through the time dependence of the coordinates) then energy is conserved.

Consider the action (3.18) where $x_p(t)$ solves the equations of motion. Now shift the coordinate t to a new coordinate $\tilde{t}$ defined by

$$\tilde{t} = t + a \quad \leftrightarrow \quad t = \tilde{t} - a \tag{3.37}$$

for some small constant a. A coordinate transformation can not, if implemented properly, change the value of the integral: the action must have the same numerical value for any given physical trajectory $x_p(t) = x_p(\tilde{t} - a)$, whether it is parameterized in terms of t or $\tilde{t}$. We wish to make the substitution $t = \tilde{t} - a$ everywhere in the action Eq. (3.15) $I[x(t)] = I[x(\tilde{t} - a)]$. Since we will now be integrating over $\tilde{t}$ instead of t, the integration endpoints are:

$$\tilde{t}_i = t_i + a \tag{3.38}$$
$$\tilde{t}_f = t_f + a \tag{3.39}$$

The action along the physical trajectory, rewritten in terms of the shifted time coordinate, is:

$$I = \int_{t_i+a}^{t_f+a} d(\tilde{t} - a) L(x_p(\tilde{t} - a), \dot{x}_p(\tilde{t} - a)) \tag{3.40}$$

Since a is constant (i.e. independent of time)

$$dt = d(\tilde{t} - a) = d\tilde{t} \tag{3.41}$$

Since a is assumed small (infinitesimal, in fact), we can do a Taylor expansion of the Lagrangian at each point along the trajectory. First, we note that under an infinitesimal shift in time $\tilde{t} \to \tilde{t} - a$, the spatial coordinate x_p and the velocity change to first order in the displacement a by:

$$\delta x_p := x_p(\tilde{t} - a) - x_p(\tilde{t}) = \frac{dx_p}{d\tilde{t}}\bigg|_{\tilde{t}} (-a) = -a\dot{x}_p(\tilde{t}) \tag{3.42}$$

$$\delta \dot{x}_p := \dot{x}_p(\tilde{t} - a) - \dot{x}_p(\tilde{t}) = \ddot{x}_p|_{\tilde{t}} (-a) \tag{3.43}$$

Treating x_p and $\dot{x}_p$ as independent variables, the corresponding change in the Lagrangian to first order in a is:

$$\begin{aligned}
\delta L &:= L(x_p(\tilde{t} - a), \dot{x}_p(\tilde{t} - a)) - L(x_p(\tilde{t}), \dot{x}_p(\tilde{t})) \\
&= \left[\frac{\partial L}{\partial x_p} dx_p + \frac{\partial L}{\partial \dot{x}_p} d\dot{x}_p \right] \\
&= \left[\frac{\partial L}{\partial x_p} \dot{x}_p(-a) + \frac{\partial L}{\partial \dot{x}_p} \frac{d\dot{x}_p}{dt}(-a) \right] \\
&= -a \left[\frac{\partial L}{\partial x_p} \dot{x}_p + \frac{\partial L}{\partial \dot{x}_p} \frac{d\dot{x}_p}{dt} \right]
\end{aligned} \tag{3.44}$$

When evaluating the variation in Eq.(3.44) above we have explicitly used the assumption that L depends on time only implicitly through its dependence on $x(t)$. The second term in square brackets in the last line of Eq. (3.44) can be rewritten to yield:

$$\delta L = -a \left[\frac{\partial L}{\partial x_p} \dot{x}_p - \frac{d}{d\tilde{t}} \left\{ \frac{\partial L}{\partial \dot{x}_p} \right\} \dot{x}_p + \frac{d}{d\tilde{t}} \left\{ \frac{\partial L}{\partial \dot{x}_p} \dot{x}_p \right\} \right] \tag{3.45}$$

The first two terms vanish by virtue of the Euler-Lagrange equations (3.31), so that δL is just a total time derivative at each point along the physical trajectory:

$$\delta L = -a \frac{d}{d\tilde{t}} \left\{ \frac{\partial L}{\partial \dot{x}_p} \dot{x}_p \right\} \tag{3.46}$$

The contribution of δL to the change in the action under the shift is therefore:

$$\begin{aligned}
\int_{t_i+a}^{t_f+a} d\tilde{t} \, \delta L &= \int_{\tilde{t}_i-a}^{\tilde{t}_f-a} (-a) d\tilde{t} \frac{d}{d\tilde{t}} \left\{ \frac{\partial L}{\partial \dot{x}_p} \dot{x}_p \right\} \\
&= -a \left[\left\{ \frac{\partial L}{\partial \dot{x}_p} \dot{x}_p \right\} \bigg|_{t_f+a} - \left\{ \frac{\partial L}{\partial \dot{x}_p} \dot{x}_p \right\} \bigg|_{t_i+a} \right] \\
&= -a \left[\left\{ \frac{\partial L}{\partial \dot{x}_p} \dot{x}_p \right\} \bigg|_{t_f} - \left\{ \frac{\partial L}{\partial \dot{x}_p} \dot{x}_p \right\} \bigg|_{t_i} \right]
\end{aligned} \tag{3.47}$$

We are able to ignore the shift by a in the two terms in the middle line because the effect of these shifts is small when a is small, and there is already a factor of a in front of the entire expression.[10]

One must now take into account the change in the action due to the shift in the end points of integration. Elementary calculus tells us that for any integrand $f(\tilde{t})$, a shift in endpoints results in the following change in the integral to first order in the shift a[11]:

$$\int_{t_i+a}^{t_f+a} d\tilde{t}\, f(\tilde{t}) = a\left[f(t_f) - f(t_i)\right] \tag{3.48}$$

This contributes to the following shift in the value of the action in Eq. (3.40):

$$\delta I = a\left[L(t_f + a) - L(t_i + a)\right]$$
$$\delta I = a\left[L(t_f) - L(t_i)\right] + \text{terms of order } a^2 \text{ and higher} \tag{3.49}$$

Putting this together with the boundary term the total change in the value of the action due to the shift in coordinates is therefore:

$$\delta I = -a\left[\left\{\frac{\partial L}{\partial \dot{x}_p}\dot{x}_p\right\}\bigg|_{t_f} - \left\{\frac{\partial L}{\partial \dot{x}_p}\dot{x}_p\right\}\bigg|_{t_i}\right]$$
$$+ a(L(x_p(t_f), \dot{x}_p(t_f)) - L(x_p(t_i), \dot{x}_p(t_i))) \tag{3.50}$$

All we have done is to evaluate the same integral in terms of shifted coordinates, while keeping the end points fixed, so δI must vanish or, rearranging terms in Eq. (3.50):

$$L(x_p(t_f), \dot{x}_p(t_f)) - \frac{\partial L}{\partial \dot{x}_p}\dot{x}_p\bigg|_{t_f} = L(x_p(t_i), \dot{x}_p(t_i)) - \frac{\partial L}{\partial \dot{x}_p}\dot{x}_p\bigg|_{t_i} \tag{3.51}$$

This must be true for any physical trajectory, whatever the starting and end points, which implies that the quantity:

$$E = -\left[L(x_p(t_f), \dot{x}_p(t)) - \frac{\partial L}{\partial \dot{x}_p}\dot{x}_p\bigg|_t\right] \tag{3.52}$$

---

[10] The corrections to the shifts are "higher order" in a and can be neglected.

[11] To convince yourself of this define $g(T) := \int^T dt f(t)$, take $T = t + a$ and Taylor expand $g(T + a) = g(T) + a\frac{dg}{dt}\big|_T$.

is constant along any physical trajectory. As the notation suggests, E is the energy of the system.[12]

Example 2 Let's verify this for the Lagrangian of a particle moving in a potential $V(x)$ as given in (3.16):

$$L(x_p(t_f), \dot{x}_p(t)) = \frac{1}{2}m(\dot{x}_p(t))^2 - V(x_p(t))$$

$$\frac{\partial L}{\partial \dot{x}_p}\dot{x}_p\bigg|_t = m\dot{x}_p(t)$$

$$\rightarrow E = -\left[\frac{1}{2}m(\dot{x}_p(t))^2 - V(x_p(t)) - m\dot{x}_p(t)\right]$$

$$= \frac{1}{2}m(\dot{x}_p(t))^2 + V(x_p(t)) \tag{3.53}$$

It must be mentioned that there are physical situations in which one wishes to describe a system that is not completely isolated, so that energy is not conserved. One example is that of a forced harmonic oscillator, namely one that is subjected to an external force in addition to the conservative restoring force. This is particularly relevant to anyone who has tried to rock a car out of a ditch. If you displace it slightly, the restoring force of gravity will cause it to oscillate around its equilibrium point, namely the force of gravity. This oscillation would continue for ever if it were not for frictional forces that cause the oscillator to lose energy and settle in the bottom again. Alternatively, by pushing the car at the right times you can increase the energy of the oscillator and hence its amplitude in the hopes of getting it out of the ditch.

To summarize, we have been able to rigorously prove Noether's theorem for a specific, but very important, family of examples. As mentioned, this is one of the most important theorems in physics. It explains and in some cases predicts the existence of fundamental conserved quantities in nature, such as electric charge, energy, momentum and angular momentum. The key message is that energy conservation follows immediately for any system whose Lagrangian does not explicitly contain time as a parameter, which in turn implies that the equations of motion are invariant under time translations.

---

[12] This explains the apparently arbitrary minus sign on the right hand side of Eq. (3.52). It is chosen to ensure that the energy as defined is positive and, more importantly, bounded below for physical systems.

Chapter 4
Symmetries and Linear Transformations

4.1 Learning Outcomes

Conceptual

- Vectors as geometrical quantities.
- Vector algebra and calculus.
- Connection between linear transformations of the plane and symmetry operations.
- Matrices as concrete mathematical representations of linear transformations and symmetry operations.
- Connection between Pythagoras theorem and Euclidean geometry.

Acquired skills

- Calculating dot products, magnitude of vectors and cross-products.
- Differentiation of time dependent vectors.
- Two dimensional matrix-vector and matrix-matrix multiplication.

4.2 Review of Vectors

4.2.1 Coordinate Free Definitions

1. Vectors and scalars

    - A *scalar* quantity is one that is specified by a single number. It has the same numerical value in any coordinate system, but the value changes depending on the units used. Examples of scalar quantities in Newtonian physics are mass, speed and energy.

© The Author(s), under exclusive license to Springer Nature Switzerland AG 2022   45
G. Kunstatter and S. Das, *A First Course on Symmetry, Special Relativity and Quantum Mechanics*, Undergraduate Lecture Notes in Physics,
https://doi.org/10.1007/978-3-030-92346-4_4

- *Vectors* have both magnitude and direction. They can be represented diagrammatically as a directed line of a specific length (i.e. a line segment with an arrowhead at one end to specify the direction). One can also think of a vector as a directed line joining two points in space, one at the tail, and the other at the head of the vector.

  In three spatial dimensions,[1] vectors are specified by three numbers: a scalar magnitude, or length whose value does not change with coordinate system, and two angles to specify direction. The latter do change as the coordinate axes change. Alternatively, one can specify a vector in terms of its projection, or components, along each of three independent coordinate axes.

  The zero vector **0** has length of zero, and hence its direction is arbitrary, or undefined. It is just a point.

2. Diagrammatical addition and subtraction.

   Scalars and vectors are geometrical quantities that have physical meaning independent of the coordinate system used to describe them. This is self evident for scalars, but not for vectors since the components change with coordinates. Nonetheless, the magnitude and direction are independent of the coordinates we use to describe the vector, and one can perform vector operations such as addition and subtraction using diagrams without reference to any specific set of coordinates.

   Vectors have magnitude and direction, but no location in that they can be moved around in space without changing their physical content. To add two vectors diagrammatically, one puts the tail of one vector, **B** to the tip of the other vector **A** and the resultant vector **C** = **A** + **B** has its tail at that of **A** and head at that of **B**, as in Fig. 4.1. It doesn't matter in what order you add vectors because **A** + **B** = **B** + **A**, as evident in Fig. 4.1.

   As illustrated in Fig. 4.2, subtraction **A** − **B** is defined as adding the negative of **B**, to **A**. That is:

$$\mathbf{A} - \mathbf{B} := \mathbf{A} + (-\mathbf{B}) \tag{4.1}$$

   where −**B** is the vector that gives the zero vector when added to **B**:

$$\mathbf{B} + (-\mathbf{B}) = \mathbf{0} \tag{4.2}$$

   The negative −**B** of **B** is therefore obtained by reversing the direction of **B** but keeping its length the same. The diagrammatic addition of −**B** to **B** gets you back to the tail, of **B**, so the sum is the zero vector.

3. Scalar product and magnitude (geometrical)

   The *scalar product* of two-vectors **A** and **B** is denoted by **A** · **B**. As the name suggests it is a way of "multiplying" one vector with another to produce a scalar quantity defined by:

$$\mathbf{A} \cdot \mathbf{B} := AB \cos(\theta) \tag{4.3}$$

---

[1] We will be considering other numbers of dimensions when we discuss special relativity.

Fig. 4.1 Geometrical definition of the vector sum $\mathbf{C} = \mathbf{A} + \mathbf{B}$. The sum is obtained by moving the tail of one vector to the tip of the other, with the resultant being the vector pointing from the tail of the first to the tip of the second, as shown

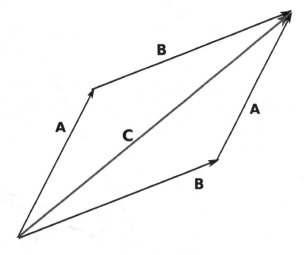

Fig. 4.2 To subtract $\mathbf{B}$ from $\mathbf{A}$, one defines $-\mathbf{B}$ by flipping $\mathbf{B}$ tip to tail and then adding this to $\mathbf{A}$

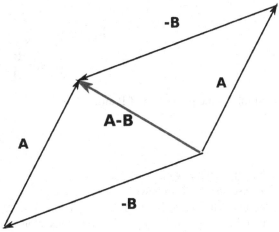

where A and B without arrows denote the length, or magnitude of the respective vectors. θ is the angle between them. Note that the definition implies that the scalar product is symmetric in the sense that:

$$\mathbf{A} \cdot \mathbf{B} = \mathbf{B} \cdot \mathbf{A} \tag{4.4}$$

The magnitude or length of a vector $\mathbf{A}$ is defined as the square root of the scalar product $\mathbf{A}$ with itself:

$$A = \sqrt{\mathbf{A} \cdot \mathbf{A}} \tag{4.5}$$

Finally, the scalar product provides the projection of $\mathbf{A}$ onto $\mathbf{B}$ (see Fig. 4.3):

Fig. 4.3 $\mathbf{A_B}$ is the projection
of the vector $\mathbf{A}$ onto $\mathbf{B}$

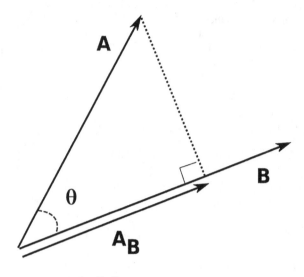

$$\mathbf{A_B} := \frac{\mathbf{A} \cdot \mathbf{B}}{B} \frac{\mathbf{B}}{B}$$

$$= A\cos(\theta)\frac{\mathbf{B}}{B} \qquad (4.6)$$

Similarly the projection of $\mathbf{B}$ along $\mathbf{A}$ is:

$$\mathbf{B_A} := \frac{\mathbf{B} \cdot \mathbf{A}}{A} \frac{\mathbf{A}}{A} \qquad (4.7)$$

Note that $\mathbf{B_A} \neq \mathbf{A_B}$.

The scalar product of two vectors is a measure of the extent to which one vector points in the same direction as the other. A negative value ($\theta > 180°$) indicates that they point more in opposite directions. Note that $\frac{\mathbf{A}}{A}$ and $\frac{\mathbf{B}}{B}$ define unit vectors parallel to $\mathbf{A}$ and $\mathbf{B}$ respectively.

Exercise 1 Draw three position vectors as follows: $\mathbf{A}$, magnitude 2 m, direction due East (E), $\mathbf{B}$ magnitude 4 m, direction North-North-East (NNE) and $\mathbf{C}$, magnitude 3 m direction South-West (SW). (SW points between South and West, whereas NNE points in the direction midway between North and North-East. North-East points mid-way between North and East).

1. Show by using diagrams, not calculating components, that

$$(\mathbf{A} + \mathbf{B}) - \mathbf{C} = \mathbf{A} + (\mathbf{B} - \mathbf{C})$$
$$= \mathbf{A} - \mathbf{C} + \mathbf{B} \qquad (4.8)$$

2. Calculate: $\mathbf{A} \cdot \mathbf{B}$ and the angle between $\mathbf{A}$ and $\mathbf{B}$.

---

4. Vector product:

The *vector product* of **A** with **B**, denoted by **A** × **B**, produces a third vector **C** whose magnitude is given by:

$$C := AB \sin(\theta), \tag{4.9}$$

where again θ is the angle between them. C is equal to the area of the parallelogram subtended by the vectors **A** and **B**, as Fig. 4.4 illustrates. In brief, the magnitude of the vector product is the area of the parallelogram subtended by the two vectors. The direction of the vector product tells us the plane in which the two vectors lie.

More specifically, the direction of **C** is defined to be perpendicular to both **A** and **B**. In three dimensions, and three dimensions only, this definition picks out a unique axis. However the sense along this axis is still ambiguous, so one uses the convention that **C** points along an axis perpendicular to **A** and **B** according to the *right hand rule*: if you curl the fingers of your right hand from **A** to **B**, then **C** points in the same direction as your thumb. A striking property of the vector product is that it changes under reflection through a mirror, which changes the right hand to the left hand. This intrinsic "handedness" in the definition is an important property of the cross product.

Fig. 4.4 Vector product
A × **B** = −**B** × **A**

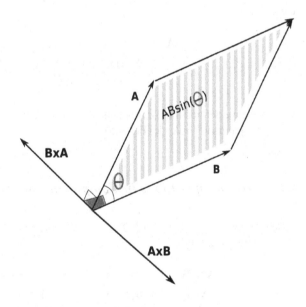

Two other important properties are:

- It is anti-symmetric, i.e. $\mathbf{A} \times \mathbf{B} = -\mathbf{B} \times \mathbf{A}$
- $\mathbf{A} \times \mathbf{B} = 0$ vanishes when $\mathbf{A}$ and $\mathbf{B}$ are parallel.

Exercise 2 For the vectors $\mathbf{A}$ and $\mathbf{B}$ in the previous exercise, draw the following:

1. $\mathbf{A} \times \mathbf{B}$,
2. $\mathbf{B} \times \mathbf{A}$,
3. $(\mathbf{A} - \mathbf{B}) \times \mathbf{B}$.

4.2.2 Cartesian Coordinates

In order to do calculations with vectors it is useful, but not altogether necessary, to go beyond the geometrical picture presented above. One must choose a coordinate system with an arbitrarily specified origin and a set of three coordinate axes. This defines a grid that allows the location of any point to be specifed by three numbers. The simplest of these are Cartesian coordinates, usually denoted as x, y, z axes, illustrated in Fig. 4.5. The Cartesian coordinate axes are straight and orthogonal. If one defines vectors of unit length $(\mathbf{i}, \mathbf{j}, \mathbf{k})$ that point along the x, y, z axes respectively, then any vector $\mathbf{A}$ can be written as the sum of its projections along each of the three axes. That is:

$$
\begin{aligned}
\mathbf{A} &= \mathbf{A}_x + \mathbf{A}_y + \mathbf{A}_z \\
&= (\mathbf{A} \cdot \mathbf{i})\,\mathbf{i} + (\mathbf{A} \cdot \mathbf{j})\,\mathbf{j} + (\mathbf{A} \cdot \mathbf{k})\,\mathbf{k} \\
&= A_x\mathbf{i} + A_y\mathbf{j} + A_z\mathbf{k}
\end{aligned}
\tag{4.10}
$$

where we have defined the components of $\mathbf{A}$ along each of the axes to be: $A_x := \mathbf{A} \cdot \mathbf{i}$, $A_y := \mathbf{A} \cdot \mathbf{j}$ and $A_z := \mathbf{A} \cdot \mathbf{k}$ (Fig. 4.5).

4.2.3 Vector Operations in Component Form

Once one has a Cartesian coordinate system with which to work, vector operations become straightforward:

$$
\mathbf{A} \pm \mathbf{B} = (A_x \pm B_x)\mathbf{i} + (A_y \pm B_y)\mathbf{j} + (A_z \pm B_z)\mathbf{k}
\tag{4.11}
$$

$$
\mathbf{A} \cdot \mathbf{B} = (A_x B_x) + (A_y B_y) + (A_z B_z)
\tag{4.12}
$$

$$
A = \sqrt{(A_x A_x) + (A_y A_y) + (A_z A_z)}
\tag{4.13}
$$

$$
\begin{aligned}
\mathbf{A} \times \mathbf{B} = {}&(A_y B_z - A_z B_y)\mathbf{i} + (A_z B_x - A_x B_z)\mathbf{j} + \\
&+ (A_x B_y - A_y B_x)\mathbf{k}
\end{aligned}
\tag{4.14}
$$

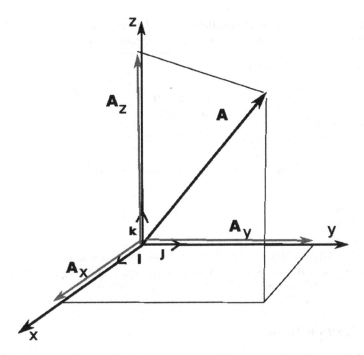

Fig. 4.5 Cartesian coordinates, showing basis vectors **i**, **j**, **k** and projections of vector **A** along each of the three coordinate axes

The expression for the cross product in three dimensions above shows that it is anti-symmetric under the interchange of A and B. Its form can be remembered by starting with one term, say $A_x B_y \mathbf{k}$, that contains all three coordinates (x, y, z) in order, then making sure the coefficient is anti-symmetric under the interchange of A and B, to give $(A_x B_y - A_y B_x)\mathbf{k}$. One then adds two more vectors that are "cyclic permutations" of this one. That is (x, y, z) is replaced by (z, x, y) in the first, and by (y, z, x) in the second. Alternatively, there is a simple expression for the cross product as a matrix determinant:

$$\mathbf{A} \times \mathbf{B} = \begin{vmatrix} \mathbf{i} & \mathbf{j} & \mathbf{k} \\ A_x & A_y & A_z \\ B_x & B_y & B_z \end{vmatrix} \tag{4.15}$$

The above prescriptions can be extended to operations involving multiple vectors, as in Exercise 3 below. In order to simplify these calculations, it is useful to know that the above rule implies:

$$\mathbf{i} \times \mathbf{j} = \mathbf{k} \tag{4.16}$$

$$\mathbf{k} \times \mathbf{i} = \mathbf{j} \tag{4.17}$$

$$\mathbf{j} \times \mathbf{k} = \mathbf{i} \tag{4.18}$$

Notice that the above three equations are cyclic permutations of each other. To obtain the remaining cross products use the anti-symmetry: $\mathbf{i} \times \mathbf{j} = -\mathbf{j} \times \mathbf{i}$ etc.

Exercise 3 1. Consider the three-vectors:

$$\mathbf{A} = \mathbf{i} - 2\mathbf{j}$$
$$\mathbf{B} = -\mathbf{i} + \mathbf{j}$$
$$\mathbf{C} = 2\mathbf{y} + \mathbf{k} \tag{4.19}$$

Calculate the following in terms of components, not diagrams:

(a) $\mathbf{A} + \mathbf{B}$
(b) $\mathbf{A} \cdot \mathbf{B}$
(c) $|\mathbf{A}|$ and $|\mathbf{B}|$
(d) $\mathbf{A} \times \mathbf{B}$
(e) $(\mathbf{A} \times \mathbf{B}) \cdot \mathbf{C}$ (*This shouldn't require a numerical calculation*)
(f) $\mathbf{D} = (\mathbf{A} \times \mathbf{B}) \times \mathbf{C}$ and $\mathbf{F} = \mathbf{A} \times (\mathbf{B} \times \mathbf{C})$. Are $\mathbf{D}$ and $\mathbf{F}$ equal?

4.2.4 Position Vector

- For simplicity we will restrict our attention to points in a two dimensional plane. As indicated above, once a set of Cartesian x and y coordinate axes are specified, including scale, one can locate any point relative to the origin by specifying two numbers, (x, y), that tell you how far along each axis you need to move to locate the point.
- Each point can also be found by specifying its position vector, $\mathbf{r}$, whose tail is placed at the origin and whose head sits on the point.
- The relationship between the coordinates and the position vector is:

$$\mathbf{r} = x\mathbf{i} + y\mathbf{j} \tag{4.20}$$

- It is sometimes convenient to denote the vector $\mathbf{r}$ by a column vector consisting of its components x and y:

$$\mathbf{r} = \begin{pmatrix} x \\ y \end{pmatrix} \tag{4.21}$$

- When we need to deal with more than one vector we will label them with a subscript (for example $\mathbf{r}_1$, $\mathbf{r}_2$, etc). If we wish to consider a set of n vectors we call the subscript i, or j,[2] and specify that i, say, takes any integer values between 1 and n. For example, the following denotes collectively the position vectors $\mathbf{r}_1, \mathbf{r}_2, \mathbf{r}_3, \ldots, \mathbf{r}_{10}$ of 10 particles:

---

[2] Not to be confused with the unit vectors $\mathbf{i}, \mathbf{j}$ which are given in bold face.

$$\{\mathbf{r}_i, i = 1, 2, \ldots 10\} \tag{4.22}$$

We also use these subscripts to distinguish components of position vectors of different particles. For example, in terms of components, the position vector of the ith particle is written:

$$\mathbf{r}_i = \begin{pmatrix} x_i \\ y_i \end{pmatrix}, \quad i = 1, 2, \ldots 10 \tag{4.23}$$

- **Relative position**:
  The position vector $\mathbf{r}_A$ as defined above gives the position of a point A on the plane relative to the origin of the coordinate system. In other words, it is the *displacement* of the point A relative to the origin. The position of the point A relative to some other point B, denoted $\mathbf{r}_{AB}$, is obtained by subtracting the position vector $\mathbf{r}_B$ of B relative to the origin from the position vector $\mathbf{r}_A$ of A relative to the origin as follows

$$\mathbf{r}_{AB} = \mathbf{r}_A - \mathbf{r}_B = \begin{pmatrix} x_A \\ y_A \end{pmatrix} - \begin{pmatrix} x_B \\ y_B \end{pmatrix} = \begin{pmatrix} x_A - x_B \\ y_A - y_B \end{pmatrix} \tag{4.24}$$

Figure 4.6 shows that $\mathbf{r}_{AB}$ as defined above is a vector that points from B to A and gives the position of the point A relative to B as expected.

The above definitions can be extended to three dimensions in which case the position vector is a column vector with 3 rows:

$$\mathbf{r} = \begin{pmatrix} x \\ y \\ z \end{pmatrix} \tag{4.25}$$

- **Distance between two points**:
  The magnitude squared of the position vector is given by:

$$r^2 := |\mathbf{r}|^2 = \mathbf{r} \cdot \mathbf{r} = x^2 + y^2 + z^2 \tag{4.26}$$

Since a position vector by definition locates the point in space with coordinates (x, y, z) relative to the origin, the expression for r^2 in Eq. (4.26) gives the distance of that point from the origin. The distance, d_{AB}, between two arbitrary points A and B is given by the magnitude of the relative position vector. From Eq. (4.24):

$$d_{AB}^2 := r_{AB}^2 = (x_B - x_A)^2 + (y_B - y_A)^2 \tag{4.27}$$

Given that the components of a vector are the magnitudes of the projections of the vector onto the x and y axes, respectively, Eqs. (4.26) and (4.27) are direct consequences of Pythagoras' theorem, the importance of which will be discussed in Sect. 4.5.

Fig. 4.6 $\mathbf{r}_{AB} = \mathbf{r}_A - \mathbf{r}_B$
points from B to A and gives
the position of the point A
relative to B

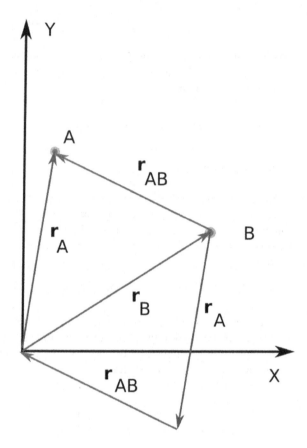

4.2.5   Velocity and Acceleration: Differentiation of Vectors

In physics we must consider vectors that change with time, such as the position
$\mathbf{r}(t)$ and velocity $\mathbf{v}(t)$ of a particle moving under the influence of a force. One also
needs to consider vector quantities that vary from point to point in space and also
depend on time. Examples include the electric field, $\mathbf{E}(x, y, z, t)$, and the magnetic
field, $\mathbf{B}(x, y, z, t)$, associated with a moving charged particle.[3] The laws of physics
necessarily involve the rates of change of physical quantities with respect to time
(and space in the case of scalar and vector fields). In this section, we focus on the
time derivatives of vectors.

---

[3] Scalar fields also exist in nature. One example is the gravitational potential $\phi(x, y, z, t)$ in Newtonian gravity, whose derivatives determine the gravitational field at each point in space and
time. In particular, at a distance r from the center of the Earth, the gravitational potential is
$\phi(x, y, z, t) = -GM/r = -GM/\sqrt{x^2 + y^2 + z^2}$.

Differentiation

In Cartesian coordinates, the time derivative of a vector is obtained by differentiating its components.

$$\frac{d\mathbf{A}(t)}{dt} := \lim_{\Delta t \to 0} \frac{\mathbf{A}(t + \Delta t) - \mathbf{A}(t)}{\Delta t}$$

$$= \frac{dA_x(t)}{dt}\mathbf{i} + \frac{dA_y(t)}{dt}\mathbf{j} + \frac{dA_z(t)}{dt}\mathbf{k} \tag{4.28}$$

In more general coordinate systems, such as spherical coordinates, the directions of the basis vectors change with location, so that if one is taking derivatives of the position of a moving particle for example, one must also differentiate the basis vectors. We will not be dealing with such curvilinear coordinates in this book.

Velocity Vector:

$$\mathbf{v} := \frac{d\mathbf{r}}{dt}$$

$$= \frac{dx}{dt}\mathbf{i} + \frac{dy}{dt}\mathbf{j} + \frac{dz}{dt}\mathbf{k}$$

$$= v_x(t)\mathbf{i} + v_y(t)\mathbf{j} + v_z(t)\mathbf{k} \tag{4.29}$$

Instantaneous acceleration:

$$\mathbf{a}(t) := \frac{d\mathbf{v}}{dt}$$

$$= \frac{dv_x}{dt}\mathbf{i} + \frac{dv_y}{dt}\mathbf{j} + \frac{dv_z}{dt}\mathbf{k}$$

$$= \frac{d^2x}{dt^2}\mathbf{i} + \frac{d^2y}{dt^2}\mathbf{j} + \frac{d^2z}{dt^2}\mathbf{k}$$

$$= a_x(t)\mathbf{i} + a_y(t)\mathbf{j} + a_z(t)\mathbf{k} \tag{4.30}$$

The magnitudes squared of the velocity and acceleration are, respectively:

$$|\mathbf{v}|^2 = \mathbf{v} \cdot \mathbf{v} = v_x^2 + v_y^2 + v_z^2$$

$$v = \sqrt{\mathbf{v} \cdot \mathbf{v}} = \sqrt{v_x^2 + v_y^2 + v_z^2}$$

$$|\mathbf{a}|^2 = \mathbf{a} \cdot \mathbf{a} = a_x^2 + a_y^2 + a_z^2$$

$$a = \sqrt{\mathbf{a} \cdot \mathbf{a}} = \sqrt{a_x^2 + a_y^2 + a_z^2} \tag{4.31}$$

Exercise 4 A ball of mass $m = 10$ gm moves in an elliptical path given by:

$$\mathbf{r}(t) = l\cos(\omega t)\mathbf{i} + 2l\sin(\omega t)\mathbf{j} \tag{4.32}$$

with $l = 2$ m and $\omega = \frac{\pi}{2}$ radians per second.

1. What is the period, T of the orbit?
2. Sketch the orbit $\mathbf{r}(t)$ for one complete cycle, i.e $0 \leq t < T$.
   Hint: Plot the points on the orbit at $t = 0$, $T/4$, $T/2$, $3T/4$ and then fill in the rest as smoothly as you can.
3. Calculate the distance of the ball from the origin as a function of t along the trajectory.
4. Find the velocity $\mathbf{v}(t) = \frac{d\mathbf{r}(t)}{dt}$ and acceleration $\mathbf{a} := \frac{d\mathbf{v(t)}}{dt}$ of the ball as a function of t.
5. Calculate $\mathbf{r} \cdot \mathbf{p}$ at $t = 1$ sec and $t = 3/2$ s.
6. Calculate the angular momentum $\mathbf{L} := \mathbf{r} \times \mathbf{p}$ at arbitrary time t.

4.3 Linear Transformations

4.3.1 Definition

We consider transformations in two dimensions. A transformation T is a rule that prescribes how to move any point (x, y) to $(x'(x, y), y'(x, y))$, where the notation $x'(x, y)$ and $y'(x, y)$ indicates that the new x' and y' coordinates can in principle be functions of both the x and y coordinates of the original point. Such transformations in general are called *non-linear* because $x'(x, y)$ and $y'(x, y)$ can be non-linear functions of their arguments. We will however focus on *linear transformations* for which the most general form is:

$$x'(x, y) = ax + by + d_x$$
$$y'(x, y) = cx + ey + d_y \qquad (4.33)$$

where a, b, c, e, d_x, d_y are constants. As the name suggests x' and y' are linear functions of the old coordinates. In mathematical notation, the transformation T acts on the coordinates (x, y) of the position vector as follows:

$$T : \mathbf{r} = \begin{pmatrix} x \\ y \end{pmatrix} \to \mathbf{r}' = \begin{pmatrix} x'(x, y) \\ y'(x, y) \end{pmatrix}$$
$$= \begin{pmatrix} ax + by + d_x \\ cx + ey + d_y \end{pmatrix} \qquad (4.34)$$

We have been using language that implies that transformations physically move the points from one place on the plane to another while keeping the coordinate axes fixed. Another viewpoint is that we are not moving anything, but instead are changing the coordinate system, and hence the numerical values of the coordinates of each point on the plane. If the transformations move the points on which they act they are called *active transformations*. If the coordinate axes move while the physical points stay put, then the transformations are called *passive*. Mathematically these two views are interchangeable, but when talking about physical consequences, especially in the context of special and general relativity, one sometimes has to be more careful.

We now discuss the three types of linear transformations.

4.3.2 Translations

One rather simple type of linear transformation is to set a, b, c and e equal to zero. This transformation takes all points in the plane and moves them over by the same constant vector, $\mathbf{d}$, where

$$\mathbf{d} = \begin{pmatrix} d_x \\ d_y \end{pmatrix}. \tag{4.35}$$

In other words, a translation D has the following action on all points (position vectors) on the plane:

$$\begin{aligned} D : \mathbf{r} \to \mathbf{r}' &= \mathbf{r} + \mathbf{d} \\ &= \begin{pmatrix} x + d_x \\ y + d_y \end{pmatrix} \end{aligned} \tag{4.36}$$

This is equivalent geometrically to taking the origin and moving it in the opposite direction by $-\mathbf{d}$.

Since physical space is three dimensional one can do independent translations in three directions, with corresponding translation vector:

$$\mathbf{d} = \begin{pmatrix} d_x \\ d_x \\ d_z \end{pmatrix} \tag{4.37}$$

Each component can be any real number, so that these are continuous transformations. It is generally assumed that the laws of physics are invariant under translations. This means that the laws of physics are the same everywhere. As long as your laboratory is isolated from its surroundings, you will get the same results for any given experiment anywhere in the Universe you do the experiment. As we learned in Sect. 2.3.2. Noether's theorem implies the existence of a conserved quantity for any continuous symmetry. In the present case, the associated conserved quantities are the components of linear momentum:

$$\mathbf{p} = \begin{pmatrix} p_x \\ p_y \\ p_z \end{pmatrix} \tag{4.38}$$

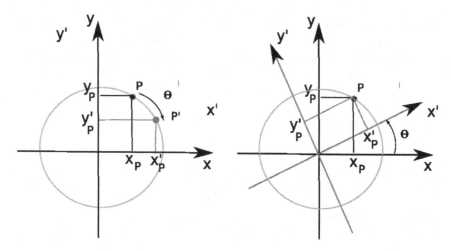

Fig. 4.7 Active rotations (shown on left) move the point P to a new point P' while passive rotations (shown on right) rotate the axes. An active rotation clockwise by angle θ yields new coordinates for P' that are the same as those obtained via a passive rotation of the axes by the same angle in the opposite direction, i.e. $-\theta$

It would be very surprising if translational invariance, and by extension momentum conservation, turned out not to be valid. In fact, if one found that the results of experiments did depend on location and that the total momentum was not conserved, the most natural explanation would be the existence of a previously unobserved field, or force, that itself depends on location and adds or carries away momentum. One would then formulate a translationally invariant theory that incorporated this extra field so that momentum conservation would be restored.

4.3.3   Rotations

A rotation is defined as a linear transformation that leaves the lengths of vectors unchanged (Fig. 4.7):

$$\mathbf{r} = \begin{pmatrix} x \\ y \end{pmatrix} \to \mathbf{r}' = \begin{pmatrix} x' \\ y' \end{pmatrix}$$
$$= \begin{pmatrix} \cos(\theta)x + \sin(\theta)y \\ -\sin(\theta)x + \cos(\theta)y \end{pmatrix} \tag{4.39}$$

Exercise 5 Verify that this transformation preserves the length of the vector, namely that:

$$|\mathbf{r}'|^2 := \mathbf{r}' \cdot \mathbf{r}' = (x')^2 + (y')^2 \tag{4.40}$$

$$= \mathbf{r} \cdot \mathbf{r} = x^2 + y^2. \tag{4.41}$$

For positive θ the linear transformation in Eq. (4.39) rotates the vector $\mathbf{r}$ by an angle θ in the **clockwise** direction about the origin. If θ is negative, then the rotation is in the counterclockwise direction. It should be obvious that $R(-\theta)$ is the inverse of $R(\theta)$, but you can verify this explicitly by performing one transformation after the other.

We are working in the $x - y$ plane for simplicity but if we think three dimensionally, we see that we are rotating about the z-axis. As with translations there are three different directions (axes in this case) about which we can rotate, each with its own independent, continuous parameter, i.e. $\theta_x, \theta_y, \theta_z$. Again, by Noether's theorem there are three associated conserved quantities, in this case the three components of the angular momentum vector:

$$\mathbf{L} = \begin{pmatrix} L_x \\ L_y \\ L_z \end{pmatrix}, \tag{4.42}$$

corresponding to rotations about the x, y and z axes respectively. As with translational invariance, we expect angular momentum to be exactly conserved and hence rotational invariance to be a good symmetry of nature.

4.3.4 Reflections

In two dimensions, reflection through the x-axis is defined by by:

$$\mathbf{r} = \begin{pmatrix} x \\ y \end{pmatrix} \rightarrow \mathbf{r}' = \begin{pmatrix} x' \\ y' \end{pmatrix}$$

$$= \begin{pmatrix} x \\ -y \end{pmatrix} \tag{4.43}$$

This also preserves lengths:

$$(x')^2 + (y')^2 = x^2 + y^2. \tag{4.44}$$

To reflect through any other axis, say one that is rotated counterclockwise by an angle θ relative to the x-axis, one needs to first rotate the vector through θ in the clockwise direction in order to line up the reflection axis with the x-axis, then reflect through the x-axis, then rotate back (counterclockwise) by θ to get the reflection axis back to where it was. As will be shown explicitly in Sect. 4.4.4, the net effect of these three transformations on a vector $\mathbf{r} = x\mathbf{i} + y\mathbf{j}$ is:

$$\mathbf{r}' = \begin{pmatrix} \left(\cos^2(\theta) - \sin^2(\theta)\right) x + 2\cos(\theta)\sin(\theta) y \\ 2\cos(\theta)\sin(\theta) x + \left(-\cos^2(\theta) + \sin^2(\theta)\right) y \end{pmatrix}$$

$$= \begin{pmatrix} \cos(2\theta)x + \sin(2\theta)y \\ \sin(2\theta)x - \cos(2\theta)y \end{pmatrix}. \tag{4.45}$$

Exercise 6 Check that the process outlined above produces Eq. (4.45) for reflections through the $x = y$ axis, i.e. $\theta = 45°$.

4.4 Linear Transformations and Matrices

4.4.1 General Discussion

A linear transformation, T, is a rule for taking any point with position vector $\mathbf{r} = \begin{pmatrix} x \\ y \end{pmatrix}$ and moving it to another point $\tilde{\mathbf{r}}$ as follows[4]:

$$T : \mathbf{r} \to \tilde{\mathbf{r}} = \begin{pmatrix} x' \\ y' \end{pmatrix}$$

$$= \begin{pmatrix} ax + by \\ cx + dy \end{pmatrix} \tag{4.46}$$

Such a transformation is characterized by four numbers (a, b, c, d) and acts in exactly the same way on all points in the plane. It is therefore useful to draw a distinction between the transformation and the points or, equivalently position vectors, on which it acts. The four numbers defining the transformation are often collected into an array, or matrix, written as follows:

$$T = \begin{bmatrix} a & b \\ c & d \end{bmatrix} \tag{4.47}$$

Using this notation the transformation above reads

$$\tilde{\mathbf{r}} = T \cdot \mathbf{r}$$

$$= \begin{bmatrix} a & b \\ c & d \end{bmatrix} \cdot \begin{pmatrix} x \\ y \end{pmatrix}$$

$$= \begin{pmatrix} ax + by \\ cx + dy \end{pmatrix} \tag{4.48}$$

---

[4] Neglecting translations.

For specific entries (a, b, c, d), Eq. (4.48) defines a linear transformation of points on a plane, something that is motivated by physical considerations of symmetry operations. It also serves to define mathematically, in the context of linear algebra, the left multiplication of a vector by a matrix. This is another example of physics and mathematics being two sides of the same coin.

So what happens if you do two linear transformations on the same point or points one after the other? Define:

$$T_1 = \begin{bmatrix} a & b \\ c & d \end{bmatrix}, \quad T_2 = \begin{bmatrix} e & f \\ g & h \end{bmatrix} \tag{4.49}$$

From the above rules we know that after the first transformation:

$$\mathbf{r} \rightarrow \tilde{\mathbf{r}} = T_1 \cdot \mathbf{r} = \begin{pmatrix} ax + by \\ cx + dy \end{pmatrix} \tag{4.50}$$

Now apply T_2 to $\tilde{\mathbf{r}}$:

$$\tilde{\mathbf{r}} \rightarrow \tilde{\tilde{\mathbf{r}}} = \begin{pmatrix} e(ax + by) + f(cx + dy) \\ g(ax + by) + h(cx + dy) \end{pmatrix}$$
$$= \begin{pmatrix} (ea + fc)x + (eb + fd)y \\ (ga + hc)x + (gb + hd)y \end{pmatrix} \tag{4.51}$$

The last line looks like a new linear transformation, T_3, with elements:

$$T_3 = \begin{bmatrix} ea + fc & eb + fd \\ ga + hc & gb + hd \end{bmatrix} \tag{4.52}$$

The above follows automatically from the expected properties of linear transformations. It also defines the product of two matrices as follows:

$$T_3 = T_2 \cdot T_1$$
$$= \begin{bmatrix} e & f \\ g & h \end{bmatrix} \begin{bmatrix} a & b \\ c & d \end{bmatrix}$$
$$= \begin{bmatrix} ea + fc & eb + fd \\ ga + hc & gb + hd \end{bmatrix} \tag{4.53}$$

From the notion of linear transformations we have derived the two basic relations of linear algebra: the multiplication of a vector by a matrix and the multiplication of a matrix by a matrix.

Exercise 7 Consider two linear transformations defined by:

$$T_1 = \begin{bmatrix} 1 & 3 \\ -2 & 2 \end{bmatrix}; \quad T_2 = \begin{bmatrix} \frac{1}{3} & 1 \\ 1 & 2 \end{bmatrix} \tag{4.54}$$

- Calculate the matrix T_3 that represents the action of first T_2 followed by T_1.
- Calculate the matrix T_4 that represents the action of first T_1 followed by T_2. Do the two operations commute?

We will now show how the matrix form of linear transformations are used to provide a concrete mathematical representation of symmetry operations. Matrices therefore provide us with tools for quantitative analysis of the effects of symmetries in the context of physical theories. For simplicity, we consider two very special types of linear transformations, namely rotations and reflections.

4.4.2 Identity Transformation and Inverse

We know that "doing nothing" to a system corresponds to a symmetry transformation called the identity, I. This too is a linear transformation:

$$I : \mathbf{r} \to \tilde{\mathbf{r}} = I \cdot \mathbf{r}$$
$$= \mathbf{r} = \begin{pmatrix} x \\ y \end{pmatrix} \tag{4.55}$$

For this transformation $a = d = 1$ and $b = c = 0$ with matrix form:

$$I = \begin{bmatrix} 1 & 0 \\ 0 & 1 \end{bmatrix} \tag{4.56}$$

In two dimensions, it is possible to calculate the inverse, T^{-1}, of any matrix, T, as given in Eq. (4.47) using the following expression:

$$T^{-1} = \frac{1}{det(T)} \begin{bmatrix} d & -b \\ -c & a \end{bmatrix} \tag{4.57}$$

where $det(T) := ad - bc$ is called the 'determinant' of the matrix T because it *determines* whether or not the matrix T has an inverse. If $det(T) = 0$, the expression in Eq. (4.57) is ill defined, and no inverse exists.

Exercise 8 Calculate the inverses of T_1 and T_2 in the previous exercise. First check if they exist. Be sure to verify in both cases that you have obtained the correct inverse.

4.4.3 Rotations

A rotation is defined as a linear transformation that leaves the lengths of vectors unchanged:

$$|\tilde{\mathbf{r}}|^2 := \tilde{\mathbf{r}} \cdot \tilde{\mathbf{r}} = \tilde{x}^2 + \tilde{y}^2 \tag{4.58}$$

$$= \mathbf{r} \cdot \mathbf{r} = x^2 + y^2 \tag{4.59}$$

The following linear transformation rotates any position vector $\mathbf{r}$ by an angle θ.

$$\mathbf{r} \to \tilde{\mathbf{r}} = R(\theta) \cdot \mathbf{r} \tag{4.60}$$

with

$$R(\theta) = \begin{bmatrix} \cos(\theta) & \sin(\theta) \\ -\sin(\theta) & \cos(\theta) \end{bmatrix} \tag{4.61}$$

If θ is positive, then the rotation is in the **clockwise** direction, otherwise the rotation is in the counterclockwise direction.

Exercise 9 A general two dimensional linear transformation takes the form:

$$T = \begin{bmatrix} a & b \\ c & d \end{bmatrix} \tag{4.62}$$

1. Show that T preserves the length of an arbitrary vector (x, y) if and only if:

$$a^2 + c^2 = 1 \tag{4.63}$$
$$b^2 + d^2 = 1 \tag{4.64}$$
$$ab + cd = 0 \tag{4.65}$$

2. Show that Eqs. (4.63)–(4.65) imply that:

$$a = \pm d \quad \text{AND} \quad c = \mp b \tag{4.66}$$

Hint: Solve Eq. (4.65) for b and then substitute this into Eq. (4.64). These conditions have two possible solutions:

$$a = d = \cos(\theta) \quad \text{AND} \quad c = -b = \sin(\theta) \tag{4.67}$$
$$a = -d = \cos(\theta) \quad \text{AND} \quad c = b = \sin(\theta) \tag{4.68}$$

Equation (4.67) corresponds to a rotation by an arbitrary angle θ as given in Eq. (4.61) while Eq. (4.68) corresponds to reflections as described in Sect. 4.4.4

Exercise 10 1. Show by explicit calculation and a diagram that when $\theta = 30°$, $R(\theta)$ as defined in Eq. (4.61) rotates the corners of a square with sides of length $2m$ centered on the origin by 30° clockwise.

2. Verify by matrix multiplication for arbitrary θ that $R(-\theta)$ is the matrix inverse of $R(\theta)$. That is:

$$R(\theta) \cdot R(-\theta) = R(-\theta) \cdot R(\theta) = I \tag{4.69}$$

where I is the unit matrix defined in Eq. (4.56).

4.4.4 Reflections

In two dimensions, reflection through the x-axis[5] is given by:

$$\mathbf{r} = \begin{pmatrix} x \\ y \end{pmatrix} \rightarrow \tilde{\mathbf{r}} = \begin{pmatrix} x' \\ y' \end{pmatrix} = \begin{pmatrix} x \\ -y \end{pmatrix} \tag{4.70}$$

This means that $a = 1, b = c = 0, d = -1$. The corresponding matrix representation is:

$$\mathcal{R} := \begin{bmatrix} 1 & 0 \\ 0 & -1 \end{bmatrix} \tag{4.71}$$

To reflect through any other axis, say one that is rotated counterclockwise by an angle θ relative to the x-axis, one simply needs to first rotate the vector through θ in the clockwise direction in order to line up the reflection axis with the x-axis, then reflect through the x-axis, then rotate back (counterclockwise) by θ to get the reflection axis back to where it was:

$$\begin{aligned} \mathcal{R}(\theta) &= \begin{bmatrix} \cos(\theta) & -\sin(\theta) \\ \sin(\theta) & \cos(\theta) \end{bmatrix} \begin{bmatrix} 1 & 0 \\ 0 & -1 \end{bmatrix} \begin{bmatrix} \cos(\theta) & \sin(\theta) \\ -\sin(\theta) & \cos(\theta) \end{bmatrix} \\ &= \begin{bmatrix} \cos^2(\theta) - \sin^2(\theta) & 2\cos(\theta)\sin(\theta) \\ 2\cos(\theta)\sin(\theta) & -\cos^2(\theta) + \sin^2(\theta) \end{bmatrix} \\ &= \begin{bmatrix} \cos(2\theta) & \sin(2\theta) \\ \sin(2\theta) & -\cos(2\theta) \end{bmatrix} \end{aligned} \tag{4.72}$$

This corresponds to Eq. (4.45). Reflections (i.e. transformations of the form in Eq. (4.72)), preserve lengths, but are distinguished from rotations by the value of the determinant of their matrix representation: it is -1 for reflections and $+1$ for rotations.

Exercise 11 1. Check that the general reflection given in Eq. (4.72) preserves lengths and that it has a determinant equal to -1.
2. Find the inverse transformation of the reflection Eq. (4.72) for arbitrary θ.

[5] Note that in three dimensions one reflects through a two dimensional mirror, or "plane". The transformation in Eq. (4.70) is a reflection through the $x - z$ plane, the two dimensional surface located at $y = 0$.

4.4.5 Matrix Representation of Permutations of Three Objects

Exercise 12 Consider the group of permutations $\{I, A, B, C, G, F\}$ of three identical objects as given in Eq. (3.7).

1. Find a 3×3 matrix that represents each operation. *Hint: the matrix contains only ones and zeros.*
2. Verify explicitly that matrix multiplication is consistent with the multiplication Table 3.3 in the following cases: $F \cdot C$, $C \cdot F$ and $F \cdot F$.

4.5 Pythagoras and Geometry

We know that for any right triangle, the square of the length of the hypotenuse is equal to the sum of the squares of the lengths of the other two sides. This is the *Pythagoras theorem*, a bit of geometry that we take for granted in our every day lives. In effect, it tells us how to calculate the distance, Δs, between any two points, A and B, in terms of the corresponding change in the Cartesian coordinates of their position vectors. Suppose $\mathbf{r}_{AB}$ is the position of point A relative to point B. We know that the distance between them, Δs, is given (implicitly) by:

$$(\Delta s)^2 := |\mathbf{r}_{AB}|^2 = |\Delta x|^2 + |\Delta y|^2 \tag{4.73}$$

where $\Delta x = x_A - x_B$ and $\Delta y = y_A - y_B$. Since the relative position vector $\mathbf{r}_{AB}$ forms a right triangle with its components along the x and y axes, the above is in essence the Pythagoras theorem. It defines the geometry of the space in which we live to be Euclidean. It is quite remarkable that from this one relationship one can derive all the other properties of Euclidean geometry that we all know and love: parallel lines never meet, the sum of the angles in a triangle is $180°$, the law of cosines and more.

It is important to note, however, that the Pythagoras theorem only applies to flat spaces, such as a blackboard, or a tabletop. Other interesting geometries are possible, each with its own version of the Pythagoras theorem. For example, if you live on the surface of a sphere of radius R (in fact we do: It's called the Earth!) you can uniquely fix your location by specifying two angles, θ, and ϕ, as shown in Fig. 4.8. These angles take the place of the Cartesian coordinates x and y that specify locations in flat space. θ specifies the line of latitude, while ϕ determines the line of longitude. The corresponding Pythagoras law that tells us the distance between two points on a sphere in terms of the angular displacements between them is:

$$|ds|^2 = R^2(d\theta^2 + \sin(\theta)^2 d\phi^2) \tag{4.74}$$

Fig. 4.8 Angles θ and ϕ
locate a point P on a sphere
of fixed radius R. The dotted
line (red) is a line of
longitude (fixed ϕ) while the
dashed line (blue) is a line of
latitude (fixed θ).
(Colour figure online)

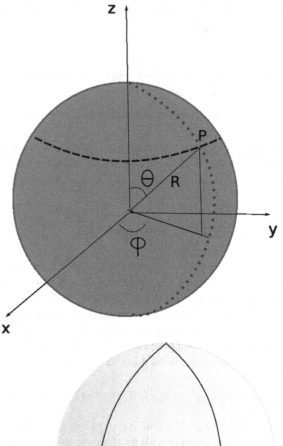

Fig. 4.9 Two airplanes head
due north from the equator
along lines of longitude and
meet at the North Pole. Since
both trajectories are
perpendicular to the equator,
the sum of the angles of the
triangle that they form with
the equator as base must be
greater than 180°. This is a
consequence of the fact that
the surface of the Earth is
spherical and obeys a
geometry that is different
from Euclid's

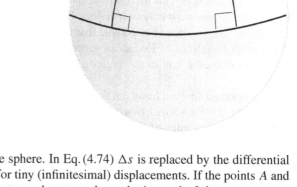

where R is the radius of the sphere. In Eq. (4.74) Δs is replaced by the differential ds because it is only valid for tiny (infinitesimal) displacements. If the points A and B on the sphere are far apart, one has to evaluate the integral of ds as one moves between them. This is because the right hand side depends explicitly on $\sin(\theta)$, that is on the line of latitude. Distances on the Earth's surface are therefore more difficult to calculate than the flat Earth society would have you believe.

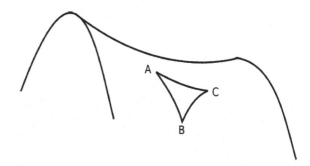

Fig. 4.10 The surface of a saddle has a different geometry from that of a tabletop or sphere. The sum of the angles in a triangle add up to <180°

We can determine some of the differences between Euclidean geometry and spherical geometry, as described by Eq. (4.74). For example, as illustrated in Fig. 4.9 the sum of the angles in a triangle is greater than 180° and parallel lines can in fact meet. Other geometries are possible as well. For example, a fly walking on the surface of a saddle would find that the sum of the angles in a triangle add up to less then 180°, as illustrated in Fig. 4.10.

Exotic geometries play a crucial role in Einstein's theory of gravitation, but even in the context of special relativity, the subject of Chap. 5, one needs to go beyond the Euclidean geometry of space and consider the geometry of the spacetime continuum. Appreciation of the role of geometry in the laws of physics provided the great paradigm shift set in motion by Einstein's special theory of relativity in 1905 and completed in 1915 when he unveiled the general theory.

Chapter 5
Special Relativity I: The Basics

5.1 Learning Outcomes

Conceptual

- Definition of inertial frame.
- The fundamental postulate as a symmetry of nature.
- Special relativity as a direct consequence of the symmetry postulate.
- The connection between Lorentz transformations in spacetime and Pythagoras' theorem.
- Special relativity as a foundation on which to construct physical theories.

Acquired Skills

- Constructing and using spacetime diagrams.
- Solving problems that involve time dilation and Lorentz contraction.

5.2 Preliminaries

5.2.1 Frames of Reference

A *frame of reference*, or *frame* for short, is simply a meter stick (or three perpendicular meter sticks in three spatial dimensions) with a clock attached. One often refers to an observer, O, attached to a particular frame of reference who measures the time and spatial coordinates of events using the corresponding clock and meter sticks. We will use the terms observer and reference frame more or less interchangeably.

Newtonian mechanics and special relativity start with the assumption that there exist a special set of reference frames in nature, known as *inertial frames*, Fig. 5.1,

© The Author(s), under exclusive license to Springer Nature Switzerland AG 2022
G. Kunstatter and S. Das, *A First Course on Symmetry, Special Relativity and Quantum Mechanics*, Undergraduate Lecture Notes in Physics,
https://doi.org/10.1007/978-3-030-92346-4_5

Fig. 5.1 Two inertial frames
O and O' moving at constant
speed relative to each other
see the same event but assign
to it different spacetime
coordinates (x, t) and
(x', t'), respectively

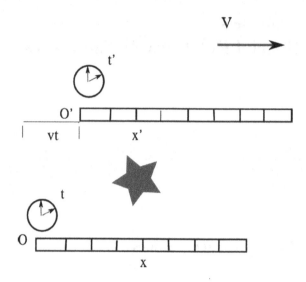

in which Newton's second law of motion $\mathbf{F} = m\mathbf{a}$ holds true, where $\mathbf{F}$ is the net
force acting on any object in that frame. Such frames move at constant velocity with
respect to each other, and also with respect to objects or other frames on which no
net force acts. An observer at rest or in uniform motion in such a frame is known as
an *inertial observer*. Inertial frames play a very important role in special relativity,
as we shall see below.

5.2.2 Spacetime Diagrams

Spacetime diagrams are essentially a series of snapshots of space at successive times
stacked on top of each other as illustrated in Fig. 5.2. Each point in the diagram
represents a location in space at some specific time that provides the possible location
in time and space of some event.[1] These points, or events, can be labelled by a set of
coordinates (t, x, y, z) in some inertial frame. Examples of possible events include
the switching on of a flashlight, collision of two billiard balls on a pool table, and a
supernova explosion in outer space.

The spatial coordinate (location) is given on the horizontal axis (or a set of horizontal axes if one considers two or three dimensional space) while the time coordinate
is specified on the vertical axis. Trajectories of moving objects map out curves on
the spacetime diagram and are called worldlines. Figure 5.3 shows the worldline of
a particle circling the origin on the x–y plane. Worldlines are essentially the same

---

[1] Some textbooks refer to the points on a spacetime diagram as events whether or not an event
in the physical sense occurs at that location and time. We may lapse into that more mathematical
terminology on occasion, but hopefully the meaning will be clear from the context.

Fig. 5.2 Successive snapshots of the motion of a particle in a single reference frame stacked on top of each other to form a spacetime diagram

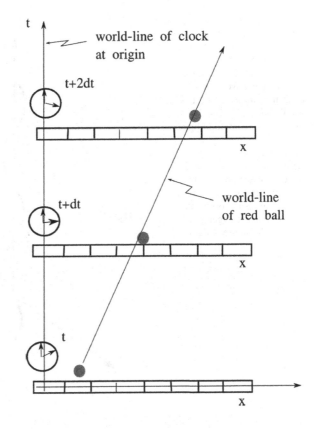

world-line of clock at origin

t+2dt

world-line of red ball

t+dt

t

as trajectories on graphs in first courses on Newtonian physics that plot the position, $x(t)$, of a particle as a function of time t. One difference and occasional source of confusion is that in spacetime diagrams time is plotted on the vertical axis in order to better illustrate the nature of spacetime as a series of successive snapshots.

As we will see, the worldlines of light rays have special significance in special relativity. In standard MKS units, the speed of light is 3×10^8 m/s so that if we used seconds on the vertical axis and meters on the horizontal axis, the trajectory of a light ray would essentially be horizontal on spacetime diagrams. This provides a graphic illustration of why we don't detect the effects of special relativity in our daily lives: the speed of light is virtually infinite relative to the speeds at which we normally travel. In order to work with situations for which special relativity is important, it is therefore useful to use units of length in which the numerical value of the speed of light is more manageable. 1 light-second (lt-sec) is defined as the distance that light travels in one second, namely 300 million meters. Similarly, 1 light-year (lt-yr) is the distance that light travels in one year, or 9×10^{15} m.

Using seconds and light-seconds as time and length units, respectively, the speed of light, by definition is precisely 1 lt-sec/s. This makes drawing spacetime diagrams involving light rays much simpler because the worldline of a light ray in these units

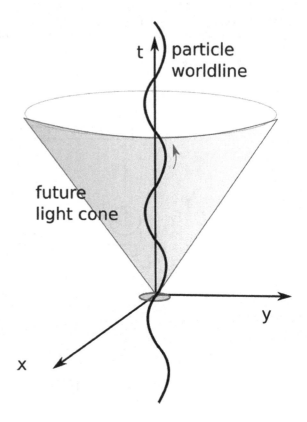

Fig. 5.3 Trajectory, or worldline, of a particle circling the origin on the x–y plane. The cone (yellow) shows the path in spacetime of a circular wave front emanating from the origin at $t = 0$ and then moving outward at the speed of light. It is the two space dimensions version of the light cone, and is discussed in detail in Sect. 6.3. In three space dimensions, the circles around the spatial origin are spheres that get larger with time. (Color figure online)

has slope equal to plus or minus one, (i.e. at 45° to both the horizontal and vertical axes). Worldlines of objects traveling less than the speed of light would have slopes whose magnitude is greater than one, while those traveling faster than the speed of light have slopes with magnitude less than one.

We will see that the speed of light is a fundamental constant in special relativity. It provides an ultimate speed limit for all forms of matter and energy. It is so important that it is given its own symbol, namely c. For the practical reasons stated above, we will often work in units in which $c = 1$, whether light-seconds per second, light-years per year, or something else.

Light years are particularly useful units for astronomy. Alpha Centauri, the nearest star to the Earth, is about 4 lt-yr from us. The centre of the Milky Way galaxy contains a supermassive black hole, Saggitarius $A^\star$, which is about 25,000 lt-yr from us. For solar system physics, light seconds are convenient. The Earth is about 8 lt-min, or 480 lt-sec from the Sun. For intergalactic scales, astronomers and cosmologists often use kiloparsecs (kpc). One kiloparsec is about 3,300 lt-yr. The galaxy is roughly in the shape of a flattened spiral disk about 30 kpc across.

A particularly useful concept in the context of spacetime diagrams is the *future light cone* associated with every point or event, P, in spacetime (see Sect. 6.3). It

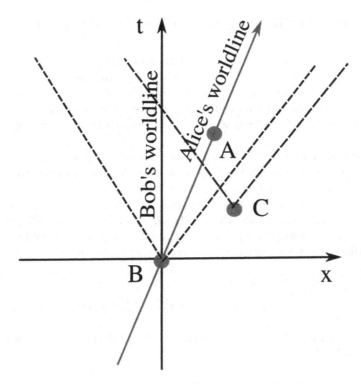

Fig. 5.4 Spacetime diagram showing two events A and B on the worldline (in blue) of an observer, Alice, moving at constant velocity relative to another observer, Bob, as seen in Bob's frame of reference. Bob is at fixed spatial coordinate. The dashed red lines are light rays emitted by two different events B and C. They mark the light cones associated with the points B and C. (Color figure online)

maps out the trajectories of all possible light rays, which together form a light front, emitted at that place and time.[2] Since as we will see in Chap. 6 special relativity implies that nothing can travel faster than light, light cones provide the boundary between events that can be influenced by an observer at P, and those that cannot. In three dimensional space, the light cone is a two dimensional sphere that grows with time. Since this is difficult to represent on a sheet of paper, Fig. 5.3 shows the light cone of the origin in two spatial dimensions. We will be dealing with only one space dimension in which case the light "cone" consists of two light rays moving in opposite directions, as in Fig. 5.4.

---

[2] The past light cone consists of light rays that could reach point P from the past. It is the boundary of events that could influence the event P.

5.2.3 Newtonian Relativity and Galilean Transformations

Newton and his contemporaries thought of the flow of time as absolute, the same everywhere in the Universe and at every epoch. Given sufficiently accurate synchronized clocks, all observers anywhere in the Universe could agree on the rate at which their respective clocks run, independent of their state of motion and when their measurements were made. Time and space were thus distinct and in no way dependent on each other.

Consider two observers, Alice and Bob, each holding a set of perpendicular rulers, i.e. Cartesian coordinates (see Fig. 5.5). We will denote the times and locations that they measure as $(t_A, \mathbf{r}_A)$ and $(t_B, \mathbf{r}_B)$, respectively. At time $t_A = t_B = 0$ they synchronize their clocks and Alice starts moving away from Bob with a constant velocity $\mathbf{v}_{AB}$. They observe two events, labeled (1) and (2), that occur at different times and different places as measured by both. According to Newton, the relationships between the coordinates of these events as measured by Alice and Bob are given by:

$$t_B^{(1)} = t_A^{(1)} =: t^{(1)} \qquad t_B^{(2)} = t_A^{(2)} =: t^{(2)}$$
$$\mathbf{r}_B^{(1)} = \mathbf{r}_A^{(1)} + \mathbf{v}_{AB} t^{(1)} \qquad \mathbf{r}_B^{(2)} = \mathbf{r}_A^{(2)} + \mathbf{v}_{AB} t^{(2)} \qquad (5.1)$$

The relative displacement, $\Delta\mathbf{r}$, and time elapsed, Δt, between the events as measured by Alice and Bob are:

$$\Delta t_B = \Delta t_A =: \Delta t \qquad (5.2)$$
$$\Delta \mathbf{r}_B = \Delta \mathbf{r}_A + \mathbf{v}_{AB} \Delta t \qquad (5.3)$$

The above transformations can also tell us how the velocity $\mathbf{v}_A$ of an object as seen by Alice is related to the velocity $\mathbf{v}_B$ of the same object as seen by Bob:

$$\begin{aligned}
\mathbf{v}_B &= \frac{\Delta \mathbf{r}_B}{\Delta t} \\
&= \frac{\Delta \mathbf{r}_A}{\Delta t} + \mathbf{v}_{AB} \\
&= \mathbf{v}_A + \mathbf{v}_{AB} \qquad (5.4)
\end{aligned}$$

Equation (5.4) matches our intuitive expectation that velocities add linearly. For example, Bob is on a train moving at 40 km/h towards Alice, who is standing at the train platform. If Bob throws a ball towards Alice at 20 km/h relative to the train (i.e. relative to his frame of reference), one expects that the ball is moving at $(40+20)$ km/h $= 60$ km/h as measured by Alice. If Bob throws the ball away from Alice at 20 km/h then one expects Alice to measure the ball's speed to be $(40-20) = 20$ km/h. This nice intuitive picture is consistent with Eqs. (5.1–5.4), which are known as *Galilean transformations*. It turns out that Galilean transformations are not consistent

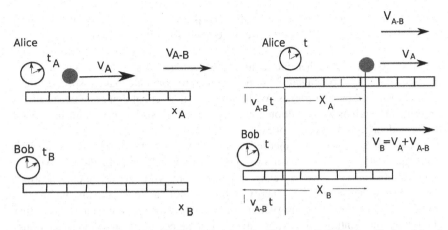

Fig. 5.5 Addition of velocities in Galilean relativity. The Figure on the left shows the location of the ball relative to both Alice's and Bob's frame at $t_A = t_B = 0$. The Figure on the right shows the location of the ball, and Alice's frame relative to Bob's at time $t > 0$

with special relativity[3] and have been proven experimentally to be incorrect when at least one of the velocities involved is close to the speed of light.

5.3 Derivation of Special Relativity

5.3.1 The Fundamental Postulate

Anyone who has taken a long, smooth airplane ride knows that in the absence of turbulence, there is no indication that one is hurtling through the atmosphere at close to 1000 km/h relative to the surface of the Earth. This observation can be elevated to the more general statement that any experiment done during such an airplane ride would yield precisely the same outcome as the corresponding experiment done on the ground,[4] or in any other inertial frame. This simple observation is, in essence, the basis for Einstein's statement of the *Fundamental Postulate* of special relativity:

The laws of physics are the same in all inertial frames.

This is a symmetry principle on par with the presumed invariance of the laws of physics under time translations, spatial translations and rotations.

Many discussions of the origin of special relativity stipulate that two postulates are required for its derivation. The first is the one that we refer to above as the

---

[3] Unless the speed of light is infinite, which it is not!.

[4] We are neglecting the rotation and gravitational pull of the Earth for the moment.

Fundamental Postulate. The second is that the speed of light is the same in all frames of reference. It is our view that the second essentially follows from the first, subject to experimental input. Sect. 5.3.2 demonstrates that given only the first postulate, one is led to a choice: either one requires the existence of an omni-present medium through which light propagates[5] or one is forced to conclude that the speed of light is the same in all inertial frames. Without further theoretical input one must rely on experiment to tell us which option is chosen by Nature. Such experiments have been performed and are the subject of Sect. 5.3.3.

In fact, further compelling theoretical input does exist. David Mermin, in a beautiful paper called "Relativity without light" [American Journal of Physics, Vol 54, p. 116 (1984)], derives special relativity purely from symmetry arguments. Mermin assumes only the following: If energy and momentum are conserved in one inertial frame, then they must be conserved in every frame. Noether's theorem tells us that energy and momentum conservation are a consequence of the time and space translation invariance of the equations of motion. Mermin shows that all one needs to assume is the Fundamental Postulate plus time and space translation invariance in one frame to derive Special Relativity. It is not necessary to consider light at all. It is amazing how far one can go starting from a basic symmetry assumption!

5.3.2 The Problem with Galilean Relativity

We will now show that Newtonian relativity and the resulting Galilean transformations are incompatible with the fundamental principle of relativity as stated above.[6] Consider the following set-up: Alice is in a spacecraft that is stationary relative to Bob. They are both in inertial frames. She sends a light ray from a mirror at the top of her spacecraft to a mirror directly below it. The mirrors are three light seconds apart (it is a big spacecraft). Both Alice and Bob measure 3 s for the light transit time between mirrors.

Alice now flies her spacecraft through the vacuum of space at speed $v = 4c/5$ (see Fig. 5.6). Bob sees the light ray move a horizontal distance of 4 lt-sec as it makes its way from the top of Alice's moving spacecraft to the bottom. Since Bob sees the light ray travel 3 lt-sec vertically and 4 lt-sec horizontally, the total displacement of the light ray, according to Pythagoras, is 5 lt-sec. At the speed of 1 lt-sec/s, the transit time of the light ray as measured by Bob must be 5 s. This suggests the following fundamental questions, the answers to which depend on whether one starts from Galilean relativity or the Fundamental Postulate of special relativity:

---

[5] Such an aether would define a preferred inertial frame, namely the one at rest with respect to the aether.

[6] GK is grateful to Richard Epp for sharing the following approach with us.

1. What time does Alice measure for the top to bottom light transit time?
   Answer(s):

   (a) Galilean physics requires that Alice measure 5 s, since the time between two events is absolute and must be the same for all observers.
   (b) The Fundamental Postulate requires that Alice still measure 3 s otherwise she would know that she had changed frames.

2. What does each of the answers in the question above imply about the speed of light that Alice measures?
   Answer(s):

   (a) If Alice measures 5 s, the velocity of light in her frame is:

   $$c_{Alice} = \frac{3 \text{ lt-sec}}{5 \text{ s}} = \frac{3c}{5} \tag{5.5}$$

   Thus, according to Galilean relativity light is moving more slowly in Alice's frame.

   (b) If Alice measures 3 s then she would conclude that the speed of light is:

   $$c_{Alice} = \frac{3 \text{ lt-sec}}{3 \text{ s}} = c = 1 \text{ lt-sec/s} \tag{5.6}$$

   This is the same as she measured when she was stationary relative to Bob and is consistent with the Fundamental Postulate. Otherwise, the different vacuum speeds of light in the two inertial frames could be used to allow Alice to deduce that she is moving, without having to refer to anything external to her spacecraft, thereby violating the Fundamental Postulate.
   On the other hand, for both Alice and Bob to measure the same speed of light when Alice is moving relative to Bob is contrary to our experience and to the Galilean law for the addition of velocities (Eq. 5.4). If in Sect. 5.2.3 Bob shines a flashlight at Alice instead of throwing a ball, then the Galilean addition of velocities (Eq. 5.4) implies that the speed measured by Alice would be: $c_A = c_B + 40 \text{ km/h}$. On the other hand, the Fundamental Postulate requires $c_A = c_B$.

To summarize the result of this thought experiment, there appears to be a choice: either Alice measures a different speed of light when she is moving and violates the Fundamental Postulate, or she measures the same speed of light and violates the principle of absolute time in Galilean relativity.

It must be said that there is in fact a loophole. If one supposes the existence of a medium permeating the vacuum of space (often referred to as an "aether") through which light propagates, then traditional wave mechanics states that the speed of the wave relative to the medium depends only on the medium's properties. This is true of the motion of sound waves through air, for example, or waves on a string. For such mechanical waves, the speed of propagation is independent of the motion of

Fig. 5.6 Alice's spacecraft
and light experiment as seen
in Bob's frame of reference.
Light ray moves straight
down as seen by Alice, while
Bob sees the vertical motion
plus a horizontal motion due
to the spacecraft's motion
relative to his frame

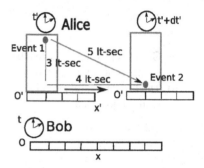

the source of the wave. The presence of such a medium would automatically imply
the existence of a preferred inertial frame, namely the one that is at rest relative to it.
Alice's ability to detect her motion relative to the aether by measuring the speed of
light would then not violate the Fundamental Postulate. For this argument to be valid,
however, the aether must be detectable by experiment. Yet, starting with American
physicists Michelson and Morley, more than a century of careful experiments have
failed to detect the existence of aether.[7]

5.3.3 Michelson-Morley Experiment

In 1887, Albert Michelson and Edward Morley performed the first careful experiment
to detect the Earth's motion through the hypothetical aether. In this interference
experiment, illustrated schematically in Fig. 5.7, a single light beam is split into two
beams that propagate to two mirrors placed at equal distances in orthogonal directions
from the beam splitter. The beams bounce back from their respective mirrors before
rejoining at the beam splitter and subsequently hitting a screen. If one of the two
beams takes longer to get back to the beam splitter, it will result in a relative phase
shift that causes the two beams to interfere destructively or constructively.[8]

Suppose that the lab is moving with the Earth at speed v to the right relative to the
aether, or equivalently that the aether is moving with a speed v to the left past the lab.
The basic assumption about the aether, as suggested by ordinary wave mechanics,
is that the speed of light c relative to the aether is independent of the motion of the
source or of the observer.

We use L to denote the distance the separated light rays moved between beam
splitter and mirror while v is the speed of the Earth through the aether. The portion

---

[7] If the aether explanation were correct, one would be tempted to ponder why Alice and Bob
happened to be in the aether's rest frame in the first place. Conversely, as Einstein supposedly asked
as a teenager, why do we never seem to find ourselves moving through the aether at close to the
speed of light, so that light would be essentially stationary in our frame.

[8] This is precisely the same type of interferometer that was used in September 2015 to detect
gravitational waves for the first time. See Sect. 7.6 and www.LIGO.org for details.

Fig. 5.7 Schematic diagram
of the Michelson-Morley
experiment

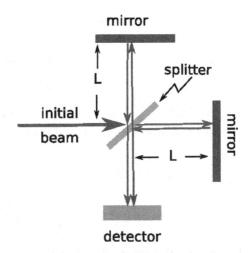

of the beam travelling parallel to the aether (to the right) has speed $c - v$ relative to the lab. On the return trip from the mirror it moves with the aether (i.e. to the left) and has speed $c + v$ relative to the lab. The time of flight to traverse a distance L to the right and a second distance L to the left is:

$$\Delta t_2 = \frac{L}{c - v} + \frac{L}{c + v} = \frac{2Lc}{c^2 - v^2} \tag{5.7}$$

The part of the beam that moves at right angles to the aether (and the lab) has speed v_{perp} given by:

$$v_{perp} = \sqrt{c^2 - v^2} \tag{5.8}$$

v_{perp} can be derived by first noting that during the time Δt the beam takes to get from the splitter to the top mirror, the lab has travelled a distance $v\Delta t$ in the horizontal direction. This is similar to the thought experiment involving Alice and Bob in Sect. 5.3.2. The total distance the light has travelled relative to the aether in getting from the splitter to the top is:

$$c\Delta t = \sqrt{v_{perp}^2 (\Delta t)^2 + v^2 (\Delta t)^2}$$
$$= \Delta t \sqrt{v_{perp}^2 + v^2} \tag{5.9}$$

Cancelling Δt from both sides and solving for v_{perp} gives the result in Eq. (5.8).

For the perpendicular beam the total time of flight from the splitter to the top mirror and back is therefore:

$$\Delta t_1 = \frac{2L}{v_{perp}}$$

$$= \frac{2L}{\sqrt{c^2 - v^2}}$$

$$= \frac{2L}{c}(1 - v^2/c^2)^{-1/2} \tag{5.10}$$

The difference in time of flight for the two perpendicular beams is

$$\Delta t_2 - \Delta t_1 = \frac{2Lc}{c^2 - v^2} - \frac{2L}{c}(1 - v^2/c^2)^{-1/2}$$

$$= \frac{2L}{c}\frac{1}{\left(1 - \frac{v^2}{c^2}\right)}\left(1 - \left(1 - v^2/c^2\right)^{1/2}\right) \tag{5.11}$$

For speeds v that are small compared to the speed of light, $(v/c) \ll 1$, we can use the Taylor approximation:

$$\left(1 - \frac{v^2}{c^2}\right)^a \sim (1 - a\frac{v^2}{c^2}) \tag{5.12}$$

for any a. Thus Eq. (5.11) reduces to:

$$\Delta t_2 - \Delta t_1 \sim \frac{2L}{c}\left(1 + \frac{v^2}{c^2}\right)\frac{1}{2}v^2/c^2$$

$$\sim \frac{Lv^2}{c^3} \tag{5.13}$$

where the v^2/c^2 term in the brackets has been dropped relative to 1. The horizontal beam therefore lags behind the perpendicular beam when they arrive at the detector by:

$$(\Delta t_2 - \Delta t_1)c = L\frac{v^2}{c^2} \tag{5.14}$$

and the fractional phase shift is:

$$\frac{\Delta\lambda}{\lambda} = \frac{(\Delta t_2 - \Delta t_1)c}{\lambda} = \frac{L}{\lambda}\frac{v^2}{c^2} \tag{5.15}$$

Since the effect is proportional to $(v/c)^2$, it is very hard to detect when $v/c \ll 1$. This is perhaps why the experiment was not attempted until the late nineteenth century. However for $v = 30$ km/s (the speed of the Earth around the Sun), $L = 10$ m (effective length between mirrors, enhanced via multiple reflections) and $\lambda = 6 \times 10^{-7}$ m, (the approximate wavelength of sodium light), this translates to $\Delta\lambda/\lambda \approx 0.4$, which is large enough to result in significant interference that would certainly have been detected had the aether existed.

Yet, meticulous experimentation stretching over day and night and various seasons of the year to produce the maximum shift and to eliminate the possibility that the Earth happens, like Bob in our thought experiment, to be in the aether's rest frame failed to produce any detectable fringe shift. The Michelson-Morley experiment has in fact been termed as the 'most famous failed experiment'. Subsequent attempts stretching over a century have reinforced this null result. The conclusion is that there is no detectable aether relative to which Alice can measure her motion. The principle of relativity therefore requires her to measure a time of 3 s, whether or not she is moving relative to Bob.

Conclusion: In the absence of a detectable aether, the speed of light is the same for all observers, irrespective of their motion with respect to the source or each other.

5.3.4 Maxwell's Equations

There exists an argument based on theory that supports the conclusion that Alice and Bob must measure the same speed of light for the Fundamental Postulate to hold. It too is related to the propagation of light. Maxwell's theory of electromagnetism is a beautiful and well-tested[9] theory that unifies electricity and magnetism within a single framework. It consists of a set of partial differential equations for the electric field, E, and the magnetic field, B, and contains two fundamental parameters. The first is the permittivity of free space, ϵ_0, that appears in Coulomb's law:

$$F_C = \frac{1}{4\pi\epsilon_0} \frac{q_1 q_2}{r^2} \qquad (5.16)$$

Its measured value is $\epsilon_0 = 8.854 \times 10^{-12}$ Coulombs squared per Newton meter squared ($C^2/(N\ m^2)$). The second is the magnetic permeability of free space, μ_0, which appears in Ampere's force law relating the magnetic force per unit length between two very long, straight parallel wires a distance r apart, each carrying a current I in the same or opposite direction. The magnitude of the force per unit length between them is:

$$\frac{F_m}{L} = \frac{\mu_0}{2\pi} \frac{I^2}{r} \qquad (5.17)$$

The force is repulsive or attractive when the currents run in the same or opposite directions, respectively. The experimental value of the permeability of free space is $\mu_0 = 4\pi \times 10^{-7}$ Newton's per Amper squared (N/A^2).

---

[9] If it did not survive the tests, the beauty of the theory would be irrelevant. Thus, in a sense, this argument based on Maxwell's theory is based on experiments, similar to those of Michelson and Morley.

A relatively straightforward manipulation of Maxwell's equations reveals that electromagnetic waves travel in vacuum at a speed determined by the fundamental constants ϵ_0 and μ_0:

$$c = \frac{1}{\sqrt{\epsilon_0 \mu_0}} = 3.0 \times 10^8 \text{ m/s}. \tag{5.18}$$

Since μ_0 and ϵ_0 are fundamental constants in Maxwell's theory, then according to Eq. (5.18), so too is c, the speed of light. As in Sect. 5.3.2, we are left with two options: Either the relationship between c, μ_0 and ϵ_0 and consequently Maxwell's equations look very different in different inertial frames, violating the Fundamental Postulate, or the speed of light must be the same in all inertial frames, independent of the motion of the emitter or the detector.

Physicists in the early 20th century preferred the first option, which would be the case if electromagnetic waves propagated through an all pervasive aether. As we saw, the experiment by Michelson and Morley to detect Earth's motion through the aether proved unsuccessful, making the second option, the one advocated by Einstein, inevitable. It is a matter of some controversy as to whether Einstein was influenced by, or even knew about, the Michelson-Morley experiment, or whether he derived special relativity, with all its wonderful consequences, purely from his conviction that all laws of physics, including Maxwell's theory, must be the same for all inertial observers.

5.4 Summary of Consequences

In the previous section we showed that the Fundamental Postulate of relativity led to the conclusion that the speed of light is the same for all inertial observers. This seemingly innocuous statement has many startling consequences. The first of these is that the more intuitive Galilean transformations in Sect. 5.2.3 cannot be right because they imply the linear addition of velocities, Eq. (5.4). If a light ray were moving at a speed, c, relative to Bob, the Galilean transformations imply that it would be travelling at speed $c - v$ relative to Alice if she were moving at speed v in the same direction as the light ray. Einstein realized that if v were close to the speed of light, then Alice would be able to see an almost stationary light ray. This is contrary to Maxwell's equations, which govern all aspects of electromagnetism and light and do not admit solutions describing nearly stationary light waves. Perhaps more importantly, slowly moving light rays have never been observed.

Other more specific and surprising consequences include:

1. *The Relativity of Simultaneity*: observers in different inertial frames do not agree on whether two events occur at the same time.
2. *Time Dilation*: observers in different inertial frames see each other's clocks ticking more slowly than their own identical clock.

3. *Length Contraction*: a moving meter stick appears to be shorter than an identical meter stick at rest.
4. The need to unify space and time into a single spacetime continuum.
5. The equivalence of mass and energy, summarized in the famous formula $E = mc^2$.

We will now elaborate on the first three consequences, while the last two will be discussed in more detail in Chap. 6.

5.5 Relativity of Simultaneity

The following thought experiment[10] illustrates that the constancy of the speed of light causes observers in different inertial frames to disagree on whether two events occurred at the same time. Consider two events, labelled P and Q, that occur at equal distances from Bob, as measured in his frame. Given the constancy of c, the light from both events reaches Bob simultaneously, forcing him to conclude that P and Q happened at the same time. Now imagine that Alice is moving relative to Bob from P towards Q, but happens to be at the same place as Bob precisely at the time in his frame that the light leaves the two events. This is illustrated in Fig. 5.8. Since Alice is moving towards the light emitted by event Q, she will necessarily be past Bob and have seen event Q before the light ray from P catches up with her. In other words, Alice will see event Q happen before she sees event P (see Fig. 5.9), and conclude that Q happened first. Bob and Alice disagree on whether or not the two events occur simultaneously. The crucial point here is that since they measure the same speed of light, there is no way to determine which of them is moving. Thus both their conclusions are equally valid in their respective frames, and we must conclude that the statement "two events happened simultaneously" has no objective meaning.

It follows directly from the above considerations that if a third observer, Charlie, were moving relative to Bob in the opposite direction to Alice, that is from Q towards P, then she would conclude that *P* happened first. Charlie and Alice disagree as to which event occurs first. This is an unavoidable consequence of the constancy of the speed of light. Thus, in some cases, two inertial observers can even disagree on which of the two events, P or Q, happened first. The *caveat* "in some cases" is necessary, because this statement only holds for events that occur so far apart in space and so close together in time that one would have to travel faster than the speed of light to be present at both events. Such events are said to be *spacelike separated*.

---

[10] Einstein came to many of his insights in both special and general relativity by performing thought (or "gedanken" after the German expression) experiments. Gedanken experiments attempt to think through the consequences of simple, conceptually possible situations that in some cases may be difficult or impossible to realize in the lab.

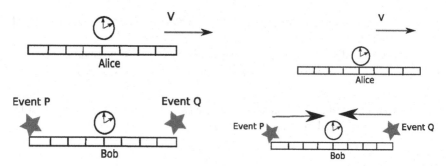

Fig. 5.8 Relativity of simultaneity. Alice intercepts light from Event Q before that of Event P. The light moves towards her at the usual speed c and she concludes that Q happened before P

Fig. 5.9 Relativity of simultaneity illustrated on a spacetime diagram. Two events P and Q that occur at equal distances from Bob emit light rays that reach him at the same time. These same light rays intersect Alice's worldline at different times because of her motion towards the location of event Q. She sees Q first followed by P. On the other hand, Charlie, who is moving towards P, sees event P as happening first

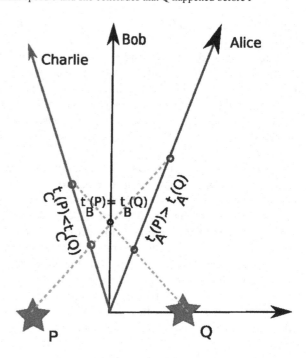

5.5.1 Surface of Simultaneity: What Time Did It Happen?

When we draw the x axis on a spacetime diagram we take for granted that the x-axis corresponds to "space" at the time $t = 0$. A horizontal line represents a snapshot of space at the time the horizontal axis crosses the time axis. For example, a horizontal line crossing the t axis at, say, $t = 3{:}00$ pm corresponds to the location in spacetime of all possible events that happen at 3:00 pm as marked on a clock sitting still at the location $x = 0$. It is said to be Bob's surface of simultaneity at 3:00 pm.

The following question arises: How would Bob determine that an event occurred at 3:00 pm in his frame? More specifically, how could it be done without moving from $x = 0$? The answer is simple, at least in principle if not in practice. Suppose Bob sends out a laser beam at 2:00 pm that causes a nearby asteroid to explode. The light from the explosion reaches him at 4:00 pm. Assuming that the speed of light is the same in both directions,[11] Bob must conclude that the explosion occurred at 3:00 pm, mid-way between 2:00 pm and 4:00 pm because the light necessarily travelled an equal distance going and coming.

This example suggests the general procedure by which any observer, including Alice, can in principle map out all the events in spacetime that coincide with a particular time on their clock. The locations in spacetime of all these events, or possible events, is called the *surface of simultaneity*.

Alice must send and receive light rays in order to observe an event. We again assume that the speed and hence the time of flight is the same going and coming. If at some time $t_A(1)$, Alice sends out a light ray that bounces off event P and returns to Alice at time $t_A(2)$, then Alice must conclude that event P occurred at time $t_A(P) = (t_A(1) + t_A(2))/2$. By sending out a series of pulses and waiting for them to come back Alice can map out her surface of simultaneity for any given time. This is illustrated in Fig. 5.10. The key thing to note here is that despite her motion relative to Bob, both Alice and Bob measure the speed of light relative to them to be the same. This leads to the conclusion that the time measured by Alice for a given event (the surface of simultaneity on which it lies) is different from that of Bob, and more importantly that both points of view are equally correct.

Note that the surface seen by Alice to be simultaneous with an event at $t = 0$, $x = 0$, subtends an angle equal and opposite to the angle of Alice's worldline relative to the path of a light ray emanating from the origin. This is a geometrical consequence of our choice of coordinates in which $c = 1$, so that light travels at $45°$ relative to the x-axis as will be explained in Sect. 6.2.3.

5.6 Time Dilation

We will now discuss the situation described in Sect. 5.3.2 more generically and derive the relationship between time intervals as measured by observers moving with respect to each other at constant velocity.

5.6.1 Derivation

Alice is in a spaceship of height L that moves at speed v relative to Bob. Their clocks were synchronized to run at the same speed before Alice took off, when they were both in the same inertial frame. A pulse of light moves from the top of the box to the bottom as shown in Fig. 5.11.

---

[11] We expect this to be true due to symmetry, but it is surprisingly difficult to prove experimentally.

Fig. 5.10 By sending out light rays, Alice observes that two events P and Q occur simultaneously at time $t_A(P) = t_A(Q)$, halfway between when she sends the light rays and receives the reflected rays. With enough light rays, Alice could map out all events that happened at $t_A(P)$, i.e. simultaneously with the event P. These events define a surface in spacetime that is called a *surface of simultaneity* for Alice. It is in effect the Universe at a particular instant in Alice's time, $t_A(P)$. If the same light rays originate and end on Bob's worldline, as shown in the Figure, the same logic forces him to conclude that the events P and Q happened at $t_B(P)$ and $t_B(Q)$, respectively, such that $t_B(P) < t_B(Q)$. Bob sees event P happening before event Q

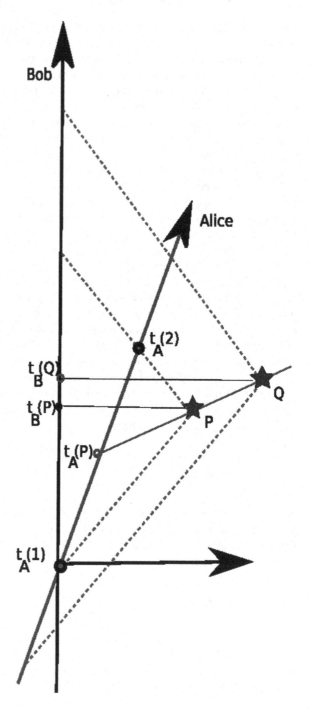

Fig. 5.11 Time dilation

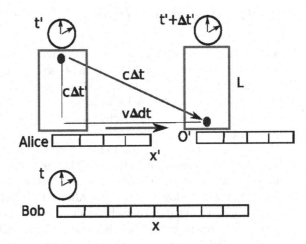

If $\Delta t'$ is the time measured by Alice between the light leaving the top of her spacecraft and reaching the bottom, then

$$L = c\Delta t' \tag{5.19}$$

Let's say that the time elapsed between these same two events as measured by Bob is Δt. During this time the spaceship has moved a horizontal distance $v\Delta t$. From Bob's point of view the light pulse has moved a total distance

$$D = \sqrt{L^2 + v^2 \Delta t^2} \tag{5.20}$$

The key point is that Bob and Alice both measure the speed of light to be the same, namely c, so that:

$$D = c\Delta t \tag{5.21}$$

Putting Eqs. (5.19), (5.20) and (5.21) together yields a relationship between Δt and $\Delta t'$:

$$c\Delta t = \sqrt{c^2 \Delta t'^2 + v^2 \Delta t^2}$$
$$\rightarrow \Delta t = \frac{1}{\sqrt{1 - \frac{v^2}{c^2}}} \Delta t'$$
$$\rightarrow \Delta t = \gamma \Delta t' , \tag{5.22}$$

where we have defined the *Lorentz factor*

$$\gamma = +(1 - v^2/c^2)^{-1/2} \tag{5.23}$$

The Lorentz factor plays a very important role in special relativity.

5.6.2 Properties of Time Dilation

1. As long as v is less than c, the Lorentz factor, γ, is a real number greater than one. This implies that from Bob's point of view, Alice's clock is running slowly. For example if $v = \sqrt{3}c/2$, then $\gamma = 2$ so it appears to Bob that it takes 2 s of his time for Alice's clock to tick 1 s. In fact, since we can just as well refer to Alice's and Bob's respective biological clocks, it is also true that Bob sees Alice as aging more slowly.

2. Alice's clock is moving relative to Bob so he sees it running slowly. On the other hand Bob's clock is moving relative to Alice and the Fundamental Postulate does not allow us to distinguish Bob's frame from that of Alice. Thus, Bob's clock must also appear to run slowly as seen by Alice. The effect is in this sense perfectly symmetrical. To convince yourself that this must be true simply redo the argument from Alice's frame. In this case, Bob must have a clock that is analogous to that of Alice's: two mirrors between which a beam of light moves.

   Figure 5.12 illustrates graphically the symmetry of time dilation, even though it does not appear symmetric at first glance when viewed in one frame or another. It also shows that the scale along Alice's time axis cannot be the same as along Bob's time axis, since $t = 0.9$ s lies along the hypotenuse of a right triangle whose opposite side, as measured in Bob's frame, is 1 lt-sec in length. This provides a hint that the geometry of spacetime is not Euclidean. This will be discussed in more detail in Chap. 6.

3. If $v/c \ll 1$, then $\gamma \approx 1 + \frac{1}{2}\frac{v^2}{c^2}$. This relation can be used to estimate time dilation for speeds that are small compared to the speed of light.

4. As v approaches c, γ approaches infinity.

5. When $v > c$, γ is imaginary (the square root of a negative number). This suggests that relative speeds greater than c are not possible in special relativity. The speed of light is the ultimate speed attainable by any form of matter or energy. There is an interesting loophole however. Mathematically, it is consistent to have an imaginary Lorentz factor if the mass of the object is also an imaginary number. Bizarre as it sounds, special relativity does allow the possible existence of *tachyons*, particles with imaginary mass that travel faster than the speed of light and cannot be slowed down below the speed of light. If tachyons existed and were able to interact with ordinary matter, it would be possible to communicate with the past. Although no evidence for the existence of tachyons exists, proposals for their experimental detection were made as recently as 2015.[12]

---

[12] Tony E. Lee, Unai Alvarez-Rodriguez, Xiao-Hang Cheng, Lucas Lamata, Enrique Solano, "Tachyon physics with trapped ions", Phys. Rev. A 92, 032129 (2015).

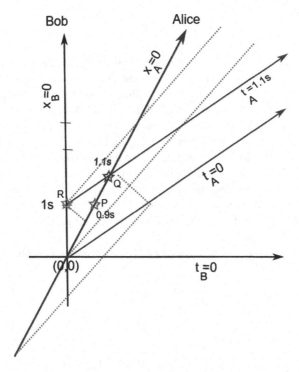

Fig. 5.12 This figure shows the symmetric effects of time dilation as seen in Bob's frame of reference. Alice moves past Bob at $v = 2c/5$, with $\gamma = 1.1$. Their worldlines intersect at the event O where they synchronize clocks. Event R corresponds to the ticking of Bob's clock at 1 s. Events P and Q correspond to Alice's clock ticking at 0.9 s and 1.1 s, respectively. Bob's surfaces of simultaneity are horizontal. Bob sees Alice's clock tick 0.9 s simultaneously with his clock ticking 1 s. He therefore concludes that her clock is slow. Alice's surfaces of simultaneity with event P and Q are represented by the two blue lines with slope less than one. Dotted (red) lines are light rays. The ones to the left of Alice's worldline represent the light rays that Alice uses to conclude that the tick of Bob's clock at 1 s (marked 1 s on Bob's worldline) is simultaneous with the tick of her clock at 1.1 secs. Alice therefore concludes that Bob's clock is slow. (Color figure online)

5.6.3 Proper Time

The time interval along the worldline of a particular observer as measured by a clock that is stationary in that observer's frame is called the observer's *proper time*. It is usually denoted by the Greek letter τ. For example, if Alice were wearing a wristwatch, it would measure her proper time as would her biological clock, (i.e. how fast she ages). A clock stationary in Bob's frame measures his proper time. As described in Sect. 5.6 their clocks appear to each other to run at different rates, so for example Alice would see Bob aging more slowly and *vice versa*.

The proper time between two nearby events can be measured in any frame O, say. It is given by:

$$\Delta\tau := \frac{1}{c}\sqrt{c^2\Delta t^2 - \Delta x^2} \tag{5.24}$$

where Δt and Δx are the time and distance between the events as measured in O. Note that expression Eq. (5.24) only yields a real number if $\Delta x \leq c\Delta t$, which means an observer can travel from one event to another at speeds less than that of light. If the proper time between two events A and B is real and non-zero, the events are said to be *timelike separated*. This will be the case if A is in B's light cone, or vice-versa. (Recall the definition of light cone in Sect. 5.2.2). If A is in B's future light cone it is said to be in the future of B, otherwise B is in the future of A. If on the other hand the proper time is imaginary, this implies that A and B cannot be connected by the worldline of an object moving at less than the speed of light, and they are said to be spacelike separated. This means that there is no definite answer to the question: which happened first, A or B? It depends on the frame of reference of the observer, as seen in Sect. 5.5.

For an observer moving at constant speed between the two events,

$$\frac{\Delta x}{\Delta t} = v = \text{constant} \tag{5.25}$$

so that

$$\begin{aligned}
\Delta\tau &:= \frac{1}{c}\sqrt{c^2\Delta t^2 - \Delta x^2} \\
&= \sqrt{1 - \left(\frac{\Delta x}{\Delta t}\right)^2}\,\Delta t \\
&= \frac{1}{\gamma}\Delta t
\end{aligned} \tag{5.26}$$

as expected from previous discussions of time dilation.

5.6.4   Experimental Confirmation of Time Dilation

Muons are a type of elementary particle created by energetic cosmic ray collisions in our upper atmosphere a few kilometers above the Earth's surface. Muons are unstable, and only live about $2.2\,\mu s$ (2.2 millionths of a second or $2.2 \times 10^{-6}\,s$) before decaying into an electron (or positron) and two neutrinos. This lifetime has been accurately measured in laboratory experiments on Earth. In these experiments, the muons are moving slowly so that they are almost at rest relative to the lab. Muons created by cosmic ray collisions in the atmosphere have much more kinetic energy than those created in the lab and travel at very high speeds, about $0.99c$ relative to the Earth. Since they only live $2.2\,\mu s$ before decaying, one might expect that between creation and decay they travel a distance:

$$L' = 0.99c \times 2.2 \times 10^{-6}s = 660 \text{ m} \qquad (5.27)$$

If this were true, we would never detect cosmic muons on Earth but in fact we do. The explanation comes from time dilation, which tells us that in the Earth's frame of reference, the muon's lifetime is measured to be: $\Delta t = \gamma \Delta t'$, where $\gamma = \frac{1}{\sqrt{1-(0.99)^2}} \sim 7$. Muons appear to live seven times as long in Earth's frame and travel 7×660 m ~ 4600 m. This is more than enough to reach the Earth's surface at their speed. The detection of cosmic muons on Earth therefore provides striking confirmation of the time dilation effect.

5.6.5 Examples

Exercise 1 A trip to the grocery store:

You travel to a store 5 km away from your home in a car moving at 100 km/h = 30 m/s.

1. What is the speed of light in km/h?
2. How long does your journey appear to take as measured by your friend sitting at home?
3. How much proper time elapses for you during the journey?

Exercise 2 Two astronauts travel to Mars at 0.001 c. The distance to Mars is about 55 million km. How long would the trip take from NASA's point of view. How much would the astronauts age? (i.e. What is the proper time elapsed during the journey?)

Example 1 The Twin Paradox

Two twins, Ashley and Mary-Kate are turning 30 years old. Ashley decides to celebrate by spending her fortune on a trip to Alpha Centauri, located four light years from Earth. Her spaceship travels at 0.8c. The other twin, Mary-Kate stays at home. It turns out that Alpha Centauri is not particularly exciting so Ashley only stays a few hours before returning to Earth. Once safely back on Earth, she attends a welcome home party.

1. What is the Lorentz factor γ at $v = 0.8c$?
   Solution:

$$\gamma := \frac{1}{\sqrt{1-\frac{v^2}{c^2}}}$$
$$= \frac{1}{\sqrt{1-(0.8)^2}}$$
$$= 1.7 \qquad (5.28)$$

2. How long does the return trip appear to take to Mary-Kate?
Solution:

$$t = \frac{\text{distance}}{\text{time}} = 2 \left[\frac{4\,\text{lt-yr}}{0.8\,\text{lt-yr/year}} \right] = 10 \text{ years} \tag{5.29}$$

3. How long does the return trip appear to take for Ashley?
Solution:

$$t' = \frac{t}{\gamma} = \frac{10}{1.7} \text{ years} = 6 \text{ years} \tag{5.30}$$

4. Which of the two twins is older when they meet up again at Ashley's welcome home party? Is there a paradox?
Solution:
Mary-Kate does indeed age 10 years to Ashley's 6 years, but there is no paradox here. The time dilation formula, and in fact all special relativistic transformation laws that we will learn, apply only to observers in inertial (non-accelerating) frames. As long as Ashley and Marie-Kate both remain in inertial frames, they cannot synchronize their clocks more than once so there is no contradiction in stating that Ashley sees Marie-Kate age more slowly and vice-versa. In order to meet up at the welcome home party, one or both of them must accelerate or decelerate for a while. In this case it is Ashley who decelerates to stop at Alpha Centauri, accelerates to return to Earth and then decelerates to be able to attend the party. She is therefore not in an inertial frame throughout the journey and her path through spacetime (worldline) is very different from that of Mary-Kate. The situation is therefore not as symmetric as it appears at first glance, so it is no surprise that one ages more. The interesting fact is that it is the twin who accelerates and decelerates who ages less. This is a generic feature for accelerating observers and will be explained further when we discuss general relativity in Chap. 7.

To reiterate, the amount one ages is determined by the proper time along the particular worldline one travels in spacetime. Ashley's trip to Alpha Centauri and back traverses a worldline for which the proper time is less than that of Mary-Kate's (even though it doesn't seem to look that way on a spacetime diagram). As long as the period of acceleration is brief so that it doesn't contribute much to the total proper time, their difference in ages is well approximated by the time dilation formula Eq. (5.22).

Example 2 Global Position (GPS) Navigation Systems:
GPS was developed by the United States Department of Defence. It consists of a network of twenty-four satellites orbiting approximately 20,000 km above the Earth's surface. They move at about 14,000 km/h, so that one orbit takes about 12 h. By triangulating your position using at least four satellites, the GPS system can locate your transmitter to an accuracy of about 10 m or less. In order to achieve this kind of accuracy, the positions of satellites must be known very accurately, and clocks on

the satellites must be synchronized to within about 30 nanoseconds (30×10^{-9} s). Atomic clocks are used because they are accurate to 1 nanosecond. To obtain this accuracy it is necessary to take into account relativistic effects, particularly time dilation.[13] At an orbital speed of 14,000 km/h, this means that the clocks on the satellites run about 7 micrometers slower per day as compared to the identical clocks at rest on Earth. This is a huge relativistic effect given the accuracy needed!

5.7 Lorentz Contraction

5.7.1 Derivation of Lorentz Contraction

As described in Sect. 5.6.4, the muon and the Earth must agree on their relative velocity. Since the muon measures its time of flight to be different than that measured on Earth, the distance it thinks it travels during that time must be different as well. This difference in distance as measured in the two frames is a special relativistic effect called *Lorentz contraction*, which can be derived generally as follows.

Alice again flies past Bob in a spaceship at speed v. The time and distance she travels as measured by Bob are Δt and L, respectively, while in Alice's frame they are $\Delta t'$ and L'. Since any two observers O and O' must agree on their relative speed

$$v = \frac{L}{\Delta t} = \frac{L'}{\Delta t'}$$
$$\rightarrow L' = \frac{\Delta t'}{\Delta t}L = \frac{1}{\gamma}L \tag{5.31}$$

5.7.2 Properties

1. Since the Lorentz factor is always greater than one, the distance, L', measured by Alice is less than the distance, L, measured by Bob. This is referred to as *Lorentz contraction*.
2. As with time dilation, the effect is completely symmetric. Lengths of objects at rest with respect to Bob appear shortened to Alice while objects at rest in Alice's frame are shortened as measured by Bob.

---

[13] In fact, corrections due to the general theory of relativity (i.e. Einstein's theory of gravity) must also be incorporated, and give a bigger effect than special relativity.

5.7.3   Proper Length and Proper Distance

In the above transformation law L is the length of an object, such as a meter stick, at rest in Bob's frame. L is then called the *proper length* of the meter stick because it is the length as measured in its own frame. As both the formula (5.31) and the name Lorentz "contraction" imply, the meter stick appears shorter to anyone flying past it.

The *proper distance* between two events is analogous to proper time, although a bit more difficult to conceptualize. It is the distance that is measured between two events by an inertial observer for whom the two events happen at the same time (that is, are simultaneous). In terms of equations, the proper distance between any two events as measured by Alice, say, is

$$\Delta D := \sqrt{(\Delta x')^2 - c^2(\Delta t')^2} \tag{5.32}$$

where $\Delta t'$ and $\Delta x'$ are the time and spatial distance between the events in Alice's frame. You will note that this definition only yields a real number if $\frac{\Delta x'}{\Delta t'} > 1$, which implies that you would have to travel faster than the speed of light to get from one event to the other. If two events are separated by a real, non-zero proper distance, then they are said to be *spacelike separated*.

The spacetime diagram illustrating Lorentz contraction is shown in Fig. 5.13. Alice's surface of simultaneity at $t' = 0$ is different from that of Bob. The meter stick as seen by Bob at $t = 0$ is depicted as the thick line along the x axis, while the same meter stick as seen by Alice at $t' = 0$ is depicted by the thick line along the x' axis. Although the length, L', as measured by Alice appears longer than the length, L, as measured by Bob, this is deceptive because our eyes don't take into account another consequence of Lorentz contraction, namely that distances along Alice's surface of simultaneity are not the same as distances along Bob's surface of simultaneity. The solid black hyperbola in the diagram is the locus of all spacetime events that would be measured by Alice to be a *proper distance*, L', from the origin ($t' = 0$, $x' = 0$) of her coordinate system. The end of Bob's meter stick, as seen in his frame, has a proper distance $L > L'$ away from Alice's origin in her frame. We will learn more about proper distance and how all this works in Chap. 6.

Example 3 How long does Earth's atmosphere appear to the muon?
Solution:

$$L' = \frac{L}{\gamma} = \frac{4600\,\text{m}}{7} = 660\,\text{m} \tag{5.33}$$

which is how far the muon can be expected to go in its frame of reference.

Example 4 At rest, a ladder 4 m long just fits into a barn that is also 4 m long in its rest frame. A sceptical physics student wants to test Lorentz contraction by running towards the open door of the barn from the left at 0.99c holding the left end of the

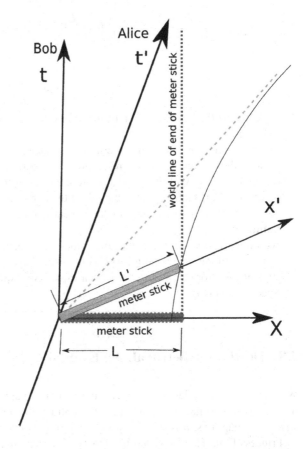

Fig. 5.13 Meter stick at $t = 0$ in Bob's frame (L). Same meter stick as seen at $t' = 0$ by Alice (L')

ladder. She knows that the barn should appear shorter as she runs so she expects the end of the ladder to hit the far end of the barn before she gets to the door. A mouse sitting in the barn is watching the proceedings. Having sat in on a couple of relativity classes, the mouse knows that the ladder appears shortened in its frame, so the student should reach the door before the end of the ladder hits the end of the barn.

1. How long is the barn in the student's frame of reference?
   Solution:
   Let L' be the length of the barn in its rest frame.

$$\gamma = \frac{1}{\sqrt{1 - \frac{v^2}{c^2}}}$$

$$= \frac{1}{\sqrt{1 - (0.99)^2}}$$

$$= 7.1 \tag{5.34}$$

$$L = \frac{L'}{\gamma}$$
$$= \frac{4}{7.1}$$
$$= 0.14\,\text{m} \qquad\qquad\qquad (5.35)$$

2. How long is the ladder in the mouse's frame of reference?
   Solution:
   Since the effect is symmetric the mouse measures the ladder to be 0.14 m, the same as the student measures the length of the barn.
3. Given the above, will the student get to the door of the barn before, or after the other end of the ladder hits the far end of the barn? In other words, is the student correct in her assessment about which event will occur first, or is the mouse?
   Solution:
   As we learned in Sect. 5.5 the time ordering of distant events depends on the frame of reference, so the answer as to which happens first is different for the student than for the mouse. This is a direct consequence of the properties of Special Relativity.

5.8 Death Star Betrayal: An Example

We now describe in detail a scenario that contains many of the key ingredients of our discussion so far, including Lorentz contraction, surfaces of simultaneity and spacetime diagrams. It is in effect a variation of the ladder in the barn problem.

Princess Leia, Han Solo and Chewbacca are planning to attack the Death Star. Solo knows from the last time he was captured that his spaceship, the Millennium Falcon, has exactly the same proper length as a hangar on the Death Star that is open at both ends. His plan is to have Chewbacca fly the Millennium Falcon through the hangar at 0.9c so that he and Princess Leia can jump onto the Death Star firing their laser weapons. He tells Leia to stand at the front of the spacecraft and fire the instant the front of the spaceship gets to the far end of the hangar. He will stand at the back of the spaceship and fire the instant he gets to the back of the hangar. Since the ship and the hangar are the same length in their respective rest frames, he tells Leia that they will fire at precisely the same instant and surprise the guards.

Unknown to Leia, Solo has taken special relativity in space cadet school and is trying to trick her. He knows that the hangar will appear shorter in the Millennium Falcon's frame because of the speed at which they are moving relative to the Death Star, so that Leia will get to the far end of the hangar before him and fire first, drawing the guards' fire and making things safer for him. Leia thinks about it for a minute and agrees. Was this a mistake?

Answer: No. Princess Leia, who also took relativity in school but got an A as opposed to Solo's C+, knows that in the Death Star frame of reference, the spaceship appears

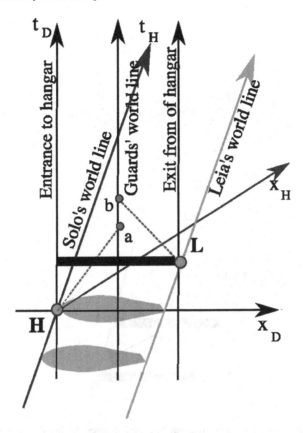

Fig. 5.14 Death Star frame of reference. The blue line denotes Han Solo's worldline, while the green line is Princess Leia's worldline. Black lines denote worldline of entrance to hangar (left), guards at center of hangar and exit to hangar (right). The letter H marks the event that Han Solo first fires his laser, while the letter L denotes the time and place at which Leia fires her laser. The Death Star hangar is shown (in black) at the instant in time (in the Death Star's frame) that Leia opens fire. Pictured in green is the Millennium Falcon when it is part way in the hangar ($t_D < 0$) and when Solo opens fire $t_D = t_H = 0$, also as seen in the Death Star frame. Solo clearly fires first in this frame. Note that in the Death Star frame the Millennium Falcon approaches from the negative x-axis. The Death Star and Millennium Falcon have been thickened along the time axis for effect. They should really be represented as one dimensional lines. (Color figure online)

shorter, so the guards will see Han Solo get to the back end of the hangar and fire first. He will therefore draw their fire, not Princess Leia.

The spacetime diagrams describing the attack on the Death Star in the Death Star frame and the Millennium Falcon frame are shown in Figs. 5.14 and 5.15, respectively. The blue line denotes Han Solo's worldline while the green is Princess Leia's worldline. Vertical (black) lines denote the worldlines of the entrance to the hangar (left), the guards in the hangar (center), and the exit to the hangar (right). The letter

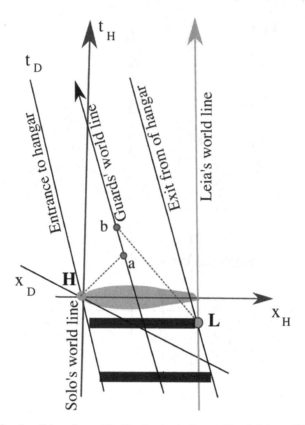

Fig. 5.15 Millennium Falcon frame. The blue line again denotes Han Solo's worldline, green line is Princess Leia's worldline. Black lines denote worldline of entrance to hangar (left), guards at center of hangar and exit to hangar (right). The letter H marks the event that Han Solo first fires his laser, while the letter L denotes the time and place at which Leia fires her laser. Pictured in green is the Millennium Falcon at the time Solo first fires his laser as seen in the Millennium Falcon frame. Also pictured (in black) is the Death Star hangar at two successive times (in the Millennium Falcon frame). The first is after Leia is already in the hangar but before she fires her laser. The second is at the instant, in the Millennium Falcon frame that Leia fires her laser. Note that as seen by Han Solo and Princess Leia the Death Star approaches them from the positive x-axis. Leia clearly fires before Solo in this frame but the guards get hit by Solo's laser first anyway because they are moving towards Solo and away from Leia so Solo's laser doesn't have as far to travel. (Color figure online)

H marks the event corresponding to Han Solo reaching the back of the hangar, while the letter L denotes the time and place at which Leia reaches the front of the hangar. The red dotted lines are the laser beams fired by Solo and Leia. The events a and b correspond to Solo's and Leia's beams, respectively, hitting the guards.

You will notice that Princess Leia arrives first in Han Solo's frame while Solo arrives first in the Death Star frame. In both frames, Solo's lasers hit the guards first. This can be understood intuitively in the Millennium Falcon frame by noting that in this frame the guards are moving away from Leia's laser beam and towards those of

Han at the moment they fire. The key is that the beams themselves move at the same speed in both frames.

We end this example and the Chapter by discussing the above result in the context of the relativity of simultaneity (see Sect. 5.5), which states that the time ordering of events could differ for different observers. In fact this is only true if the events are spacelike separated, as defined above. The events H and L on the spacetime diagrams Figs. 5.14 and 5.15 are spacelike separated, so the guards on the Death Star disagree with Han Solo as to which happens first. On the other hand, the events corresponding to Solo's and Leia's laser beams hitting the guards are timelike separated. This is evident from the fact that the guards' worldlines intersect both. Everyone must agree which happens first. This illustrates that special relativity is not a purely academic endeavour, but can have dire consequences!

Chapter 6
Special Relativity II: In Depth

6.1 Learning Outcomes

Conceptual

- Properties of general Lorentz transformations.
- Physical interpretation of light cones.
- Definition and physical interpretation of timelike, spacelike and null separated events.
- Lorentz transformations as a symmetry of Nature.

Acquired Skills

- Drawing spacetime diagrams and using them to solve problems.
- Use of full Lorentz transformations (in one space dimension) to solve problems.
- Solving simple relativistic scattering problems.

6.2 Lorentz Transformations

6.2.1 Derivation of General Form

Time dilation relates the proper time of a moving clock to the time between the same two events as measured by a stationary observer. Length contraction, on the other hand, relates the proper length of a meter stick (the length measured in its rest frame) to its length as measured by a moving observer. We would like to be able to deal with the more general case that relates the space and time coordinates of general events as measured in two different inertial frames. Time dilation and length contraction should follow as direct consequences of these more general transformations.

© The Author(s), under exclusive license to Springer Nature Switzerland AG 2022   101
G. Kunstatter and S. Das, *A First Course on Symmetry, Special Relativity and Quantum Mechanics*, Undergraduate Lecture Notes in Physics,
https://doi.org/10.1007/978-3-030-92346-4_6

Two observers O and O' moving at relative speed v watch a clock (called C, for short) fly by.[1] Let τ denote the proper time of the clock. That is, τ is the time elapsed between ticks as measured in the clock's frame of reference. The speed of the clock, C, relative to O in terms of coordinates (t, x) measured by O is

$$u = \frac{dx}{dt} \tag{6.1}$$

The speed of C relative to O' in terms of coordinates (t', x') measured by O' is

$$u' = \frac{dx'}{dt'} \tag{6.2}$$

The time dilation effect derived in Sect. 5.6 implies that both O and O' see the moving clock running slowly compared to their respective clocks. For an infinitesimal proper time interval $d\tau$ of the moving clock, the corresponding time intervals measured by O and O' are, respectively:

$$dt = \gamma(u)d\tau = \frac{1}{\sqrt{1 - u^2/c^2}}d\tau \tag{6.3}$$

$$dt' = \gamma(u')d\tau = \frac{1}{\sqrt{1 - (u'^2)/c^2}}d\tau \tag{6.4}$$

By squaring both sides of Eqs. (6.3) and (6.4) we can solve for $d\tau^2$ in both, to get, respectively:

$$c^2 d\tau^2 = (1 - \frac{u^2}{c^2})c^2 dt^2 = c^2 dt^2 - u^2 dt^2$$
$$= c^2 dt^2 - dx^2 \tag{6.5}$$
$$c^2 d\tau^2 = \left(1 - \frac{u'^2}{c^2}\right)c^2 dt'^2 = c^2 dt'^2 - u'^2 dt'^2$$
$$= c^2 dt'^2 - dx'^2 \tag{6.6}$$

where to obtain the last line in each of Eqs. (6.5) and (6.6) we have used the following relations (see Eqs. (6.1) and (6.2)):

$$udt = dx$$
$$u'dt' = dx' \tag{6.7}$$

Since the change in proper time of the clock $d\tau$ is a property of the clock, it must be the same in both the O and O' frames. Equations (6.5) and (6.6) therefore imply

---

[1] We have now switched from giving our observers proper names (Alice and Bob) to this more traditional (and less interesting) designation O for one observer and O' for the other.

the following relationship between infinitesimal changes in the coordinates of the two events as measured by O and by O':

$$c^2 d\tau^2 := c^2 dt^2 - dx^2 = c^2 dt'^2 - dx'^2 \qquad (6.8)$$

Apart from the minus signs, Eq. (6.8) is very similar to the condition defining rotations in Eq. (4.59) of Sect. 4.4.3. As will be explained in more detail in Sect. 6.2.3, this is no coincidence. It suggests that in the context of Lorentz transformations from one inertial frame to another, we need to consider a non-Pythagorian definition of lengths, and hence an interesting new form of non-Euclidean geometry.

Notice that the Galilean transformations given in Eq. (5.3) do not satisfy Eq. (6.8).

Exercise 1 Use the Galilean transformations in Eq. (5.3) to substitute for (t', x') on the right hand side of Eq. (6.8), thereby proving explicitly that Eq. (5.3) does not satisfy Eq. (6.8) in general.

We would like to find a modified set of linear transformations that relate the coordinates (t, x) of an event as measured by O to the coordinates (t', x') of the same event as measured by O' in a way that respects Eq. (6.8). An important constraint in our search for these transformations is that they must agree, approximately at least, with Galilean transformations for speeds much less than that of light, for which we know the latter are valid. A rigorous derivation will be presented in Sect. 6.10.1. The derivation is very similar to that of the general form of rotations presented in Sect. 4.4.3 so here we will just state the transformations.

Both O and O' measure coordinates of an event relative to the origins of their respective spacetime coordinates. We assume that they have both set up these origins at the same event, specifically at the time and location that their worldlines intersect. More simply, they set their clocks and meter sticks to be zero when they are in the same place at the same time. The Lorentz transformations are then:

$$
\begin{aligned}
ct' &= \gamma(v)ct - \frac{v}{c}\gamma(v)x \\
&= \gamma(v)\left(ct - \frac{v}{c}x\right) \qquad (6.9)
\end{aligned}
$$

$$x' = \gamma(v)\left(-\frac{v}{c}ct + x\right) \qquad (6.10)$$

In the case we are considering, the time elapsed and distance between two events as measured in the two different frames, the Lorentz transformations in Eqs. (6.9) and (6.10) become:

$$c\Delta t' = \gamma(v)\left(c\Delta t - \frac{v}{c}\Delta x\right) \qquad (6.11)$$

$$\Delta x' = \gamma(v)\left(-\frac{v}{c}c\Delta t + \Delta x\right) \qquad (6.12)$$

Fig. 6.1 Coordinates
(t_p, x_p) and (t'_p, x'_p) of a
spacetime event P as
measured by two different
observers O and O',
respectively. The t' axis is
the worldline of observer O',
and the x' axis is the surface
that is simultaneous with
$t' = 0$ of O'

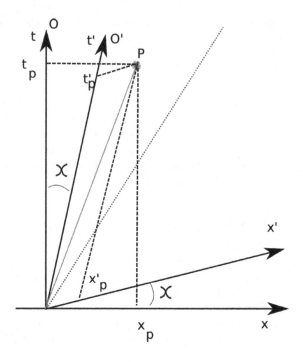

When the two events occur infinitesimally close in time and space:

$$cdt' = \gamma(v) \left(cdt - \frac{v}{c}dx \right) \qquad (6.13)$$

$$dx' = \gamma(v) \left(-\frac{v}{c}cdt + dx \right) \qquad (6.14)$$

Exercise 2 Verify by direct substitution of dt' and dx' in terms of dx and dt as given by Eqs. (6.13) and (6.14) that Eq. (6.8) is satisfied.

The Lorentz transformations of Special Relativity in Eqs. (6.9) and (6.10) replace the Galilean transformations Eqs. (5.2) and (5.3) of Newtonian physics. The relationship between the coordinate axes of O and O' is represented graphically in Fig. 6.1. The t' coordinate of an event P as measured by O' is obtained by projecting onto the t' axis along a surface of simultaneity, i.e. a line parallel to the x' axis. The x' coordinate of **P** is obtained by projecting onto the x' axis along the worldline of a potential observer who is moving at the same speed as O', i.e. parallel to the t' axis.

6.2.2 Properties of Lorentz Transformations

The Lorentz transformations have several important properties.

1. **Linearity**: The Lorentz transformations relate the elapsed time and distance in one inertial frame to time and distance as measured in another inertial frame. The coordinates t' and x' in one frame are linear functions of the coordinates x and t in the second frame. The Lorentz transformations therefore constitute linear transformations of spacetime that are completely analogous to the spatial transformations considered in Sect. 4.3.

2. **Completeness**: A Lorentz transformation with speed v_1, followed by another with speed v_2 corresponds to a third Lorentz transformation, although **not** with the speed $v_1 + v_2$ as one might expect from Galilean transformations. (see Sect. 6.5).

3. **Identity**: When $v = 0$, one has the identity Lorentz transformation which does not change one's reference frame.

4. **Inverse**: The inverse to the Lorentz transformation represented by Eqs. (6.9) and (6.10) is obtained by replacing v with $-v$ everywhere in the transformation. This can be deduced from two basic observations: first, the fundamental postulate of special relativity requires that there are no preferred inertial frames, and second, if the speed of O' relative to O is v, then the speed of O relative to O' is $-v$.

Exercise 3 Verify explicitly that a Lorentz transformation with speed v followed by a Lorentz transformation with speed $-v$ takes one back to the original coordinates.

---

5. **Associativity**: One can verify by a somewhat tedious calculation that the product of three Lorentz transformations satisfy the associativity condition Eq. (3.4)

6. **The group of Lorentz transformations**: The previous five properties verify what we might have suspected from the considerations in Chap. 3, namely that Lorentz transformations are a group of symmetry operations that satisfy all the properties of groups, as listed in Sect. 3.2.2.

7. **Galilean limit**: For small velocities relative to the speed of light, $\frac{v}{c} \ll 1$, a Taylor expansion reveals that the Lorentz transformations Eqs. (6.9) and (6.10) agree approximately with Galilean transformations. In particular,

$$ct' \rightarrow ct + O\left(\frac{v}{c}\right) \tag{6.15}$$

$$x' \rightarrow x - vt + O\left(\frac{v}{c}\right)^2 \tag{6.16}$$

where $O(\frac{v}{c})^n$ in general means that terms in the Taylor expansion containing the nth power or higher of (v/c) are left out.

Exercise 4 Taylor expand Eqs. (6.9) and (6.10) to obtain Eqs. (6.15) and (6.16). Calculate the next term in the Taylor expansion to obtain the $(v/c)^2$ corrections to the Galilean transformations.

---

8. **Time dilation**: Time dilation compares the time $\Delta t'$ elapsed between two ticks of a clock carried by an observer O' who is moving at speed v relative to O as measured by O', to the time Δt between the same two ticks as measured by O. Since the clock is stationary with respect to O', one must set $\Delta x' = 0$ in Eq. (6.12) when applying the full Lorentz transformations. This implies, as expected, that $\Delta x/\Delta t = v$, which, when substituted into the other Lorentz transformation, Eq. (6.11), gives:

$$c\Delta t' = c\gamma(v)\Delta t - \frac{v}{c}\gamma(v)v\Delta t \qquad (6.17)$$

Solving for Δt, we obtain the expected result:

$$\Delta t = \gamma(v)\Delta t' = \gamma(v)\Delta \tau \qquad (6.18)$$

9. **Length contraction**: Consider now a meter stick of length L at rest in the frame O. O' flies past the front end of the meter stick at speed v, at time $t' = 0$. In order to determine the instantaneous length of the meter stick in her frame O' must locate the distance to the far end of the meter stick also at $t' = 0$, so that it constitutes a measurement of the instantaneous length in her frame. See Fig. 5.13. Thus, when implementing the full Lorentz transformations, we need to set $\Delta t' = 0$ in Eq. (6.11), which again implies that $\Delta x = v\Delta t$. Solving Eq. (6.12) for $\Delta x'$ we find that

$$L' := \Delta x' = \frac{1}{\gamma(v)}\Delta x = \frac{1}{\gamma(v)}L \qquad (6.19)$$

in agreement with the Lorentz contraction formula of Sect. 5.7.

6.2.3 Lorentzian Geometry

Rotations preserve Pythagorian lengths, which means that after any point (x, y) is rotated by any angle around the origin to $(\tilde{x}, \tilde{y})$ the new coordinates satisfy:

$$x^2 + y^2 = (\tilde{x})^2 + (\tilde{y})^2 = a^2 \qquad (6.20)$$

This implies that a circle of radius a centered on the origin gets rotated onto a circle of the same radius. Rotations leave circles invariant. This terminology applies to so called *active transformations*. As described in Sect. 4.3.1, this means that the transformations physically move a given point in space to a new point while the coordinate axes stay fixed.

Henceforth we will adopt the view that the transformations we consider are *passive*: the physical points stay in the same place, but the coordinate axes, and the coordinates of the points, are transformed. This is represented graphically in

Fig. 6.2 Points at constant
radial distance from the
origin form circles that are
left invariant under rotations

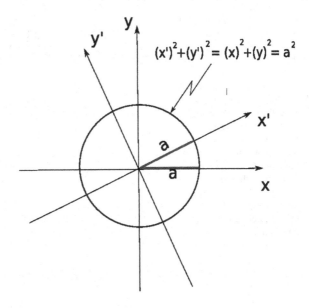

Fig. 6.2. Rotations preserve lengths: the distance from any point to the origin, in fact the distance between any two points remains unchanged by the rotation. This in turn implies that the scales on both the rotated and unrotated axes are the same. If a centimeter along any axis of the graph represents one meter before the rotation, then a centimeter also represents one meter after the rotation.

Lorentz transformations are linear transformations of spacetime coordinates that describe the change in coordinates when going from one inertial reference frame to another inertial frame. Such transformations are also called *boosts*. Lorentz transformations preserve magnitudes of spacetime vectors, but in terms of a different, non-Pythagorean notion of length, namely proper time τ or proper length L depending on whether the displacement from the origin is timelike or spacelike.

$$\text{Timelike} \quad c^2\tau^2 = c^2 t^2 - x^2 = (ct')^2 - (x')^2 \tag{6.21}$$
$$\text{Spacelike} \quad L^2 = -c^2 t^2 + x^2 = -(ct')^2 + (x')^2 \tag{6.22}$$

In order to understand the significance of the new geometry defined by these non-Pythagorean lengths[2] recall that the equation

$$c^2 t^2 - x^2 = c^2\tau^2 = \text{constant} \tag{6.23}$$

describes a hyperbola in the upper or lower quadrant of the (ct, x) plane, as shown on the right hand side of Fig. 6.3.

Similarly,

---

[2] This geometry is not suprisingly called *Lorentzian geometry*.

Fig. 6.3 All points located on the hyperbola in top quadrant are separated from the origin by the same proper time. All points on the hyperbola in right quadrant are the same proper distance from the origin. Because we are trying to represent Lorentzian geometry on a flat diagram, the scales along the individual worldlines connecting the origin to the hyperbolas must be different, just as the map of the Earth must be deformed to represent it on a flat map. (Color figure online)

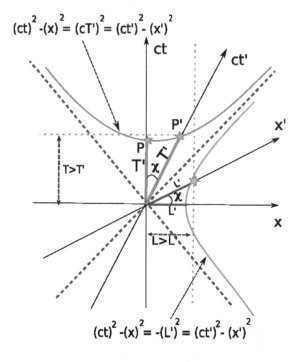

$$- c^2 t^2 + x^2 = L'^2 = \text{constant} \tag{6.24}$$

defines a hyperbola in the left or right quadrants.[3] If one considers the more physical case of two spatial dimensions, Fig. 6.3 must be rotated about the time axis so the hyperbolas form a two-dimensional surfaces called a *hyperboloid*, and there is no distinction between left quadrant and right quadrant.[4]

Equations (6.21) and (6.22) imply that boosts, i.e. Lorentz transformations from one frame to another, leave hyberbolae such as those in Eqs. (6.23) and (6.24) unchanged. If the events with coordinates (ct, x) lie on a particular hyperbola in the original frame, then the same events with transformed coordinates in the boosted frame also lie on the same hyperbola.

Physically, the hyperbolas defined by Eqs. (6.23) and (6.24) consist of all the points (events) in spacetime that lie at fixed proper time τ from the origin (if timelike sparated), or fixed proper length L from the origin (if spacelike separated). The hyperbolas at fixed proper time are horizonal (spacelike), while those at fixed proper distance lie vertically (are timelike).

---

[3] Since one can change $x \to -x$ and $t \to -t$ without affecting either equation, they define similar hyperbolas in the upper and lower quadrants, and right and left quadrants, respectively.

[4] This hyperboloid is a two dimensional timelike surface because at each point there exists a timelike vector tangent to it. The hyperbolas in the upper and lower quadrants, on the other hand, form a spacelike hyperboloid when rotated about the time axis.

The only way these proper times and lengths in Lorentzian spacetime can be consistently represented on a piece of paper whose geometry is Euclidean is to adjust the scales on all the timelike axes (worldlines) and spacelike axes (surfaces of simultaneity) accordingly, as explained below.

6.2.3.1 Graphing Lorentz Transformations

Figure 6.3 represents the effects of Lorentz transformations on a diagram drawn on a flat blackboard or piece of paper whose geometry necessarily obeys Pythogoras, i.e. is Euclidean. Such drawings are sometimes difficult to interpret but are unavoidable. The following discussion is meant to minimize the potential confusion as much as possible.

As mentioned above the scales along the two sets of axes in different frames are different. The main point is that a point, **P**, on the axis of O that is proper time T' from the origin must transform onto a point **P**$'$ on the axis of O' that is also a proper time T' from the origin. Given that we are using Lorentzian geometry to determine lengths, this means that **P** must stay on the same hyperbola when it is transformed to **P**$'$, as shown in Fig. 6.3. This also tells us how length scales along the axis of O are related in the diagram to length scales along the axis of O', as follows.

We can compare the relative lengths in the diagram of the (green) line segment along O's worldline and that of O' using Euclidean geometry and the definition of a hyperbola. Suppose the physical length of the (green) line segment on the diagram between the origin and **P** is s and the length of the line segment between the origin and **P**$'$ is called s'. Since the geometry of the page (or computer screen) on which the diagram is drawn is Euclidean, we can use Pythagoras to determine that

$$(s')^2 = (ct)^2 + x^2 \tag{6.25}$$

where ct is the length along O's time axis and x is the length along O's space axis locating the point **P**$'$. On the other hand we know that the point **P** and **P**$'$ are on the same hyperbola, so the length s' of the segment between the origin and **P** along $O's$ axis is

$$s^2 = (ct)^2 - x^2 \tag{6.26}$$

where again ct is the time and x is the length measured on the diagram specifying spacetime coordinates of the point **P**$'$ on O's frame. Combining Eqs. (6.25) and (6.26) we find the ratio:

$$\frac{s^2}{(s')^2} = \frac{(ct)^2 - x^2}{(ct)^2 + x^2} \tag{6.27}$$

Using that $x/(ct) = (v/c)$, we find

$$\frac{s}{s'} = \sqrt{\frac{1 - (v/c)^2}{1 + (v/c)^2}} \tag{6.28}$$

For $(v/c) < 1, s < s'$ as expected from the figure. In order to dispel any confusion, we emphasize that s is the length along the O time axis on the diagram, not a physically measurable proper time in Minkowski spacetime.

We can also work out the angle χ on the diagram between the t axis of O and the t' axis of O'. We just need to know the time t measured by O for the event P'. It is given by time dilation to be:

$$ct = \gamma s \tag{6.29}$$

Thus we have:

$$\begin{aligned}
\cos(\chi) &= \frac{ct}{s'} \\
&= \gamma \frac{s}{s'} \\
&= \frac{1}{\sqrt{1 - (v/c)^2}} \sqrt{\frac{1 - (v/c)^2}{1 + (v/c)^2}} \\
&= \frac{1}{\sqrt{1 + (v/c)^2}} \tag{6.30}
\end{aligned}$$

As expected $\chi = 0$ for $v = 0$ and $\chi = 45°$ when $v = c$.

It is very important to note that the angle χ does not correspond to a physical rotation angle. It is instead the angle required in the diagram to accurately represent a boost or Lorentz transformation with speed v.

A similar analysis will confirm that the angle that the spatial axis of O' makes in the diagram with that of O is the same angle χ.

The overall message that one can take away from Fig. 6.3 is that spacetime diagrams are very useful in solving problems qualitatively and getting some intuition as to what is happening as long as one is aware of the different length scales required along different axes due to the Lorentzian geometry of spacetime.

6.3 The Light Cone

The future light cone at a point, or event, **P** in four dimensional spacetime is formally defined as the surface of all possible light rays, or light fronts, that could emanate from that point. It consists of a two dimensional sphere that grows in time at the speed of light. Thus it is a three dimensional *cone* in spacetime. The past light cone at **P** consists of all possible light rays that could converge on that point from the past. Both the past and future light cone of **P** is illustrated in Fig. 6.4 for the case of three

Fig. 6.4 Spacetime diagram showing future and past light cones in two spacial dimensions of an event **P**

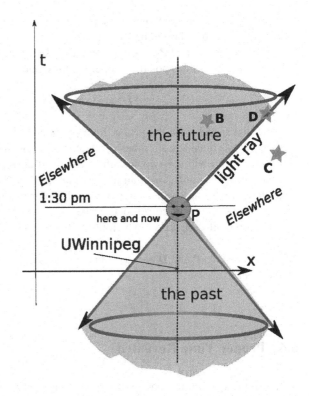

spacetime dimensions instead of four. In this case light fronts emitted at **P** are circles that grow with the speed of light, mapping out a cone in spacetime.

The future light cone provides the boundary between all those events that can be reached from **P** without going faster than the speed of light and those that cannot. Similarly, the past light cone separates those events that can communicate or affect the event at **P** without exceeding the speed of light from those that cannot. All events outside the light cone of **P** are "elsewhere", because there can be no communication between them and **P** without faster than light propagation.

Some definitions are useful at this stage. Please refer to the events/points labelled **P**, **B**, **C** and **D** in Fig. 6.4.

1. The two events **P** and **B** are said to be *timelike separated* because the spacetime vector $(c\Delta t, \Delta x)$ joining them has positive proper time squared:

$$c^2 \Delta \tau^2 = c^2 \Delta t^2 - \Delta x^2 > 0 \tag{6.31}$$

This implies that **P** and **B** are within each other's light cones. Since **B** is in **P**'s future light cone, **P** is in **B**'s past light cone. That is, if **B** is to the future of **P** then **P** is in the past of **B**. The relative time ordering of two timelike separated events is invariant under Lorentz transformations. This means that all inertial observers agree on which of the two events happened first, even if they do not necessarily agree on the time elapsed between them.

2. The two events **C** and **P** are said to be *spacelike separated* because the spacetime vector $(c\Delta t, \Delta x)$ has positive proper length squared:

$$\Delta L^2 = \Delta x^2 - c^2 \Delta t^2 > 0 \tag{6.32}$$

Note that in this case the proper time squared $\Delta \tau^2$ is negative, signifying that **C** and **P** are outside each other's light cones. There will always exist an inertial observer who will see two spacelike separated events as occurring at the same time. In other words, one can always find a Lorentz transformation, or boost, to the frame of an inertial, timelike observer for whom the straight line joining these two events will correspond to their surface of simultaneity. Other inertial observers will disagree on which of the two events occurs first, as described in Sect. 5.5.

3. The two events **P** and **D** are said to be *null separated* if the proper time between them is zero:

$$c^2 \Delta \tau^2 = c^2 \Delta t^2 - \Delta x^2 = 0 \tag{6.33}$$

6.4 Proper Time Revisited

Recall from Sect. 5.6 that the proper time along a particular worldline joining two events is defined as the time measured by an observer moving along that worldline. Although we have been restricting attention to inertial observers, we can now generalize the calculation of proper time to accelerating worldlines. Consider an observer O' moving along an accelerating trajectory $x(t)$ as measured in the rest frame of another observer O. One can calculate the proper time of O' along this trajectory by splitting it up into tiny (infinitesimal) sections, of duration dt and displacement dx. Along such an infinitesimal section, the speed $v(t) = \frac{dx(t)}{dt}$ is approximately constant, so the proper time can be calculated using the usual formula:

$$d\tau = \frac{dt}{\gamma(v(t))} = \sqrt{1 - \frac{v(t)^2}{c^2}} dt \tag{6.34}$$

where the notation $v(t)$ indicates that the speed is not constant along the trajectory.

The proper time along any finite segment connecting the points A and B, say, of the trajectory is obtained by adding up (integrating) the infinitesimal proper time (length) at each point along the trajectory:

$$c\Delta\tau = \int_A^B d\tau$$

$$= c\int_A^B \frac{dt}{\gamma(t)}$$

$$= c\int_A^B dt\sqrt{1 - \frac{v(t)}{c^2}} \tag{6.35}$$

We emphasize that this equals the time as measured by a clock carried by an observer moving along this worldline. Different observers, travelling along different world-lines between the same two events can measure different proper times along their particular trajectories. This fact is key to the resolution of the twin paradox (See Example 1, Sect. 5.6.5), for example.

6.5 Relativistic Addition of Velocities

Suppose a rocket is moving to the right relative to O' at speed u', and O' is moving to the right with speed v relative to O, as shown in Fig. 6.5. The Lorentz transformations Equations (6.11) and (6.12) imply, after a bit of algebra that:

$$\begin{aligned} u' &:= \frac{\Delta x'}{\Delta t'} \\ &= \frac{-\frac{v}{c}\gamma(v)c\Delta t + \gamma(v)\Delta x}{\gamma(v)c\Delta t - \frac{v}{c}\gamma(v)\Delta x} \\ &= \frac{u - v}{1 - \frac{uv}{c^2}}, \end{aligned} \tag{6.36}$$

where $u := \Delta x/\Delta t$.

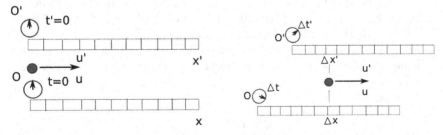

Fig. 6.5 A rocket ship (red dot) moving at constant speed relative to two inertial frames O and O'

If O' is moving to the left instead of the right, then simply change $v \rightarrow -v$ in Eq. (6.36). One can invert Eq. (6.36) to express the velocity u as measured by O in terms of u':

$$u = \frac{u' + v}{1 + \frac{u'v}{c^2}}. \tag{6.37}$$

An important property of Eq. (6.36) is that whenever u/c or v/c are much less than one, it reduces to the usual Galilean (intuitive) addition of velocities Eq. (5.4) as required by the correspondence principle.

Example 1 Suppose that an observer O' moves at speed $v = c/2$ relative to O. She then fires a rocket R_1 at speed $u' = c/2$ relative to her frame, in the direction of motion.

1. What is the speed u of the rocket as observed by O?
   Solution:
   Using the addition of velocities in Eq. (6.36)

   $$u = \frac{u' + v}{1 + \frac{u'v}{c^2}} = \frac{c/2 + c/2}{1 + c^2/4c^2} = \frac{4}{5}c. \tag{6.38}$$

2. Now suppose the rocket R_1 fired by O' fires a smaller rocket R_2 moving at $c/2$ relative to R_1. What is the speed u_2 of R_2 relative to O?
   Solution:
   The speed of R_2 relative to R_1 is u' and the speed of R_2 relative to O is u, so that:

   $$u_2 = \frac{\frac{4c}{5} + \frac{c}{2}}{1 + \frac{4}{5 \cdot 2}} = \frac{13}{14}c. \tag{6.39}$$

   Note: If you keep repeating this process the speed relative to O gets closer and closer to the speed of light, but can never reach it. Thus special relativity, specifically the Lorentz transformation law, prevents a massive object from being accelerated to the speed of light. Another way to say this is that as an object moves faster relative to a given frame O, it gets more difficult to accelerate. Recall that inertia, or inertial mass, is the resistance of an object to a change in motion. This suggests that the faster an object moves relative to O, the greater its inertial mass. This argument is somewhat heuristic, but we will see in Sect. 6.7.2 that it is rigorously born out by the equations of special relativity.
3. Suppose O' fires a laser beam, moving at the speed of light, c, relative to her frame. Use the velocity transformation law to find the speed of the laser beam with respect to O.
   Solution:

   $$u = \frac{u' + v}{1 + \frac{u'v}{c^2}} = \frac{c/2 + c}{1 + c^2/2c^2} = c. \tag{6.40}$$

Thus, as expected, the speed of light is the same for O as it is for O'.

4. Use Eq. (6.37) to show that if u' or v is less than c then $|u| \leq c$.

Solution:

We only need to consider u' and v. Write $u'/c = 1 - a$ and $v/c = 1 - b$. From Eq. (6.37), we need to show that:

$$\frac{u' + v}{1 + \frac{u'+v}{c^2}} \leq c$$

Using the definitions of a and b, a bit of algebra yields:

$$(1 - a) + (1 - b) \leq 1 + (1 - a)(1 - b)$$
$$2 - (a + b) \leq 2 - (a + b) + ab, \tag{6.41}$$

which is true for all non-negative a and b.

6.6 Relativistic Doppler Shift

6.6.1 Non-relativistic Doppler Shift Review

Consider a source S' that is emitting sound waves with frequency f. The distance between the crests of the waves is called the *wavelength* λ and is given by

$$\lambda = \frac{v_s}{f}. \tag{6.42}$$

where v_s is the speed of sound relative to the air. An observer O who runs at speed v towards the source, and hence towards the oncoming waves, measures the speed of the sound relative to him to be $v_s + v$. Since we are ignoring special relativity, the distance between the crests of the wave is not affected by the motion of O, so the rate at which O encounters crests of the wave (the frequency) is:

$$f_1 = \frac{v_s + v}{\lambda} = \left(1 + \frac{v}{v_s}\right) f. \tag{6.43}$$

On the other hand if the source S' is moving towards O then the crests do not get as far from S' before the next crest is emitted, so the wavelength of the emitted waves is shortened to

$$\lambda' = (v_s - v)\Delta t \tag{6.44}$$

The frequency measured by O is:

$$f_2 = \frac{v_s}{\lambda'} = \left(\frac{v_s}{v_s - v}\right)^{-1} f = \frac{f}{1 - v/v_s} \qquad (6.45)$$

Note that $f_1 \neq f_2$ so that one can distinguish via the non-relativistic Doppler shift whether it is the source or the observer that is moving. This is consistent with the fundamental postulate because the speed of sound is only a constant relative to the medium through which it moves, namely the air. The air therefore provides a preferred frame of reference. The speed of the sound waves relative to the air does not change if the source moves, but it is different for an observer in motion relative to the air.

6.6.2   Relativistic Doppler Shift

Consider a source S' that emits light waves with frequency f' while moving towards an observer O as shown in Fig. 6.6.

The time between emitted crests as measured by S' is $\Delta t' = 1/f'$ where f' is the frequency as measured in the frame of the source. The wavelength as measured by S' is $\lambda' = c\Delta t'$. Because of the relative motion between observer O and the source, the distance between successive crests that are emitted is decreased by $v\Delta t$, so the time between crests as measured by O is:

$$\Delta t = \gamma \Delta t' - \frac{v\Delta t}{c}. \qquad (6.46)$$

The first term in the above is the time between emissions of crests as measured by O, taking into account time dilation. The second term specifies the decrease in distance $v\Delta t$ that successive crests have to travel due to the motion of O towards them. Using the time dilation formula $\Delta t = \gamma \Delta t'$ in the second term of Eq. (6.46) and doing a bit of algebra:

$$\Delta t = \gamma \left(1 - \frac{v}{c}\right) \Delta t'$$
$$= \frac{1 - v/c}{\sqrt{1 - \frac{v^2}{c^2}}} \Delta t'$$
$$= \sqrt{\frac{1 - v/c}{1 + v/c}} \Delta t' \qquad (6.47)$$

Fig. 6.6 A source S' moves towards observer O emitting light at intervals of $\Delta t'$ as measured in the source's frame

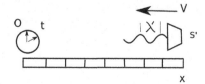

where we have used $1 - v^2/c^2 = (1 - v/c)(1 + v/c)$ to get the last line.
    The frequency observed by O is

$$f = \frac{1}{\Delta t}$$

$$= \sqrt{\frac{1 + v/c}{1 - v/c}}\, f' \tag{6.48}$$

and the corresponding wavelengths $\lambda = c/f$, $\lambda' = c/f'$ are related by:

$$\lambda = \sqrt{\frac{1 - v/c}{1 + v/c}}\, \lambda'. \tag{6.49}$$

If O is moving away from the source at speed v, then the successive crests have
further to travel, so that one needs to change $v \rightarrow -v$ in the above formulas. The most
important thing to remember, besides the fact that the numerator and denominator
must contain opposite signs, is that if the observer and source are moving towards each
other, the wavelength decreases and the frequency increases, while if the observer
and source are moving away from each other then the wavelength must increase and
the frequency decreases. These qualitative features agree with the non-relativistic
case, but there is an important difference, as described directly below.

Properties of the relativistic Doppler shift:

1. The effect is completely symmetric: The formulas depend only on the relative
   motion of source and observer because in special relativity there is no preferred
   inertial frame. This is not the case for the usual non-relativistic Doppler shift
   because the medium through which the wave moves defines a preferred rest frame.
2. If $v/c \ll 1$ then a Taylor expansion in v/c shows that all three expressions, f, f_1
   and f_2 in Eqs. (6.48), (6.43) and (6.45), respectively, are approximately the same.

Exercise 5 Prove that for speeds much less than the speed of light the relativistic
and non-relativistic expressions for Doppler shift agree.

Example 2 The most distant quasar has a red-shift of $z = 6.43$, where the red-shift
is defined as:

$$z := \frac{\lambda' - \lambda}{\lambda}. \tag{6.50}$$

If this red-shift were due to a proper motion of the quasar,[5] it would imply

---

[5] Strictly speaking, it is not due to proper motion. It is really due to the expansion of the Universe.
See Sect. 7.8.

$$z := \frac{\lambda' - \lambda}{\lambda}$$

$$= \frac{f}{f'} - 1$$

$$= \sqrt{\frac{1 + v/c}{1 - v/c}} - 1. \tag{6.51}$$

After a bit of algebra, the speed of the quasar away from Earth would be:

$$\frac{v}{c} = \frac{(1+z)^2 - 1}{(1+z)^2 + 1} = 0.964. \tag{6.52}$$

6.7 Relativistic Energy and Momentum

6.7.1 Relativistic Energy-Momentum Conservation

The Newtonian expressions for energy and momentum are useful because in the absence of external forces or changes in internal energy due to deformations (i.e. inelastic collisions), the total kinetic energy and total momentum are conserved in the collision of two (or more) particles. Moreover, if energy and momentum are conserved in a collision in one inertial frame, they are conserved in any other inertial frame. This latter fact is a direct consequence of the form of the Galilean transformation laws in Sect. 5.2.3 and the resulting expression for the addition of velocities in Eq. (5.4).

The velocity addition law in Newtonian mechanics in Eq. (5.4) is linear, whereas the corresponding law in special relativity is non-linear (See Eq. (6.36)). The non-linear velocity addition law has the consequence that the usual expressions for total energy and momentum are not conserved when transforming from one inertial frame to another. One must find expressions for relativistic energy and momentum possessing the property that if they are conserved in one frame, then after a Lorentz transformation to a different frame they are still conserved.

6.7.2 Relativistic Inertia

We saw in Sect. 6.5 that the addition law for velocities implies that the faster an object is moving relative to a given frame, the more difficult it is to speed it up further relative to that frame. Since *inertia* (or *inertial mass*, or just mass) is essentially the resistance of an object to a change in motion, this immediately implies that the inertia of a moving object must increase with speed. We first define the *rest mass*, m_0 of an object to be the inertial mass as measured in its rest frame. The corresponding relativistic inertial mass, $M(\gamma)$ of the same object as observed in a frame moving at speed v relative to it is defined as:

$$M(\gamma) := m_0 \gamma(v) = \frac{m_0}{\sqrt{1 - \left(\frac{v}{c}\right)^2}}. \tag{6.53}$$

We cannot provide a rigorous derivation of this formula at this point, but we can verify that Eq. (6.53) has several desirable properties.

1. It increases with speed as required by the discussion in Sect. 6.5.
2. For non-relativistic velocities, when $v/c \ll 1$, Eq. (6.53) reduces via the usual Taylor approximation to:

$$M(\gamma) \sim m_0 + \frac{1}{2} m_0 \left(\frac{v}{c}\right)^2. \tag{6.54}$$

This implies two things. First, for low speeds compared to the speed of light, the inertial mass is approximately constant to high accuracy, as required by the correspondence principle. Secondly Eq. (6.54) shows that in the limit $v/c \ll 1$, the increase in inertia is proportional to the kinetic energy of the object:

$$\Delta M(\gamma) = M(\gamma) - m_0 \sim \frac{KE}{c^2} \tag{6.55}$$

where :

$$KE := \frac{1}{2} m_0 v^2. \tag{6.56}$$

The fundamental postulate has now led us, via the special theory of relativity, to a remarkable realization. Kinetic energy contributes to mass, and by extension, any mass must contain energy. Thus for low velocities, we have shown that kinetic energy has a "mass equivalent" equal to KE/c^2.

6.7.3 Relativistic Energy

We have argued that kinetic energy contributes to the inertial mass of an object, which suggests that mass and energy are in some ways equivalent to each other. This experimentally confirmed equivalence is made explicit in the relativistic expression for the total energy, E, of an object moving at speed u relative to some observer O:

$$E = M(\gamma)c^2 = m_0 \gamma(u)c^2. \tag{6.57}$$

A direct consequence of this formula is that even a stationary object with rest mass m_0 has non-zero total energy, which we call the *rest energy*:

$$E_0 = m_0 c^2. \tag{6.58}$$

Equation (6.58) is Einstein's famous, and to some extent infamous, formula the consequence of which is that mass and energy are equivalent and can be converted into each other in physical processes. This equivalence is inevitable in light of the Fundamental Postulate of special relativity. It was discovered by Einstein as part of his curiosity driven quest to understand the fundamental workings of the Universe, specifically the behaviour of light. Eventually, it led to the development of nuclear power and nuclear weapons. While the former holds the potential to benefit the human race, the latter have immense destructive power: the speed of light is so large when expressed in terms of laboratory scale units of meters and seconds that a small amount of mass contains a large amount of rest energy.

Exercise 6 Calculate the energy equivalent in Joules of 1 g of fissionable material (assumed to be at rest). How many tons of TNT are required to produce this much energy in an explosion? (1 ton of TNT can produce 4×10^9 J.)

6.7.4 Relativistic Three-Momentum

The spatial momentum, or three-momentum,[6] of an object is defined as the inertial mass times the speed. The increase of the inertia of an object with speed should be reflected in the expression for the relativistic momentum. In one spatial dimension, the expression is:

$$p = M(\gamma)u. \tag{6.59}$$

In three dimensions, u becomes the velocity vector $\mathbf{u}$.

$$\mathbf{p} = M(\gamma)\mathbf{u} = m_0 \gamma(u)\mathbf{u}. \tag{6.60}$$

6.7.5 Relationship Between Relativistic Energy and Momentum

Using $p^2 := \mathbf{p} \cdot \mathbf{p} = m_0^2 \gamma^2 u^2$, where $u^2 := \mathbf{u} \cdot \mathbf{u}$, Eqs. (6.57) and (6.60) imply the following energy-momentum relation:

---

[6] It is necessary to specify that we are dealing with the momentum three-vector (three-momentum), since in the next section we will introduce the concept of a four component momentum, or four-momentum.

$$m_0^2 c^4 = E^2 - p^2 c^2$$
$$\rightarrow E = \sqrt{m_0^2 c^4 + p^2 c^2}, \tag{6.61}$$

where $m_0 := (\sqrt{E^2/c^2 - p^2})/c$ is the invariant rest mass. As the name suggests, it is invariant under Lorentz transformations, i.e the same in every inertial frame of reference. The energy E and momentum p, on the other hand, are not invariant under Lorentz transformations.

Note that when $p = 0$, Eq. (6.61) yields the familiar expression for the rest energy of the particle:

$$E_0 = m_0 c^2. \tag{6.62}$$

6.7.6 Kinetic Energy

The energy of a moving object as defined in Eq. (6.57) contains contributions from the rest energy of the particle and the energy of motion. We wish to isolate the latter since this is the kinetic energy of the particle. We therefore define the kinetic energy as the difference between the total energy and the rest energy:

$$T(u) = m_0 c^2 \gamma(u) - m_0 c^2. \tag{6.63}$$

By doing the Taylor expansion of $\gamma(u)$, one finds that for small speeds compared to the speed of light, $u/c \ll 1$:

$$T(u) \sim m_0 c^2 (1 + \frac{1}{2}\frac{u^2}{c^2}) - m_0 c^2$$
$$= \frac{1}{2} m_0 u^2. \tag{6.64}$$

Thus for low speeds one recovers the usual expression for the kinetic energy of a non-relativistic particle, as required by the correspondence principle.

Exercise 7 1. Taylor expand the expression for $T(u)$ in Eq. (6.63) in a power series in u/c to order $(u/c)^6$.
2. For a 1 kg soccer ball moving at $u = 50$m/s, what is the percentage error made by calculating the kinetic energy using the last line of Eq. (6.64) and neglecting the extra terms you calculated in the first part of this exercise.

6.7.7 Massless Particles

Since $\gamma(u) \to \infty$ as $u \to c$, Eq. (6.57) for the total energy of a moving particle seems to imply that a particle moving at the speed of light has an infinite amount of energy. This is true as long as the rest mass m_0 of the particle is not zero. The conclusion then is that it would require an infinite amount of energy to accelerate a massive particle to the speed of light.

On the other hand we know that light travels at the speed of light. This is possible mathematically and physically because light has zero rest mass. It does not imply, however that it has zero energy or zero momentum. The energy-momentum relation Eq. (6.61) implies that the energy and momentum of a photon are related by:

$$E = pc. \tag{6.65}$$

We have seen from Eq. (6.36) that anything travelling at the speed of light in one frame necessarily travels at the speed of light with respect to every frame. What does it mean to say that a particle has zero rest mass? Physically, it means that the particle can never be observed at rest. More precisely, a zero rest mass particle travels at the speed of light in all frames of reference and its energy/momentum relation is given by Eq. (6.65). [7]

Example 3 We would like to send a large spaceship to Mars in order to explore it. The propulsion method involves ejecting a large quantity of radiation. Consider an approximation in which the spaceship of mass M_i is initially at rest. After the propulsion phase is finished, the ejected radiation has energy E_r and moves to the left while the spaceship, with remaining mass M_f, moves to the right at speed $v = 0.8c$.

1. What fraction of the initial mass M_i remains as "payload" M_f?
2. Given that the spaceship must come to a stop once it reaches Mars, what is the final payload, as a fraction of the initial mass once it arrives?

Solution: This requires the use of relativistic energy and momentum conservation.

$$\begin{aligned} \text{Initial energy:} \quad & E_i = M_i c^2 \\ \text{Final energy:} \quad & E_f = M_f \gamma(v) c^2 + E_r \end{aligned}$$

$$\tag{6.66}$$

1. Energy conservation requires:

$$E_i = E_f \implies M_i c^2 = M_f \gamma(v) c^2 + E_r \tag{6.67}$$

---

[7] In Sect. 6.8.2 the rest mass is defined as a frame independent quantity in terms of the invariant magnitude of a particle's energy-momentum vector. For light, the magnitude of this vector is zero.

Similarly, the final momentum is the sum of the momentum of the radiation and the remaining mass, and since the initial momentum is zero, momentum conservation requires:

$$p_i = p_f \;\Rightarrow\; 0 = -\frac{E_r}{c} + M_f \gamma(v) v \tag{6.68}$$

Note that the minus sign in the first term tells us that the radiation moves to the left and the remaining mass moves to the right. Given the speed v and the initial mass M_i, we now have two equations and two unknowns. Since we were asked for the fraction of the mass lost, we can divide the above equations by M_i, to make the two unkowns M_f/M_i and E_r/E_i. First solve for E_r/M_i using Eq. (6.68):

$$\frac{E_r}{M_i c^2} = \frac{M_f}{M_i} \gamma(v) \frac{v}{c} \tag{6.69}$$

Then use Eq. (6.67) to solve for M_f/M_i:

$$1 = \frac{M_f}{M_i} \gamma(v) + \frac{E_r}{M_i c^2} \tag{6.70}$$

$$= \frac{M_f}{M_i} \gamma(v) + \frac{M_f}{M_i} \gamma(v) \frac{v}{c}$$

$$\rightarrow \frac{M_f}{M_i} = \frac{1}{\gamma(v) \left(1 + \frac{v}{c}\right)}$$

$$= \frac{\sqrt{1 - (v/c)^2}}{1 + (v/c)}$$

$$= \sqrt{\frac{1 - (v/c)}{1 + (v/c)}} \tag{6.71}$$

For $v = 0.8 = 4/5$, $\gamma = 5/3$:

$$\frac{M_f}{M_i} = \frac{1}{3} \tag{6.72}$$

Using Eq. (6.70) this implies that:

$$\frac{E_r}{M_i c^2} = 1 - \gamma \frac{M_f}{M_i} = 1 - \frac{1}{3} \times \frac{5}{3} = \frac{4}{9} \tag{6.73}$$

2. In order to come to a complete halt from $v = 0.8c$, symmetry of the situation suggests that the rocket ship must again eject 2/3 of its mass as fuel. It is starting with a mass of $M_f = M_i/3$, so after it comes to a stop, the remaining mass is

$$M_{\text{final}} = \frac{M_f}{3} = \frac{M_i}{9} \tag{6.74}$$

It is interesting that the final payload after the rocket ship comes to a rest again at its destination, is only 1/9 of the original mass. This is a problem for this type of space travel: most of the spaceship needs to be made up of fuel. A better option would be to harness the solar wind, which consists of a stream of high energy particles flowing out of the sun.

Note that the ratio of the payload to initial mass in Eq. (6.71) was obtained without specifying the details of the propulsion phase, that is whether the fuel is ejected all at once, continuously or in several larger portions. The answer depends only on the speed of the ejected fuel, in this case the speed of light, and the final speed of the payload.

6.8 Spacetime Vectors

Spacetime vectors are, as the name suggests, vectors in spacetime. They have a time component in addition to the usual three spatial components. Just like spatial vectors, they are geometrical and physical quantities whose components transform non-trivially under changes of inertial frame, or boosts, in much the same way that the three components of an ordinary vector transform under rotations. Because of the close analogy with spatial three component vectors, spacetime vectors are often called *four-vectors*. We will use the two expressions more or less interchangeably even though we consider mostly one spatial dimension so that our spacetime vectors have only two components.

6.8.1 Position Four-Vector

As discussed in Sect. 6.2.3, Lorentz transformations are linear transformations of the spacetime coordinates that preserve the proper time or proper length between two timelike or spacelike separated events, respectively. It is therefore natural to think of the spacetime coordinates of an event relative to the origin as a vector in spacetime[8]:

$$\mathbf{X} = \begin{pmatrix} ct \\ x \end{pmatrix} \tag{6.75}$$

Just as the length of a three-vector is preserved by rotations, Lorentz transformations preserve the "length" of the position vector, as defined by the Lorentzian version of the Pythagoras theorem. In the Lorentzian version, the (length)^2 is given by the **difference** (as opposed to sum) of the lengths squared of the time component and spatial component. For a timelike position four-vector $\mathbf{X}$:

---

[8] We henceforth denote vectors in spacetime by bold faced upper case letters.

$$\mathbf{X}^2 = \mathbf{X} \cdot \mathbf{X} := c^2 t^2 - x^2 = c^2 \tau^2 . \tag{6.76}$$

This Lorentzian length of $\mathbf{X}$ is just c times the proper time along the constant velocity worldline from the origin to the event.

If the position four-vector $\mathbf{Y}$ is spacelike:

$$\mathbf{Y}^2 = \mathbf{Y} \cdot \mathbf{Y} := c^2 t^2 - y^2 = -D^2 . \tag{6.77}$$

where D is the proper distance from the event to the origin

In Sect. 4.4 we found it useful to express rotations, in fact any linear transformation, in matrix form. Since Lorentz transformations are also linear transformations, they too can be written in matrix form:

$$
\begin{aligned}
L(u) &= \begin{bmatrix} \gamma(u) & -\gamma(u)u/c \\ -\gamma(u)u/c & \gamma(u) \end{bmatrix} \\
&= \gamma(u) \begin{bmatrix} 1 & -u/c \\ -u/c & 1 \end{bmatrix}
\end{aligned}
\tag{6.78}
$$

The action of $L(u)$ on a spacetime vector $\mathbf{X}$ is:

$$
\begin{aligned}
\mathbf{X} \to \mathbf{X}' &= \begin{pmatrix} ct' \\ x' \end{pmatrix} \\
&= L(u)\mathbf{X} \\
&= \begin{bmatrix} \gamma(u) & -\gamma(u)u/c \\ -\gamma(u)u/c & \gamma(u) \end{bmatrix} \begin{pmatrix} ct \\ x \end{pmatrix}
\end{aligned}
\tag{6.79}
$$

where $\mathbf{X}'$ are the spacetime coordinates in a frame moving at speed u in the positive x-direction with respect to a rest frame in which the coordinates are given by $\mathbf{X}$. Writing Lorentz transformations in matrix form often simplifies calculations. In Sect. 6.10, we will take advantage of the matrix form of Lorentz transformations to highlight their connection to symmetry. In four spacetime dimensions:

$$\mathbf{X} = \begin{pmatrix} ct \\ \mathbf{x} \end{pmatrix} = \begin{pmatrix} ct \\ x \\ y \\ z \end{pmatrix} \tag{6.80}$$

Given the above interpretation of the spacetime position vector, it follows that, there are other spacetime vectors that have physical significance, analogous to three-vectors in Newtonian physics. Four-vectors, and their generalizations called tensors are discussed in the supplementary Sect. 6.11.

6.8.2 Momentum Four-Vector

The spacetime extension of momentum that transforms linearly under Lorentz transformations is called the four-momentum $\mathbf{P}$. It is defined by:

$$\mathbf{P} := \begin{pmatrix} \frac{E}{c} \\ p \end{pmatrix}$$

$$= \begin{pmatrix} m_0 \gamma(u)c \\ m_0 \gamma(u)u \end{pmatrix} \tag{6.81}$$

Note that E/c has the same dimensions, or units, as momentum (kg-m/s) so that the time component of both components of $\mathbf{P}$ have the same dimensions as momentum, namely kg-m/s in S.I. units. If we are in the same frame of reference as the particle:

$$\mathbf{P} = \begin{pmatrix} m_0 c \\ 0 \end{pmatrix} \tag{6.82}$$

Under a Lorentz transformation, i.e. change of frame to one that is moving to the right with speed u:

$$\mathbf{P} \rightarrow \mathbf{P}' = \begin{pmatrix} \frac{E'}{c} \\ p' \end{pmatrix}$$

$$= L(u) \begin{pmatrix} m_0 c \\ 0 \end{pmatrix}$$

$$= \begin{bmatrix} \gamma(u) & -\gamma(u)u/c \\ -\gamma(u)u/c & \gamma(u) \end{bmatrix} \begin{pmatrix} m_0 c \\ 0 \end{pmatrix}$$

$$= \begin{pmatrix} m_0 \gamma(u)c \\ -m_0 \gamma(u)u \end{pmatrix} \tag{6.83}$$

as expected, since in this frame the particle is moving to the left with speed u (velocity $-u\mathbf{i}$).

Exercise 8 You are in a frame O in which you see a particle of rest mass m_0, moving to the right with speed v, so that its momentum four-vector is given by Eq. (6.81). Now boost to a Lorentz frame, O', moving with speed v in the same direction as the particle. O' is the rest frame of the particle. Do the appropriate Lorentz transformation to verify that $E' = m_0 c^2$ and $p' = 0$ in the boosted frame O'.

In four dimensions:

$$\mathbf{P} := \begin{pmatrix} \frac{E}{c} \\ \mathbf{p} \end{pmatrix} = \begin{pmatrix} \frac{E}{c} \\ p_x \\ p_y \\ p_z \end{pmatrix} \tag{6.84}$$

As well you can prove that under a Lorentz transformation:

$$\mathbf{P}' \cdot \mathbf{P}' = \mathbf{P} \cdot \mathbf{P} = m_0^2 c^2 \tag{6.85}$$

which shows that $m_0 c = \sqrt{\mathbf{P}' \cdot \mathbf{P}'}$ is the invariant magnitude of the four-vector $\mathbf{P}'$, using the Lorentzian as opposed to Euclidean definition of lengths.[9]

In Galilean collisions, the total kinetic energy and total three-momentum are separately conserved in the absence of external forces. This is also the case for relativistic collisions: they must conserve all four components of the total momentum four-vector in a given frame.

6.8.3 Null Four-Vectors

Light travels at the speed of light (not surprisingly), (i.e. $dx/dt = c$). For a light ray emanating from the origin:

$$c^2 \Delta t^2 - \Delta x^2 = c^2 \Delta t^2 - c^2 \Delta t^2 = 0 . \tag{6.86}$$

Eq. (6.86) is valid in all inertial frames. Such paths are called null worldlines. The proper time along a light ray is zero, and the position four-vector that describes the propagation of a null ray has zero magnitude. It is therefore called a null spacetime vector or null four-vector.

Classically we think of light as a wave, with no definite position. As we will learn in Chap. 8, quantum mechanics suggests that waves sometimes behave as particles, and conversely, particles sometime behave as waves. This "particle-wave duality" is a key attribute of the microscopic world as described by quantum mechanics. The quantum particle associated with light is called a photon. Photons travel at the speed of light. They necessarily have zero rest mass, otherwise one could find a reference frame in which the photon is at rest, something that is forbidden in special relativity. This implies that the four momentum, $\mathbf{P}_\gamma$ of anything moving at the speed of light, including photons, must be a null vector, by which we mean:

$$\mathbf{P}_\gamma \cdot \mathbf{P}_\gamma = (E_\gamma/c)^2 - \mathbf{p} \cdot \mathbf{p} = 0$$
$$\rightarrow E_\gamma = pc \tag{6.87}$$

where $p = \sqrt{\mathbf{p} \cdot \mathbf{p}}$ is the magnitude of the photon's three-momentum. Remarkably, we recovered Eq. (6.65) from these general considerations.

---

[9] Note that it has units kg·m/s.

6.8.4   Relativistic Scattering

In the absence of external forces, the total momentum and total energy of a system of particles is conserved. This is true in Newtonian mechanics and in relativistic mechanics. Consider a collision between two particles, A and B. Before the collision, their respective four-momenta are:

$$\mathbf{P}_A^{(i)} = \begin{pmatrix} E_A^{(i)} \\ \mathbf{p}_A^{(i)} \end{pmatrix}$$

$$\mathbf{P}_B^{(i)} = \begin{pmatrix} E_B^{(i)} \\ \mathbf{p}_B^{(i)} \end{pmatrix} \tag{6.88}$$

After the collision:

$$\mathbf{P}_A^{(f)} = \begin{pmatrix} E_A^{(f)} \\ \mathbf{p}_A^{(f)} \end{pmatrix}$$

$$\mathbf{P}_B^{(f)} = \begin{pmatrix} E_B^{(f)} \\ \mathbf{p}_B^{(f)} \end{pmatrix} \tag{6.89}$$

Energy-momentum conservation requires:

$$\begin{aligned} \mathbf{P}_{\text{TOT}}^{(i)} &= \mathbf{P}_A^{(i)} + \mathbf{P}_B^{(i)} \\ &= \mathbf{P}_{\text{TOT}}^{(f)} \\ &= \mathbf{P}_A^{(f)} + \mathbf{P}_B^{(f)} \end{aligned} \tag{6.90}$$

Eq. (6.90) can be applied to the relativistic collisions of particles, including massless particles such as photons, as long as one uses the relativistic expressions for energy and momentum. We study a particularly important example of such a collision, Compton scattering, in Sect. 8.2.3 but for the moment we consider a simpler example (Fig. 6.7).

Exercise 9   A pion, rest mass $m_\pi = 270 m_e$, decays from rest into a muon ($m_\mu = 205 m_e$) and an anti-neutrino, $\bar{\nu}$, ($m_{\bar{\nu}} \sim 0$). The mass of an electron is approximately 10^{-30} gms.

1. Show that after the collision the speed v_μ of the muon is given by:

$$\frac{v_\mu}{c} = \frac{m_\pi^2 - m_\mu^2}{m_\pi^2 + m_\mu^2} \tag{6.91}$$

Fig. 6.7 Pion decays into a muon moving to the right and a neutrino moving to the left

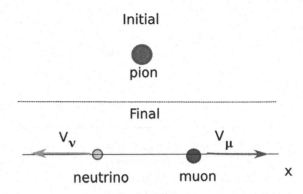

2. What is the energy of the muon after the collision, expressed in terms of m_π and m_μ? What is it in Joules?
3. What is the momentum of the neutrino after the collusion?

Note that working with MKS units to analyze the collisions of elementary particles is somewhat tedious. In Sect. 6.9 we will introduce units that are much more convenient for this purpose.

6.9 Relativistic Units

For discussions of Newtonian mechanics, we normally use MKS units, because our day to day experience involves distances, masses and time intervals that are most easily measured in meters, kilograms and seconds. The corresponding unit of energy that emerges is $1 \, \text{kg} \, \text{m}^2/\text{s}^2 = 1 \, \text{J}$. Particle physicists find it more useful to work with a much smaller unit of energy. An electron volt is the amount of kinetic energy gained by an electron that is accelerated across a potential difference of one volt. This is too small a unit for modern accelerators, so the following are used:

Unit	Joules	Mass equivalent (energy/c^2)	Particle
1 eV	1.6×10^{-19}J	1.7×10^{-36}kg	Neutrino mass
1 keV=10^3 eV	1.6×10^{-16}J	1.7×10^{-33}kg	Energy of x-rays
1 MeV=10^6 eV	1.6×10^{-13}J	1.7×10^{-30}kg	Electron mass (1/2 MeV)
1 GeV=10^9 eV	1.6×10^{-10}J	1.7×10^{-27}kg	Proton/neutron mass
1 TeV=10^{12} eV	1.6×10^{-7}J	1.7×10^{-24}kg	Higgs mass (0.1 TeV)

Example 4 The Large Hadron Collider (LHC) in Geneva, Switzerland has produced protons with total energy of $E = 7$ TeV. What is the corresponding speed of the protons?

$$E = m_0\gamma(v)c^2 = 7\,TeV$$
$$\rightarrow \gamma(v) = \frac{7\,TeV}{m_0c^2} = \frac{7\,TeV}{1\,GeV}$$
$$= 7000$$
$$\rightarrow \frac{v}{c} \sim 1 - \frac{1}{2}\left(\frac{1}{7000}\right)^2 \sim 1 - 10^{-8} \qquad (6.92)$$

Example 5 Fusion: In a typical fusion reaction a deuterium atom (^2_1D), (one proton and one neutron) and tritium atom (^3_1H), (one proton and two neutrons) combine at rest to form Helium 4 (two protons and two neutrons) plus a single neutron. How much energy is released per reaction?

The mass of ^2_1D is 2.01410178 u, where $1u = 1.661 \times 10^{-27}kg = 931.5MeV/c^2$ is an atomic mass unit, defined to be precisely one twelfth of the mass of a ^{12}C atom. Because of the binding energy of the carbon atom, 1 u is less than the mass of a proton or neutron. The mass of $^3_1H = 3.0160293u$, while the mass of $^4_2He = 4.002602$ u, and the mass of a proton is 1.007276467u:

Solution: Net change in mass $= (3.016 + 2.014 - 1.007 - 4.003)$ u $= 0.02$ u $\sim$ 0.02GeV $\sim 10^{-12}$ J.

Note: This is about 1% of the initial mass. So 1000 kg of nuclear fusion matter would yield $0.01 \times 1000\,kg \times c^2 = 10^{18}$ J. According to Statistics Canada, this is about 5% of Canada's total 2013 energy production of 17,912 petaJoules $= 18 \times 10^{18}$ J (http://www.statcan.gc.ca/daily-quotidien/141212/dq141212c-eng.htm).

6.10   Symmetry Redux

6.10.1   Matrix Form of Lorentz Transformations

We have seen in Sect. 6.8 that we can think of the coordinates (ct, x) of an event as defining a vector, denoted by **X**, in spacetime (with tail at origin $(0, 0)$). We can write Eqs. (6.9) and (6.10) as the operation of a matrix $L(v)$ on the vector **X**, just as we did for spatial rotations:

$$\mathbf{X}' = L(v)\mathbf{X}$$
$$\begin{pmatrix} ct' \\ x' \end{pmatrix} = \begin{bmatrix} \gamma(v) & -\frac{v}{c}\gamma(v) \\ -\frac{v}{c}\gamma(v) & \gamma(v) \end{bmatrix} \begin{pmatrix} ct \\ x \end{pmatrix}$$
$$= \gamma(v)\begin{bmatrix} 1 & -\frac{v}{c} \\ -\frac{v}{c} & 1 \end{bmatrix} \begin{pmatrix} ct \\ x \end{pmatrix} \qquad (6.93)$$

In order to emphasize the analogy between Lorentz transformations and rotations, it is convenient to write $L(v)$ in terms of hyperbolic functions:

$$L(v) = \begin{bmatrix} \cosh(\omega) & -\sinh(\omega) \\ -\sinh(\omega) & \cosh(\omega) \end{bmatrix} \qquad (6.94)$$

The relationship between the argument of the hyperbolic functions and the speed is:

$$\omega = \tanh^{-1}(v/c) \qquad (6.95)$$

The parameter ω is called the "rapidity". A key property of the hyperbolic functions is that:

$$\cosh^2(\omega) - \sinh^2(\omega) = 1 \qquad (6.96)$$

in contrast to the ordinary geometric functions:

$$\cos^2(\theta) + \sin^2(\theta) = 1 \qquad (6.97)$$

The appearance of hyperbolic functions is a consequence of the Lorentzian geometry discussed in Sect. 6.2.3.

Exercise 10 1. The speed of the Earth is 30 km/s. What is its rapidity?
2. Show that for small speeds $\omega \sim \frac{v}{c} + O(v^2/c^2)$.
3. Prove the following relationship between the rapidity ω and the Lorentz factor γ:

$$\cosh(\omega) = \gamma(v) \qquad (6.98)$$

4. Use the relationship in Eq. (6.98) to prove that the expressions for $L(v)$ and $L(\omega)$ in Eqs. (6.93) and (6.94) are equivalent.

Given the matrix form of the Lorentz transformations in two spacetime dimensions, it is straightforward to generalize them to three spatial dimensions. Equation (6.93) corresponds to a boost in the x-direction. Adding the y and z directions requires increasing the size, (i.e. number of rows), of our position vector and hence the Lorentz transformation matrix. The following corresponds to a boost by speed v_x in the x direction. It leaves the y and z components of the position vector unchanged.

$$L_x(v_x)\mathbf{X} = \begin{bmatrix} \gamma(v_x) & -\gamma(v_x)\frac{v_x}{c} & 0 & 0 \\ -\gamma(v_x)\frac{v_x}{c} & \gamma(v_x) & 0 & 0 \\ 0 & 0 & 1 & 0 \\ 0 & 0 & 0 & 1 \end{bmatrix} \begin{pmatrix} ct \\ x \\ y \\ z \end{pmatrix} \qquad (6.99)$$

Similarly, a boost in the y direction by a speed v_y is given by:

$$L_y(v_y)\mathbf{X} = \begin{bmatrix} \gamma(v_y) & 0 & -\gamma(v_y)\frac{v_y}{c} & 0 \\ 0 & 1 & 0 & 0 \\ -\gamma(v_y)\frac{v_y}{c} & 0 & \gamma(v_y) & 0 \\ 0 & 0 & 0 & 1 \end{bmatrix} \begin{pmatrix} ct \\ x \\ y \\ z \end{pmatrix} \qquad (6.100)$$

and similarly for a boost in the z direction by v_z.

The most general boost in three directions can therefore be written as the product of boosts in the x, y and z directions:

$$L(v_x \, v_y \, v_z) = L_x(v_x)L_y(v_y)L_z(v_z) \qquad (6.101)$$

which corresponds to first a boost in the z direction followed by a boost in the y direction followed by a boost in the x direction. If three boosts are done in a different order they will result in a different transformation because Lorentz transformations in three dimensions, just like rotations, are not commutative. Nonetheless any three dimensional Lorentz transformation can be written as a product of the form Eq. (6.101).

Exercise 11 Show that $L_x(v_x)L_y(v_y) \neq L_y(v_y)L_x(v_x)$ for general v_x and v_y.

Exercise 12 Suppose a particle of rest mass m_0 and three-momentum $\mathbf{p} = (p_x, 0, 0)$ is acted on by a Lorentz transformation with speed v_y in the y-direction. What is its new four-momentum vector $\mathbf{P'}$? Show that the $\mathbf{P} \cdot \mathbf{P} = \mathbf{P'} \cdot \mathbf{P'} = m_0^2 c^2$.

6.10.2 Lorentz Transformations as a Symmetry Group

In Sect. 3.3.4 we verified that the set of rotations form a symmetry group. We argued the same for Lorentz transformations in Sect. 6.2.2. This must be true because the fundamental postulate states that the laws of physics are the same in all inertial frames. Thus, Lorentz transformations, which transform coordinates from one inertial frame to another, are symmetries of the laws of physics, and we know from Sect. 3.2.4 that the complete set of symmetries of an object or system of equations forms a group. We now verify that Lorentz transformations obey the four defining properties of a group of symmetry transformations using the matrix representation Eq. (6.93).

1. **Identity** A boost by $v = 0$ corresponds to the identity matrix in Eq. (6.93) and as expected leaves all spacetime vectors invariant.
2. **Inverse** The matrix inverse of $L(v)$ is:

$$L^{-1}(v) = L(-v) \qquad (6.102)$$

Exercise 13 Verify Eq. (6.102) for the Lorentz transformation given in Eq. (6.78).

---

3. **Completeness** One can verify by explicit matrix multiplication that a boost by v_1 followed by a boost by v_2 is equivalent to a boost by v_3:

$$v_3 = \frac{v_1 + v_2}{1 + \frac{v_1 v_2}{c^2}} \qquad (6.103)$$

This is the same velocity addition rule as derived using a different method in Sect. 6.5.

Exercise 14 Use matrix multiplication to verify that $L(v_2)L(v_1) = L(v_3)$ where v_3 is given in Eq. (6.103).

---

4. **Associativity** this implies that for any three boosts v_1, v_2, and v_3:

$$[L(v_3)L(v_2)] L(v_1) = L(v_3) [L(v_2)L(v_1)] \qquad (6.104)$$

Exercise 15 Prove that Lorentz transformations obey associativity, namely Eq. (6.104).

6.11 Supplementary: Four-Vectors and Tensors in Covariant Form

We saw in this chapter that the three-dimensional spatial vectors (*three-vectors*) of non-relativistic mechanics can be generalized to four-dimensional spacetime vectors (*four-vectors*) in relativistic mechanics that specify the locations in spacetime of specific events. In the limit $c \rightarrow \infty$ or $v/c \rightarrow 0$, each four-vector reduces to a three-vector plus a scalar. For example, the *event four-vector*

$$X^\mu = \begin{pmatrix} ct \\ x \\ y \\ z \end{pmatrix} \qquad (6.105)$$

reduces to the *position three-vector*

$$x^i = \begin{pmatrix} x \\ y \\ z \end{pmatrix} \qquad (6.106)$$

and scalar time t, which transform independently under Galilean transformations in the limit $c \rightarrow \infty$. Here we have denoted the above four-vector using the *covariant* notation X^μ, where the 'index' or superscript μ runs from $0, 1, 2, 3$, with $x^0 = ct, x^1 = x, x^2 = y, x^3 = z$. Similarly, the components of the three-vector $\mathbf{x}$ is denoted by x^i, where the index i runs from $1, 2, 3$, with $x^1 = x, x^2 = y, x^3 = z$. We will follow this commonly adopted convention in this book: a vector with a Greek index will signify a four-vector, with the index running over time and space, while a vector with a Latin index will signify a three-vector, with the index running over space alone. Further, to facilitate composition of four-vectors to form 'four scalars', 'world scalars' or 'Lorentz scalars', which are invariant under Lorentz transformations, we will call the vector X^μ a 'contravariant vector'. Corresponding to each such contravariant vector, we will define a 'covariant vector' X_μ (i.e. now μ is a subscript) in the following way:

$$X_\mu = \begin{pmatrix} ct \\ -x \\ -y \\ -z \end{pmatrix}, \tag{6.107}$$

i.e. with the sign of its spatial components reversed. In other words,

$$X_0 = X^0, \ X_1 = -X^1, \ X_2 = -X^2, \ X_3 = -X^3 . \tag{6.108}$$

With the above, one can define the four-dimensional 'length' or norm of the four-vector as follows:

$$\mathbf{X}^2 = \sum_{\mu=0}^{3} X^\mu X_\mu = (ct)^2 - x^2 - y^2 - z^2 = c^2 \tau^2 , \tag{6.109}$$

which as we showed earlier is a Lorentz invariant quantity, a Lorentz scalar or a space-time scalar. Note that $\mathbf{X}^2$ as defined above can be positive (timelike), negative (space-like) or zero (null). The above notation lends itself nicely to the four-dimensional generalization of the Pythagoras theorem. We will now go one step further and drop the summation symbol $\sum_{\mu=0}^{3}$ altogether, and write Eq. (6.109) as

$$\mathbf{X}^2 = X^\mu X_\mu = (ct)^2 - x^2 - y^2 - z^2 = c^2 \tau^2 . \tag{6.110}$$

In this convention, whenever one sees a *repeated* Greek index on four-vectors, with one up (contravariant) and one down (covariant), it will be assumed that there is an implicit summation $\sum_{\mu=0}^{3}$ to the left of it. Einstein introduced this convention to simplify calculations with multiple vectors and tensors, which appear frequently in special and general relativity. It is therefore referred to as the *Einstein Summation Convention*.

The above generalizes to any four-vector A^μ, with

$$A^\mu = \begin{pmatrix} A^0 \\ A^1 \\ A^2 \\ A^3 \end{pmatrix} \tag{6.111}$$

$$A_\mu = \begin{pmatrix} A^0 \\ -A^1 \\ -A^2 \\ -A^3 \end{pmatrix} . \tag{6.112}$$

For example, the four-vector A^μ can be the four velocity vector $V^\mu = dx^\mu/d\tau$ of a particle, or the covariant version of the electromagnetic vector potential:

$$A^\mu \propto \begin{pmatrix} \phi \\ A^1 \\ A^2 \\ A^3 \end{pmatrix}. \tag{6.113}$$

where ϕ is the electrostatic potential whose gradient yields the electric field and $\mathbf{A}$ is the magnetic vector potential whose curl gives the magnetic field.

We will return to other examples momentarily, but first recall the definition of the three-dimensional gradient operator

$$\nabla = \begin{pmatrix} \frac{\partial}{\partial x} \\ \frac{\partial}{\partial y} \\ \frac{\partial}{\partial z} \end{pmatrix} =: \begin{pmatrix} \partial_1 \\ \partial_2 \\ \partial_3 \end{pmatrix}. \tag{6.114}$$

How do we generalize this to four-dimensions? It turns out that one can define

$$\partial_\mu = \begin{pmatrix} \partial_0 \\ \partial_1 \\ \partial_2 \\ \partial_3 \end{pmatrix} \tag{6.115}$$

$$\partial^\mu = \begin{pmatrix} \partial_0 \\ -\partial_1 \\ -\partial_2 \\ -\partial_3 \end{pmatrix}, \tag{6.116}$$

where $\partial_\mu = \partial/\partial x^\mu$. Note that in this case the covariant form with all plus signs is defined first, and subsequently its contravariant form with the spatial components reversed. This ensures that the four-dimensional divergence $\partial_\mu A^\mu = \partial_0 A^0 + \nabla \cdot \mathbf{A}$ of a four-vector is Lorentz invariant.

Another example of a four-vector and its norm is the 'Energy-Momentum' four-vector

$$P^\mu = \begin{pmatrix} E/c \\ p_x \\ p_y \\ p_z \end{pmatrix} \tag{6.117}$$

$$P_\mu = \begin{pmatrix} E/c \\ -p_x \\ -p_y \\ -p_z \end{pmatrix} \tag{6.118}$$

$$P^2 = P^\mu P_\mu = (E/c)^2 - p_x^2 - p_y^2 - p_z^2 = m_0^2 c^2. \tag{6.119}$$

Once again, the quantity m_0 in the last line is a Lorentz invariant or four scalar, and as we have seen is the rest mass, or invariant mass of the particle. When the three-momentum of the particle (components p_x, p_y, p_z and hence magnitude $|\mathbf{p}|$) is negligible, one obtains the celebrated relation $E = m_0 c^2$. This is the non-relativistic limit. In the opposite, ultra-relativistic limit, when $|\mathbf{p}| \sim \mathbf{E}/c$, one has

$$P^\mu = \begin{pmatrix} |\mathbf{p}| \\ p_x \\ p_y \\ p_z \end{pmatrix} \tag{6.120}$$

$$P^2 = P^\mu P_\mu = |\mathbf{p}|^2 - \mathbf{p_x^2} - \mathbf{p_y^2} - \mathbf{p_z^2} = \mathbf{0} \tag{6.121}$$

The summation convention can be used to define scalar products of two four-vectors, for example X^μ and P^μ as follows

$$\mathbf{X} \cdot \mathbf{P} = X^\mu P_\mu = x^0 p_0 + x^1 p_1 + x^2 p_2 + x^3 p_3 \tag{6.122}$$
$$= x^0 p^0 - x^1 p^1 - x^2 p^2 - x^3 p^3 = Et - \mathbf{r} \cdot \mathbf{p} . \tag{6.123}$$

When expressed as quantum mechanical relations from wave-particle duality, $E = \hbar\omega$, $\mathbf{p} = \hbar\mathbf{k}$, where ω and $|\mathbf{k}| = 2\pi/\lambda$ are the angular frequency and wave number of the corresponding wave, respectively, while λ is the wavelength. ω and $k = |\mathbf{k}|$ are related by $\omega/k = v$.

$x \cdot p = \hbar(\omega t - \mathbf{k} \cdot \mathbf{r}) = \hbar\phi$ where ϕ is the phase of the wave and $\hbar$ is the reduced Planck constant (See Chap. 8). The amplitude of the wave in spacetime can be written as $\psi = e^{i\phi} = e^{i(\omega t - \mathbf{k} \cdot \mathbf{r})}$. For real waves (such as sound or light waves), one takes the real part of ψ at the end of the calculation. As we have seen earlier, the frequency and wavelength of the wave changes under a Lorentz transformation (they are Doppler shifted), but the phase remains unchanged. In fact if it is a light wave, then its speed c remains unchanged as well.

One can now generalize the definition of four-vectors to four tensors of arbitrary rank. The simplest way to construct these is by using two or more four-vectors. For example, one can define the four angular momentum tensor as follows:

$$L^{\mu\nu} = x^\mu p^\nu - x^\nu p^\mu . \tag{6.124}$$

This is a tensor of rank two. One can easily verify the following:

- $L^{\mu\nu}$ is anti-symmetric in the indices μ and ν. That is, $L^{\mu\nu} = -L^{\nu\mu}$ under the exchange of any pair of indices $\{\mu, \nu\}$.
- $L^{ij} = x^i p^j - x^j p^i$ are the components of three angular momentum $\mathbf{L} = \mathbf{r} \times \mathbf{p}$.
- $L^{0j} = ct\, p^j - (E/c)x^j$ are known as the 'boost generators'. They are the space-time generalizations of the angular momentum operators, conjugate to Lorentz boosts, which can be thought of as rotations in spacetime, but with imaginary or hyperbolic angles[10]

---

[10] Recall one can relate the sine of an imaginary angle to the hyperbolic sine of a real angle using the expression for the sine function written in terms of complex exponentials:

• A particularly important tensor is the completely anti-symmetric fourth rank four tensor known as the *Levi-Civita tensor* $\epsilon^{\mu\nu\lambda\sigma}$. It is defined by $\epsilon^{0123} = 1$. All symmetric permutations of the indices also give 1 while anti-symmetric permutations give -1. Thus for example $\epsilon^{1230} = 1$, while $\epsilon^{0132} = -1$. This requires that the Levi-Civita tensor with one or more repeated indices, such as ϵ^{0113}, are equal to zero. For example, in terms of this tensor one can write: $L^{\mu\nu} = \epsilon^{\mu\nu\lambda\sigma} x_\lambda p_\sigma$, where the summation convention is assumed.

One can define other four tensors such as the stress-energy tensor $T^{\mu\nu}$, whose (0, 0) component, T^{00}, give the energy density, while the components $T^{0j} = T^{j0}$ correspond to the components of the momentum density along each of the three spatial directions. The components T^{ij} give the stress and shear stress of the matter. These are analogous to the stresses that occur in ordinary materials due to forces that try to compress or twist them. For classical fields, such as a scalar or electromagnetic field, these components normally depend on the derivatives of the fields. The conservation of energy and momenta is expressed as the partial differential equation

$$\partial_\mu T^{\mu\nu} = 0 \qquad (6.126)$$

again with the summation convention over repeated index μ assumed.

An important feature of tensor equations such as Eq. (6.126) is that the spacetime indices μ, ν, ... obey simple rules. Specifically, it is possible to sum over a pair of indices provided one is up and the other is down. In addition, excluding summed indices there must be an equal number of up and down indices on both sides of the equations. Assuming the quantities in the equations are tensors, these rules guarantee their covariance under Lorentz transformation.

Four-vectors and tensors are especially useful in the General Theory of Relativity, a relativistic theory of gravitation, the discovery of which in 1915 provided Einstein's crowning achievement. These tensors are more general than the ones we have discussed so far, since the equation must be covariant under more general coordinate transformations than the Lorentz transformations. General Relativity is the subject of the next chapter.

$$\sin(i\theta) = \frac{e^{i(i\theta)} - e^{-i(i\theta)}}{2i} = -\frac{e^\theta - e^{-\theta}}{2i} \qquad (6.125)$$
$$= i \sinh(\theta)$$

Chapter 7
General Relativity

7.1 Learning Outcomes

Conceptual:

- The problems with Newtonian gravity.
- The weak and strong equivalence principles and their significance in Einstein's formulation of gravity.
- Geometrical structure and some basic consequences.
- Gravitational force due to non-constant gravitational acceleration.
- Black holes: properties, observational evidence and theoretical importance.
- Gravitational waves: what they are and recent observational evidence.
- Cosmology: a general description.

7.2 Problems with Newtonian Gravity

7.2.1 Review of Newtonian Gravity

In 1687 Newton revolutionized the world of science with the publication of *Principia*.[1] It detailed three laws[2]:

1. The first law postulates that the velocity of an object stays constant in the absence of external forces.

---

[1] The full name of the work is *Philosophiæ Naturalis Principia Mathematica* which is Latin for *Mathematical Principles of Natural Philosophy*.

[2] There is some debate as to whether there are three laws or two. The first law follows from the second, but in some sense the first is the starting point.

© The Author(s), under exclusive license to Springer Nature Switzerland AG 2022
G. Kunstatter and S. Das, *A First Course on Symmetry, Special Relativity and Quantum Mechanics*, Undergraduate Lecture Notes in Physics,
https://doi.org/10.1007/978-3-030-92346-4_7

2. The second specifies that the acceleration of an object is proportional to the net force acting on it. The constant of proportionality is a property of the object, namely the inverse of its inertial mass.
3. The third states that every action has an equal and opposite reaction.

Newton also described the gravitational force responsible for the motions of projectiles, satellites and planets. According to Newton, the gravitational force between two masses that are small compared to their separation is given by:

$$\mathbf{F}_{2on1} = -\frac{Gm_1^{(g)}m_2^{(g)}}{r_{12}^3}\mathbf{r}_{12}, \tag{7.1}$$

where the subscript $2on1$ denotes that this is the force of mass 2 on mass 1. $\mathbf{r}_{12}$ is the position vector (not a unit vector) of mass 1 relative to mass 2, while G is a fundamental constant of nature, called Newton's constant. $m_1^{(g)}$ and $m_2^{(g)}$ are the gravitational masses of the two objects that determine the strength of the force that each exerts on the other. By Newton's third law the force on particle 1 due to particle 2 is of equal magnitude and opposite direction to the force on particle 2 due to particle 1, as required by the assumption that the net force on the two particles is zero.

The constant G was not prescribed by the theory, but had to be measured. It is now known to be:

$$G = 6.67408 \times 10^{-11} \text{ N m}^2/\text{kg}^2. \tag{7.2}$$

Assuming no other forces act on it, the motion of mass 1 is determined from Newton's second law for the acceleration $\mathbf{a}_1$:

$$m_1^{(i)}\mathbf{a}_1 = -\frac{Gm_1^{(g)}m_2^{(g)}}{r^3}\mathbf{r_{12}}. \tag{7.3}$$

The mass $m_1^{(i)}$ on the left hand side of Eq. (7.3) is the inertial mass that, by definition, determines the resistance of mass 1 to a change in motion, whatever the force. The gravitational masses $m_1^{(g)}$ and $m_2^{(g)}$ on the right hand side determine the strength of the gravitational force between the two objects. Gravitational mass plays the same role in Newtonian gravity as electric charge does in Maxwell's theory. Gravitational mass is therefore conceptually and physically distinct from inertial mass, despite the fact that both are given the same units. These units do not in principle have to be kilograms. They could be something new, let's say "Einsteins", in which case Newton's constant would be measured in N · m^2/(Einstein^2).

It turns out that Newtonian gravity reproduces observed gravitational phenomena if one takes the gravitational mass of any object to be proportional to its inertial mass: $m^{(g)} \propto m^{(i)}$. Without loss of generality one can take the two to have the same units and set the constant of proportionality to be 1. In this case, Newton's constant takes the value given in Eq. (7.2). The equality (proportionality) of gravitational and inertial mass has been verified to a high degree of accuracy. One well known manifestation of the proportionality of gravitational and inertial mass is that in the absence of air resistance all objects fall towards Earth at the same rate. In brief $m_1^{(i)}$ and $m_1^{(g)}$ cancel out of Eq. (7.3). More will be said about this in Sect. 7.2.4.

Newtonian gravity is a central force, in that it points directly towards the source of the gravitational field. This is the same as the electric force in Maxwell's theory of electromagnetism. The difference is that Newtonian gravity is always attractive, unless of course one were to find objects with negative mass, something that has not been encountered to date. There are in fact theorems in general relativity proving that negative mass objects cannot be constructed from matter that obeys certain generic and quite reasonable conditions.

Newton's universal law of gravitation is incredibly successful in describing a large range of phenomena, both terrestrial and astronomical. It does, however, have some fundamental problems associated with it, which we will outline in the subsequent subsections.

7.2.2 Perihelion Precession of Mercury

Newtonian gravity predicts stable elliptical planar orbits for all satellites orbiting a central object that is small compared to the size of the orbit and much more massive than the satellite. An *elliptical* orbit is one that takes the shape of an elongated circle, while *planar* means that the line joining the central object to the satellite remains in the same plane as the satellite orbits. The term satellite includes satellites orbiting the Earth, moons orbiting their respective planets and of course the planets themselves orbiting the Sun. According to Newton's laws, the elliptical orbit of a satellite is affected by other massive objects in the vicinity as well as the finite size of the central object that it orbits. If these other effects are small but not zero, which is the case for the planets, they cause the orbits to *precess* slightly. The elliptical shape and planarity are not affected, but the line through the central object and satellite rotates slowly as the satellite orbits. This is shown in Fig. 7.1.

Fig. 7.1 Newtonian gravity predicts the orbit of Mercury to be an ellipse whose perihelion (and aphelion) precesses around the Sun at about 500 s of arc per century, as shown. This differs from Einstein's prediction by 43 s of arc per century. It takes Mercury's orbit about 100,000 Earth years to shift the 45° shown in the figure

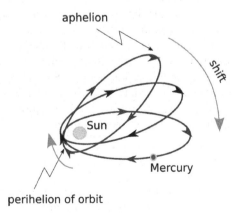

Another way to describe the precession is to say that the point of closest approach to the central object, called the *perihelion*, does not stay in the same place. It shifts very slightly with every successive orbit, so that this shift or precession, as it is called, builds up over time and is measurable. In the late 1800s, Le Verrier, the director of the Royal Observatory in Paris at the time, calculated from Newton's theory the expected precession of Mercury's orbit taking into account the finite size of the Sun and perturbations due to other planets in the solar system. He compared these to the observed value obtained from 150 years of astronomical data and found that Mercury was precessing more quickly than Newton's theory predicted. The discrepancy was about 40 s of arc per century.[3] A second of arc is an angle that corresponds to one sixtieth of one sixtieth of a degree, just as a second of time is one sixtieth of one sixtieth of an hour. In order to visualize how small an angle this is, note that it corresponds to the angle subtended by a quarter seen from a distance of 5 km. Now imagine the quarter orbiting you so slowly that after a year it has moved about half of its own diameter. A discrepancy of 43 s of arc per century may seem incredibly small, but it was a big source of concern for astronomers of the day. In order to explain this anomaly, Le Verrier postulated the existence of an unseen planet, which he called Vulcan, on the far side of the Sun from the Earth.[4]

7.2.3 Action at a Distance

Once Einstein convinced the world Newtonian mechanics must be replaced by its relativistic version, it quickly became apparent that there was a fundamental inconsistency between Newtonian gravity and special relativity. The gravitational force in Eq. (7.1) depends on the instantaneous distance between the two masses. Imagine that at precisely noon in your time zone, a technologically superior alien race makes the Sun disappear by moving it into a parallel Universe. According to Newton's theory of gravity, the Earth would immediately feel the absence of the Sun's gravitational pull and go flying off into space in a straight line. This seems a bit odd, to say the least, because the light that you see at noon was emitted by the Sun at around 11:52 am,[5] so that the Sun would still appear to be shining in the sky. More significantly, the Newtonian description of this event violates the theory of special relativity. We would learn instantaneously by the change in the Earth's trajectory that the Sun had disappeared. This information would have to travel to us outside the future light cone of the disappearance of the Sun, a blatant violation of the laws of relativity. To emphasize this point further, we recall that in special relativity "instantaneous" is in the eye of the observer, so the formulation of this entire process is ambiguous at best in the context of Newtonian gravity.

---

[3] More accurate calculations and observations have determined that the discrepancy is 43 s of arc per century.

[4] Vulcan is now known not to exist, except on re-runs of the old television series Star Trek.

[5] It takes light about 8 min to get to us from the Sun.

7.2.4 The Puzzle of Inertial Versus Gravitational Mass

As mentioned in Sect. 7.2.1, it has been verified experimentally to a high degree of accuracy that the inertial mass of any object is equal to its gravitational mass[6]:

$$m^{(i)} = m^{(g)} \tag{7.4}$$

The equality of gravitational and inertial masses for all forms of matter is dubbed the *Weak Equivalence Principle* (WEP). One physical manifestation of this equality is that, when air resistance is negligible, everything falls towards the Earth at the same rate, irrespective of its shape, mass or constitution. You can verify this for yourself by holding your keys and your wallet at the same height above the floor before releasing them. As is well known, they will accelerate downwards at the same rate and hit the floor simultaneously.

In the context of Newtonian gravity this is a complete mystery, one that played a major role in the discovery of the General Theory of Relativity. Einstein understood that if all massive objects obey the *Weak Equivalence Principle*, then experiments done in free fall in a constant gravitational field would give indistinguishable results from those done in an inertial frame in the absence of gravity.[7] Imagine, for example, that Alice and fellow astronaut Bob are tossing a tennis ball back and forth in a non-accelerating space ship far from the effects of any gravitating objects or external forces. We know from Newton's laws that the ball will travel at constant velocity from Alice to Bob and back again. Now imagine that they are tossing a tennis ball back and forth in an elevator on Earth that is in free fall. Alice, Bob and the tennis ball are all accelerating downward at the same rate (9.8 m/s^2), so from their point of view the tennis ball again travels in a straight line back and forth between them. Neither Alice nor Bob would be able to tell from this, or any other experiment involving projectiles that they were in an elevator in free fall, until of course they either hit the ground or are rescued by super heroes.

The above argument follows from the WEP but only applies to the mechanics of massive objects. Einstein realized that some form of the WEP must be valid for all interactions and therefore formulated a new and powerful symmetry principle: the *Strong Equivalence Principle*.

---

[6] We emphasize again that the correct statement is that experiment verifies that they are proportional. Without loss of generality one can take them to be equal, thereby fixing the numerical value of Newton's constant uniquely.

[7] If the gravitational field were not constant, different parts of the lab would fall at different rates or in different directions, something that would be readily detectable.

7.3 Strong Equivalence Principle

Statement of the Strong Equivalence Principle (SEP): All laws of physics have
the same form locally in all frames of reference, including non-inertial frames (i.e.
ones not in uniform motion).

"*Locally*" in this context means in a sufficiently small volume. It suggests that if
the lab is too big, things would be a bit different. This is indeed the case. As will be
highlighted in Sect. 7.4, when the gravitational field is not constant the acceleration
it causes is not constant and introduces interesting, and potentially fatal, effects.
The lab must be small enough so that the gravitational acceleration is approximately
constant throughout.

The SEP is a symmetry principle that generalizes the fundamental postulate of
special relativity from just inertial frames to all frames of reference. This is why
Einstein's theory of gravity is called "general relativity". Given that all frames of
reference are equivalent, one can now do transformations of coordinates, including
boosts, that vary from point to point. Recall that special relativity is based on a global
symmetry in that Lorentz transformations are specified by three boost parameters that
are the same everywhere. One cannot do an independent boost at each point in space.
Symmetries whose parameters can be functions of the spatial coordinates are called
local symmetries. In this sense, general relativity extends the global symmetries of
special relativity to local symmetries.[8]

Einstein realized that experiments done in a lab in free fall in a gravitational field
that is constant everywhere within the lab would give results that were indistinguishable from what would be obtained if the lab were in an inertial frame (i.e. subjected
to no external forces, including gravity).

His next intuitive leap was the realization that if the laws of physics are the same
in the inertial frame of the spaceship as in the freely falling frame of the elevator and
the results of all experiments are the same in both cases, then it must follow that if
there is no measurable gravity in the spaceship there can be no measurable gravity
in the freely falling elevator. As with Einstein's previous insights, this seems at first
glance to be paradoxical. How can one eliminate gravity by going to a freely falling,
accelerated frame? Surely the elevator is accelerating because gravity is pulling down
on it!

Before answering this question, we emphasize that the elevator is accelerating
relative to the surface of the Earth, just as the surface of the Earth is accelerating
relative to the elevator. It perhaps is reasonable to ask which of the two frames is

---

[8] For future reference, not in this course but in more advanced courses, whenever a global symmetry
is extended to a local symmetry, it is necessary to introduce a new dynamical field, sometimes called
a *compensating field*. Maxwell's theory of electromagnetism extends the global phase symmetry
of quantum mechanics (see Sect. 9.2.2) to a local symmetry, and the electromagnetic field is the
(derivative of) the corresponding compensating field. In general relativity the compensating field is
essentially the metric, or geometry of spacetime, which becomes dynamical to preserve the local
symmetry in the equations.

actually accelerating. Although the answer may seem obvious, consider the following situation.[9]

You are on a ride in an amusement park in which you are standing against the wall of a circular cage that is being rotated rapidly at constant angular velocity. Now hold your keys and your wallet directly overhead, above the cage wall and release them.[10] You know from Newton's laws that once released, the keys and wallet will move outward in a straight line while you continue to be whipped around in a circle by the walls of the cage. From your point of view, the keys and wallet will appear to accelerate away from you at exactly the same rate. In your frame, they both appear to be acted on by a mysterious force that is proportional to their inertial mass. In this case, however, you know that you are being acted on by the force of the cage on your back, and that the keys and wallet are in an inertial frame. You are in a non-inertial frame and this is the reason that the keys and wallet appear to be accelerating away from you at the same rate. It is you who are accelerating away from them. The force that you think is pulling the objects directly away from you is *fictitious*. [11]

Since, according to the Fundamental Postulate, the laws of physics are the same when you are standing stationary on the ground as they are in the accelerated frame on the amusement park ride, this *gedanken* experiment should suggest to you, if not convince you outright, that in both cases the keys and wallet that you release are in inertial frames, and it is you who is accelerating. But to be in an accelerating frame, you must be subject to a net external force. On the ride the force is provided by the wall of the cage. When you are standing on the ground, it is the floor that provides the electromagnetic forces keeping you in your accelerated frame. Why then do you not feel this force? Upon reflection, you will realize that you do feel a force on the soles of your feet, which you have been mistakingly interpreting as the force of gravity pulling you down. In fact, the only force acting on you is caused by the floor that is keeping you in an accelerated frame. The force of gravity that makes things fall to Earth at the same rate is a fictitious force! Two important questions should come to mind:

Question: How can acceleration towards the center of the Earth and straight line motion in the absence of gravity both be described as inertial motion using equations that have the same form for all observers?

Answer: Both types of motion follow the straightest line possible in the curved spacetime near them. At the surface of the Earth, the curvature of space and spacetime cause a frame that is stationary with respect to the center of the Earth to be non-inertial. The force term appears in our usual description of acceleration in a constant gravitational field because the equations are written in the non-inertial frame. The

---

[9] This and the elevator experiment are examples of the thought experiments (*gedanken* experiment in German) often used by Einstein to clarify concepts. In other words, they are not to be tried at home!

[10] Remember, this is a *gedanken* experiment! We recommend that you do not try it in real life.

[11] We stress that although this force is indeed fictitious, real gravitational forces do exist whenever the gravitational field is not constant. This will be discussed further in Sect. 7.4.

description of the keys on the amusement park ride would also contain such a constant acceleration term in your rotating frame. If one were to describes the motion of the object falling towards the earth in an inertial (freely falling) frame, there would be no fictitious force term and the equations would look the same as that for the spaceship moving at constant velocity far from Earth.

Question: What causes the geometry of spacetime to curve?

Answer: Einstein's equations tell us that the presence of matter and energy causes spacetime to curve.

These considerations lead us to the following beautiful two line summary of Einstein's theory of gravity proposed by John Wheeler,[12] the late Princeton physicist who also first coined the phrase "Black Hole".

1. Matter tells spacetime how to curve.
2. Curved spacetime tells matter how to move.

Matter at any point A in spacetime can create curvature at any other point B. Any change in the location or motion of A will cause the curvature at B to change. Special relativity requires that this change not manifest itself at B instantaneously. It must somehow propagate towards B at a finite speed that is no greater than the speed of light.

7.4   Geometry of Spacetime

Euclidean geometry is defined by the expression for the length Δs of the line segment joining any two points. In Cartesian coordinates and three dimensions the famous Pythagoras theorem states that:

$$\Delta s^2 = \Delta x^2 + \Delta y^2 + \Delta z^2. \tag{7.5}$$

Equation (7.5) determines the geometry of three-space. You know many aspects of the geometry already: the sum of the angles in a triangle is 180°, parallel lines never meet and the areas of all parallelograms of the same height are equal, to name a few.

---

[12] It is perhaps an apocryphal tale that John Wheeler wrote this two line summary as a Japanese form of poetry called Haiku, which consists of 17 syllables in three lines with 5-7-5 syllables each. In this case it should be written:

Matter tells spacetime
how to curve. Curved spacetime tells
matter how to move.

Whether or not Wheeler intended it as Haiku, it is a beautiful way to recall the essence and elegance of general relativity. We are grateful to Jack Gegenberg for bringing this to our attention.

Fig. 7.2 Lines of longitude
on the Earth or on a balloon
cross the equator at 90° but
still meet at the poles.
Airplanes or ants that wish to
minimize time and energy
will travel along great circles
and therefore be able to
distinguish the geometry
from that of Euclid

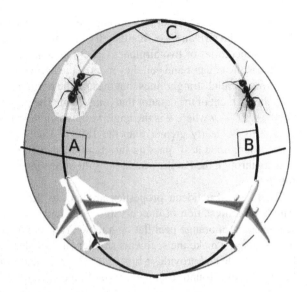

In Chap. 6 we saw that special relativity unifies space and time into a single space-time continuum. An important quantity is the proper time $d\tau^2$ along any segment of the worldline of an observer:

$$c^2 d\tau^2 = c^2 dt^2 - dx^2 - dy^2 - dz^2. \tag{7.6}$$

Equation (7.6) defines the geometry of spacetime in exactly the same way that Eq.(7.5) defines the geometry of three dimensional space.[13] In special relativity this geometry is rigid, unchanging. It is the same at all places and for all time. This geometry is known as flat spacetime, to reflect the absence of curvature. It is also called Minkowski spacetime after Hermann Minkowski, the German mathematician (and Einstein's teacher) who was the first to explain the underlying geometrical structure of special relativity.

The simplest non-Euclidean geometry we can consider is that of the surface of a sphere. On the surface of the Earth, which is roughly a sphere, parallel lines do meet, and the sum of the angles in a triangle is greater than 180°, in contrast to what Euclid tells us. To convince yourself that this is true, consider two lines of longitude on the Earth that run from the south pole to the north pole. They are both the straightest lines you can draw on the the surface of the Earth, and form right angles with the equator, where they are all parallel to each other (Fig. 7.2).

Any airplane pilot flying from the south pole to the north pole who wishes to minimize the distance travelled, and hence the fuel consumption, will fly along a line of longitude. This does not require complicated navigational calculations because in

---

[13] We have switched to infinitesimal line segments because the notion of a long straight line gets complicated once the geometry is neither Euclidean nor Lorentzian.

the absence of heavy winds and turbulence, the pilots need only make sure that they do not turn the airplane, i.e. that they fly in a straight line.

The trajectories of two airplanes travelling along different lines of longitude are parallel at the equator and yet, they meet at the north pole. Two lines of longitude are therefore parallel straight lines that meet at the north pole. They also form a triangle with the segment of the equator that joins them. The sum of the angles of this triangle is $90° + 90° + \theta$ where θ is the angle with which their trajectories meet at the north pole. This is clearly greater than the 180° of Euclidean geometry. In fact for the lines of longitude at 0° passing through Greenwich, UK and at 180° East or West passing through the Pacific ocean, the sum of the angles of the resulting triangle is a full 360°.

These non-Euclidean properties are due to the curved geometry of spheres. Another manifestation of the curvature of spheres appears when you try to lay the segments of an orange peel flat on a table. You cannot do it without deforming the pieces. As you make the segments smaller and smaller this effect becomes less and less noticeable and provides a hint as to why the ancients found it difficult to believe that the Earth is round from their rather limited perspective on its surface. However, the effects of curvature can never be completely removed by going to smaller and smaller scales. No matter how small you make a triangle on a curved surface, the sum of the angles will not be 180°. It is only when the triangle is reduced to a single point with zero extension that it is impossible to detect the curvature of the surface.

The geometry of the sphere is completely encoded in how the length of any line segment is related to the change in coordinates needed to get from one end of the line segment to the other. Suppose you use the usual angles θ, ϕ to navigate the surface of a sphere of radius r. The length of the line segment between two infinitesimally close points on a sphere is:

$$ds^2 = r^2(d\theta^2 + \sin^2(\theta)d\phi^2) \tag{7.7}$$

There is no way to choose coordinates on the sphere to make Eq. (7.7) look like Eq. (7.5).

Now consider a stretched balloon. Its surface has roughly the same spherical geometry as that of the Earth. The straightest possible lines are again great circles that are parallel at the middle, or "equator" and meet at the "poles". A tiny ant walking on the surface of the balloon towards a piece of food will naturally follow a great circle, simply by making sure it does not turn: the legs on both sides travel exactly the same distance each step. Now imagine that the surface of the balloon, if it is not too tightly stretched, can deform. In this case the geometry of the surface changes near the deformation. The ant walking on the balloon's surface will still go along the straightest possible line that it can, but its trajectory will change as it follows the curved contours of the balloon. Despite the different trajectory, the ant will not feel any force causing it to turn as it walks the straightest line possible. If the ant were a physics major, it would consider itself to be moving at constant speed in an inertial frame.

To reiterate, the airplanes and the ant follow the straightest possible lines on the curved spaces to which they are confined to move. In John Wheeler's words, the geometry of space (and spacetime) tells them how to move. In addition a heavy ant sitting on the top of the balloon will change the geometry of the balloon in its neighbourhood, changing the straightest possible lines in its vicinity. Again, in Wheeler's words, matter tells space and spacetime how to curve. You will note that we have been a bit cagey about the distinction between the curvature of space and that of spacetime. They are closely related, but the latter is much harder to visualize, so we have focussed the discussion on the former.

Finally, we stress that even though the ants and airplanes travel along straightest possible lines along the curved surface of the Earth as long as they do not experience external (non-gravitational) forces that would veer them away from the straight line, true non-fictitious gravitational forces nonetheless do exist. Suppose you are in free fall towards the center of a very massive star, or even a black hole. The gravitational acceleration you experience is not the same everywhere. Specifically, it is somewhat stronger at your feet, which are slightly closer to the center of the star than your head. As well, your left and right shoulder are accelerating in slightly different directions because they move along slightly different radial lines. The net affect is that you feel both stretched and squeezed as you fall because you are of finite size. These finite size affects in a non-constant gravitational field correspond to forces that are very real: if you were unlucky enough to be falling into a black hole, these *tidal forces*, as they are called, would rip you apart as you got too close. For example, a person of height $h = 2$ m in free fall just outside the horizon of a solar mass black hole would feel a tidal acceleration, a_t, of their feet relative to their head of about 10^{10} m/s^2. Ouch!

In general relativity spacetime is no longer a rigid stage on which actors move. Instead it is an active participant whose shape is influenced by and also influences the play as it unfolds. The precise nature of the geometry of spacetime in the presence of matter and/or energy is governed by the equations of general relativity as written down by Einstein in 1915, ten years after his discovery of special relativity in 1905.

7.5 Some Consequences of General Relativity

- General relativity solves Newtonian gravity's problem of action at a distance. Objects respond to deformations of the geometry of spacetime in their immediate vicinity. These deformations are caused by the motion of distant objects and travel at finite speed along the fabric of spacetime, like ripples on the surface of a balloon. It turns out that these deformations travel at the same speed as light, so that, as expected from special relativity, if the Sun were removed from our Universe, the Earth would deviate from its orbit at exactly the same moment that its rays of sunshine ceased to reach us.

- The deformations or ripples in the fabric of spacetime are called *gravitational waves*. Gravitational waves were first observed in 2015 by the Light Interferometry Gravitational Wave Observatory (LIGO), which detected gravitational waves emitted by the merger of two massive black holes 1.3 billion light-years away. We will say more about this in Sect. 7.6.
- The equations of general relativity yield the correct observed value of the perihelion shift of Mercury. It is worth emphasizing the magnitude of this achievement. Einstein started from a symmetry principle (the SEP) and deduced its consequences using pure reason. He then showed that the resulting theory of general relativity gave precisely the right correction to Mercury's motion, thereby solving a long standing problem in astronomy. Einstein deduced something both fundamental and verifiable about the workings of the Universe.
- General relativity correctly predicted the amount that starlight bends as it passes near the Sun on its way to the Earth. This prediction was a factor of two greater than that of Newton's theory. Unlike the perihelion shift, this effect had not as yet been measured in 1915, so it was a true prediction (as opposed to retro-diction) of Einstein's theory. The value obtained from general relativity was confirmed by Sir Arthur Eddington in 1919 during a solar eclipse. This made Einstein a household name overnight.
- Spacetime is curved so that the distances between events in spacetime are described by a generalization of the Pythagoras theorem. The straightest possible lines can look curved and can meet even if they start off parallel.
- Time slows down in strong gravitational fields. This means that time runs more slowly for an observer on the Earth as compared to a clock on a high altitude plane or satellite. In fact it is vital to incorporate this effect when calculating the very accurate timing and location required for the Global Positioning System (GPS) to function. If this correction were not taken into account, GPS would very rapidly stop working.
- The equations of general relativity are able to describe with remarkable accuracy the time evolution of the Universe as a whole. In Sect. 7.8 we will see that recent observations of certain types of supernovae (exploding stars), the Cosmic Microwave Background Radiation (CMBR) etc. have shown beyond doubt that the Universe is not just expanding, but doing so at an ever accelerating rate!
- General relativity predicts the existence of black holes, exotic objects that are, roughly speaking, regions of very strong gravitational fields from which nothing, not even light, can escape. The boundary of such a region is called an event horizon. In 2019, the Event Horizon Observatory produced for the first time a remarkable image of the event horizon at the center of a distant galaxy (see Sect. 7.7).

7.6 Gravitational Waves

7.6.1 Introduction

As mentioned in the previous section, general relativity implies the existence of *gravitational waves* or "ripples in the fabric of spacetime" that are similar to electromagnetic waves in that they are also waves travelling at the speed of light. In order to emit electromagnetic waves, you need to "jiggle" (i.e. accelerate) an electric charge, or pair of electric charges. In order to generate gravitational waves, you need to jiggle a collection of masses. The effect of the resulting gravitational wave as it passes through a ring of masses is shown in Fig. 7.3. The wave in the figure is propagating perpendicular to the page, i.e. coming at you directly through the ring of masses.

Gravitational waves from distant events such as black hole or neutron star mergers are very weak by the time they reach the Earth. The oscillation of masses that such events produce are so minute, that their recent detection was a technological *tour de force*. The first such observation was made by the LIGO (Light Interferometry Gravitational Wave Observatory) in 2015, for which Rainer Weiss, Kip Thorne and Barry C. Barish, the scientists instrumental in creating the detector, received the 2017 Nobel prize in Physics.

Gravitational wave

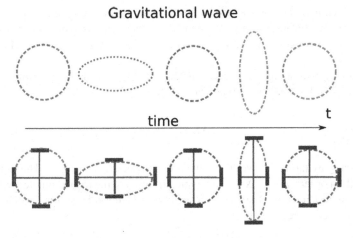

time   t

Response of LIGO's Mirrors

Fig. 7.3 Illustration of a gravitational wave that causes alternating expansion and contraction of the space perpendicular to its motion, as shown in the top half of this figure. As the wave passes through the LIGO detector along a path perpendicular to both mirrors (idealization), the mirrors separate and draw together as shown in the bottom half. The figure is not to scale. The actual movement of the mirrors is about one part in 10^{21} of the original distance between them

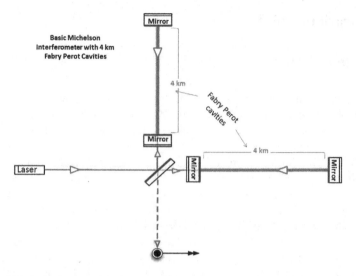

Fig. 7.4 Schematic diagram of interferometer used in the LIGO gravitational wave detections

7.6.2 Detection

Gravitational waves were detected by LIGO using an interferometer, similar to that used by Michelson and Morley. A schematic of the set-up is given in Fig. 7.4.

For more information about the interferometer see: https://www.ligo.caltech.edu/page/ligos-ifo.

The fractional change ΔL in the distance L between the interferometer's two mirrors caused by the gravitational wave of amplitude A is:

$$\Delta L = \frac{A}{2}L \qquad (7.8)$$

For typical gravitational waves that LIGO tries to detect, $A \sim 10^{-21}$. This is incredibly small. In order to make ΔL measurable, the arms of the interferometer separating the mirrors at LIGO are 4 km long. The corresponding relative displacements that were measured were:

$$\Delta L \sim 2 \times 10^{-18}\,\text{m}. \qquad (7.9)$$

This distance is about one thousandth of the diameter of a proton, which is about 1 fermi or 10^{-15} m.

7.6.3 Early Observations

We now briefly describe some of the early observations. Their technical names quite unimaginatively start with GW, for "gravitational wave" followed by the date, e.g. GW150914 was detected September 14, 2015. The links in the headings take you to more detailed descriptions that are available on the LIGO website.

1. http://www.ligo.org/detections/GW150914.php
   The observations in September 2015 of gravitational waves by the LIGO collaboration provided the first direct confirmation of the existence of black holes. The experiment detected the ripples in the fabric of spacetime that were caused by the merger of two massive black holes 1 billion years ago. Remarkably, the frequencies of the mirror oscillations caused by the passing gravitational waves were in the audible range.
2. http://www.ligo.org/detections/GW170814.php
   On August 14, 2017, another gravitational detector, called VIRGO, joined LIGO in the almost simultaneous observation of gravitational waves that were also produced by a pair of merging black holes.
3. http://www.ligo.org/detections/GW170817.php
   On August 17, 2017, the first observation was made of a gravitational wave emitted by the merger of two neutron stars. Optical observatories were able to observe the electromagnetic signals emitted by this event. It was the first time that a cosmic event was observed using two very different types of signals: electromagnetic and gravitational waves.

There have been many further confirmed gravitational wave detections since then. These observations show that gravitational waves will open a new observational window on the Universe. It will be known as gravitational wave astronomy. Exciting stuff!

7.7 Black Holes

Black holes are among the more bizarre predictions of Einstein's theory of general relativity. A black hole is defined as a region of space into which so much matter has fallen that nothing, not even light can escape. The *event horizon* of the black hole is the boundary of the region of no return. Outside the event horizon it is possible to escape to infinity as long as you have sufficient rocket power. This gets more and more difficult as you get closer to the horizon. At the horizon itself you would need to travel at the speed of light just to stay at a fixed distance from the center of the black hole. Once you cross the horizon, you are doomed. You can no more avoid arriving at the very center than you can avoid moving forward in time. The trip from the horizon to the center of a solar mass black hole would only take one millionth of a second, during which time you would join the matter that formed the black hole in the first place. If you had not already been ripped apart by the tidal forces before entering the horizon, this final event would be even more cataclysmic.

7.7.1 Properties of Black Holes

- As implied above, Einstein's equations predict that all the matter and energy that fall through the horizon of a black hole invariably end up at the very center. The density of matter therefore seems to become infinite. Of course, Nature and physicists abhor infinities because they imply, by definition, the existence of physical quantities that are larger than anything that can even in principle be measured. A region of spacetime in which a theory predicts infinities is called a *singularity*. The singularity implies a limit to the realm of validity of the theory. Einstein's equations cease to be valid near the center of black holes and must be replaced by a more complete theory. It is thought that this can only be accomplished by unifying gravity with quantum mechanics, a very difficult and as yet unsolved problem in theoretical physics.
- As light tries to escape from just outside the event horizon of a black hole it loses energy and gets shifted towards the long wavelength, red end of the electromagnetic spectrum. It is said to be *red-shifted*. This effect also occurs at the surface of ordinary planets and stars, but the black hole's red shift goes to infinity as the point of emission gets closer to the horizon. Thus, to a distant stationary observer, a collapsing star would effectively go black even before the event horizon was reached (which would in any case take forever).
- A clock falling towards the event horizon of a black hole from the outside will be seen by a distant stationary observer to tick more and more slowly as it approaches the horizon. This is due to the gravitational time dilation effect that occurs as an object gets close to any massive object. For example gravity causes time to pass more slowly (by a very small amount) on the ground than at the top of a 100 storey sky scraper. Time dilation near a black hole is distinguished by the fact that it gets infinitely large at the horizon. The ticking of the falling clock appears to a distant observer to stop completely at the horizon, as would its motion. This means that it would appear to take an infinite amount of time for the clock to reach the horizon. By extension, it would appear from outside to take an infinite amount of time for a star to collapse to a black hole.[14] On the other hand, for an observer falling with the surface of the star, time passes at the normal rate, and the formation of the black hole is rather quick. If the in-falling observer is small enough to make the tidal forces negligible, she will not feel anything unusual since she is in free-fall and hence in an inertial frame. The tidal forces are strong but not infinite. For this reason, the apparent infinite red-shift and time-dilation detected by a distant stationary observer do not constitute singularities. They are just manifestations

---

[14] For this reason Russian astrophysicists dubbed black holes to be "frozen stars".

Table 7.1 The masses, radii and densities of the black holes that would be formed by typical astrophysical objects

Object	Mass (kg)	Radius (km)	Density (kg/m^3)
Earth	10^{24}	5×10^{-5}	10^{31}
Sun	10^{30}	3	10^{19}
Sagitarius A	5×10^{36}	10^7	10^6

of the highly non-inertial nature of the observer's frame in the curved spacetime around the black hole.

- The radius R_H of the event horizon radius of a black hole is proportional to its gravitational mass M:

$$R_H = \frac{2GM}{c^2} \tag{7.10}$$

Therefore the minimum density ρ required to form the event horizon of a black hole of mass M is:

$$\rho = \frac{M}{4\pi R_H^3/3} = \frac{3c^6}{32\pi G^3 M^2} . \tag{7.11}$$

After the horizon forms all the matter piles up at the center, where the density goes to infinity, so that the rest of the spacetime inside the horizon is essentially vacuum.

For a solar mass black hole the minimum density predicted by Eq. (7.11) is 6×10^{18} kg/m^3. The densest element found on Earth, Osmium, has a density of about 22,000 kg/m^3. The difference is literally astronomical.

A very important feature of Eq. (7.11) is that the more massive the black hole, the **lower** the density required. Some typical black hole parameters (order of magnitude) are found in Table 7.1.

7.7.2 Observational Evidence for Black Holes

General relativity makes testable predictions in regions of weak gravity, such as the perihelion precession of Mercury, and regions of strong gravity, such as black holes. Until recently, tests of the theory involved mostly weak gravitational fields, so that one might be forgiven if they doubted that event horizons actually existed. Recent observations, however, have put the concerns of all but the hardiest sceptics to rest.

Prior to the LIGO observations, the evidence for the existence of black holes was circumstantial. By this we mean that the properties of the black hole were deduced from observations of physical processes that occurred well outside the event horizon in relatively weak field regions. The strong gravitational fields near the horizon itself were not measured. For example, the first black hole candidate, Cygnus X-1 appeared in 1976 in the form of a dark compact object that was observed (or rather not observed) orbiting an ordinary star (in a so-called binary system) in the constellation Cygnus. Its presence was detected by the "wobble" of its companion visible star. The mass of the black hole was first estimated from the orbital motion of the binary pair, and also the observed luminosity of the black hole's accretion disk, which is a flat disk-shaped concentration of matter outside the event horizon. If the predictions of general relativity and certain fairly well established astrophysics are valid, then the observed mass of 15 solar masses was too large for the compact object in the binary to be anything other than a black hole; circumstantial evidence at best.

This type of indirect evidence has by now convinced astronomers that black holes with mass a few times that of the Sun exist in many binary systems within our galaxy and in other galaxies. In addition, supermassive black holes whose mass is millions to billions times that of our Sun exist at the centre of most if not all galaxies, including our own (known as Sagittarius A^*). Observations show that the stars near Sagittarius A^* are apparently orbiting a tiny object whose mass, as deduced by Newtonian mechanics, is a few million times that of our Sun. Although the observed stars are quite far from the event horizon of the central object, it is hard to imagine that so much mass confined to such a relatively small volume does not constitute a black hole.

Recent observations have moved beyond circumstantial to provide direct evidence for the existence of black holes. As mentioned in Sect. 7.6, the first detected gravity wave and several subsequent ones were created by the merger of two black holes. Numerical simulations using Einstein's equations to model the gravity waves that should reach us from such events match incredibly well with those observed by LIGO (the waveforms from exotic compact objects which are not black holes would have been quite different). More recently (2019), the Event Horizon Telescope (EHT) imaged the event horizon of a supermassive black hole at the centre of the galaxy Messier 87 using an array of radio telescopes working in tandem around the world. This arrangement gave rise to a huge telescope whose effective aperture (area that collects signals) was that of the entire Earth, giving it extremely high resolution.[15] The EHT was able to image the shape and size of the black hole, with results that again agreed almost exactly with predictions of general relativity (see Fig. 7.5). Future observations will give us more detailed knowledge of black holes and their interactions with other astrophysical objects. However, there is by now little doubt that black holes exist in abundance in Nature, and that they come in a large variety of shapes, sizes and ages.

---

[15] Recall that size of the smallest object a telescope can resolve varies inversely with aperture size.

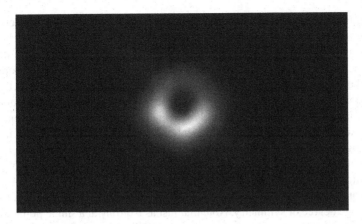

Fig. 7.5 Black hole at the centre of galaxy Messier 87 as imaged by the Event Horizon Telescope. *Picture Credit:* Event Horizon Telescope

7.8 Cosmology

Cosmology is the study of the evolution of the Universe as a whole. It is a field in which general relativity plays a dominant role, although the differences between Einstein's theory and Newtonian gravity already start to become significant within our solar system.

At astronomical scales, and even larger cosmological scales, every object is in free fall in the curved spacetime created by all other objects in their vicinity. Such large distances are usually measured in *parsecs* or *kiloparsecs*, where 1 kiloparsec $\sim$ 1,000 parsecs $\sim$ 3,000 light years. As one examines space at greater and greater scales, interesting structures appear. For example, 16 kiloparsecs (kpc) or about 50, 000 light years is the size of our own Milky Way galaxy, which contains around 250 billion stars, many with their own planetary system. At about 1 Megaparsec (Mpc) and larger one begins to see clusters of galaxies. Finally, as one approaches 300 Mpc, the so-called 'cosmological' scale, the galactic clusters appear to be distributed more or less homogeneously (i.e. the density of clusters is roughly the same everywhere) and isotropically (i.e. the same in all directions). This homogeneity is not perfect however, as one occasionally encounters large *voids*, regions of space that appear to be empty of galactic clusters.

As seen from Earth, the Universe looks pretty much the same in all directions. The distant galaxies are more or less uniformly distributed across the sky, and the *Cosmic Microwave Background*, which is a bath of radiation that comes to us from the beginning of the Universe 14 billion years ago, is the same everywhere we look to 1 part in 10^5.[16] This *isotropy* as it is called, could have only one of two possible explanations. Either we occupy a special place in the Universe, namely the center, or the Universe looks the same in all directions everywhere. It would be somewhat

---

[16] The Cosmic Microwave Background will be discussed in more detail in Sect. 8.3.5.

ironic if five hundred years after Copernicus removed the Earth from the center of the Universe we were to find ourselves back there again.

The notion that we occupy such a special place in the Universe is so disliked by physicists that they have elevated the opposite view to a principle, appropriately called the *Copernican principle*. If one adopts the Copernican principle, the only explanation for the observed isotropy is that the Universe looks isotropic from every location in space. This in turn can only be true if the Universe is homogeneous as well as isotropic. Deep space astronomical observations have been able to measure the average density of the Universe at great distances from the Earth. The fact that this density does not depend on the radial distance to the Earth provides more direct observational evidence that space is indeed homogeneous.

It is quite amazing that observations have revealed to us that the Universe has such a huge degree of symmetry. Note that just as isotropy at a single point does not imply homogeneity neither does homogeneity of the Universe imply isotropy. It is in principle possible for example, to have a non-zero electric field that has the same magnitude and direction everywhere in space. Such an electric field would pick out a preferred direction in space without destroying the homogeneity. In fact, the presence of both homogeneity and isotropy makes our Universe the most symmetrical that any space can be.

The high degree of symmetry of our Universe has useful implications. First of all it restricts the geometry of space to just one of three possibilities, the so called *constant curvature* geometries. We have discussed spherical geometry in detail in Sect. 7.4. Spheres have constant positive curvature and are *compact*: if you head off in a straight line in any direction, you always end up back where you started. The sum of the angles in a space of constant positive curvature will always be greater than 180°.

The second possibility is the geometry of a saddle (see Fig. 4.10), which has constant negative curvature. The sum of the angles in a triangle drawn on a saddle is always less than 180°. Spaces of constant negative curvature are *non-compact*: if you head off in a straight line in any direction, you will keep on going forever.

Observations suggest that our space (but not spacetime as a whole) has chosen the third possibility, namely zero curvature. On cosmological scales space appears to be very close to flat, like a blackboard, implying that its geometry is on average Euclidean. On such scales the Pythagoras theorem holds for any right angled triangle, parallel lines never meet, and the sum of the angles in a triangle is 180°.

In the context of general relativity, the homogeneity and isotropy of space implies that there exists a special frame of reference, known as the *comoving frame*. Using spherical coordinates, in this frame the metric of spacetime for cosmological scales, assuming it is flat, can be written everywhere in space and for all time as[17]:

$$c^2 d\tau^2 = c^2 dt^2 - a^2(t) \left[dr^2 + r^2(d\theta^2 + \sin^2\theta d\phi^2) \right] , \qquad (7.12)$$

---

[17] If the Universe were either spherical or saddle-shaped, the metric would take a slightly different form, but would still contain only one free function of time.

The simple form of metric provided by the symmetry of space makes it surprisingly easy to work out the physical properties and dynamics of the Universe as a whole. The symmetry of the cosmos is yet another unexpected gift from Nature to physicists!

The physical consequences of a homogeneous and isotropic Universe were first worked out by the Soviet mathematician Alexander Friedmann in 1924. Further analysis was done in the 1930s by Howard P. Robertson and Arthur Geoffrey Walker from the United States and United Kingdom, respectively. The important contributions of George Lemaitre mentioned in Footnote 19 suggest the commonly used acronym, FLRW, for Friedman-Lemaitre-Robertson-Walker.

The only variable in the FLRW metric Eq. (7.12) is the time-dependent *scale factor* $a(t)$, whose value at any time determines how far two given astronomical objects are from each other. A scale factor that increases with time implies that these distances are growing while if $a(t)$ is decreasing, cosmological objects are getting closer. It is important to note that this expansion or contraction is cosmological in origin and does not represent proper motion of the galaxies in the usual sense. To understand this distinction better, consider the galaxies to be heuristically represented as a pattern of evenly spaced dots painted on a flat rubber sheet. If the rubber sheet is stretched equally in all directions, then every dot will grow further and further apart from every other dot, even though none of them is moving relative to the rubber sheet.

The time evolution of the scale factor $a(t)$ is governed by Einstein's equations. Since the geometry of spacetime is determined by its matter content, one cannot make any predictions without making assumptions about the nature of the large scale matter in the Universe. It turns out the observed Universe can be well described by one of the simplest forms of matter, namely a perfect fluid. The nature of such a fluid is completely specified by only two parameters, the pressure p and the density ρ which are related to each other by an equation of state. Again, it is sufficient to consider the simplest equation of state, namely:

$$p = w\rho c^2 \qquad\qquad (7.13)$$

where w is a constant that needs to be determined via experiment.[18] For example, baryonic matter (the ordinary matter that we can directly observe, such as stars, planets and dust) has $w = 0$, as does the invisible dark matter that is conjectured to exist due to its gravitational effects. For fast relativistic particles on the other hand, such as neutrinos (with a tiny mass) and photons, $w = 1/3$. Finally, there appears to exist in the Universe dark energy, that has $w = -1$ and therefore negative pressure, $p = -\rho c^2$. This negative pressure is thought to be the source of the astounding relatively recent observation that the Universe is not only expanding, but its expansion is accelerating.

By looking deeper and deeper into space using the Hubble space telescope as well as other powerful orbiting and land-based telescopes, astronomers have been

---

[18] You can confirm for yourself that the factor c^2 in Eq. (7.13) is necessary for consistency with respect to units.

able to get a fairly good handle on how the scale factor $a(t)$ has depended on time over the last 14 billion years. Theoretical predictions can model the observed time dependence of $a(t)$ accurately using just three kinds of fluid to account for the entire matter content of our Universe: In these models, 5% of the matter in the Universe consists of ordinary matter, 25% is unobserved dark matter clustered around galaxy cores and 70% is dark energy ($w = -1$). Many theories exist for the true nature of this unobserved dark energy, so it is currently a subject of hot debate.

By observing redshifts of distant galaxies and interpreting them as relativistic Doppler shifts, Edwin Hubble discovered in 1928 that the Universe appears to be expanding. Moreover, astronomical objects at greater distances from the Earth were receding at greater velocities than closer objects.[19]

Hubble found that objects twice as far from the earth were receding approximately twice as fast. This is consistent with the rubber sheet model of the Universe described earlier. If the sheet doubles in size in 1 s, say, then dots initially 10 cm apart will be 20 cm apart after the expansion. Dots that were initially twice as far apart before the expansion, namely 20 cm initially, will also double their distance and end up 40 cm apart. Thus the closer dots will have an apparent speed with respect to each other of 10 cm/s, while the dots initially twice as far apart will have an apparent speed relative to each other of 20 cm/s. If the distance is doubled then so is the apparent velocity, in agreement with Hubble's observations of the red-shifts of galaxies. In the context of general relativity and the FLRW metric in Eq. (7.12), Hubble discovered that the scale factor is an increasing function of time, i.e. $da/dt > 0$ and the Universe is expanding.

It appears that the Universe at large scales could hardly be simpler! Have we then completely solved all the mysteries of the observable gravitational Universe? The answer is unfortunately no,[20] as there are still many puzzles to consider, including the following:

• **The Cosmic Acceleration**

Given that gravity is an attractive force, the expectation was that the expansion would eventually slow down and perhaps even stop altogether at which point the Universe would start to contract. Remarkably, supernovae data showed that the Universe is not only expanding, but also accelerating,[21] i.e. $d^2a/dt^2 > 0$. The expansion rate in the past as observed for distant objects was slower than it is today, as seen by observing closer objects. The conjectured explanation for this expansion was the presence of an unseen *dark energy* with negative pressure that makes up 70% of the energy/matter content of the Universe.

The most promising candidate for dark energy is simply a constant parameter that can be inserted either on the left hand, geometrical side of Einstein's equations or

---

[19] Georges Lemaitre was the first to point out that the apparent recession of the galaxies away from the Earth could be explained by the expansion of the Universe. Lemaitre [1].

[20] Or fortunately if you make your living as a cosmologist.

[21] Adam Riess, Brian Schmidt, Saul Perlmutter shared the 2011 Nobel Prize in physics for this astounding discovery.

the right side describing the matter content. Where it is placed is determined by its physical interpretation as either a form of pure geometry, in which case it is named a cosmological constant, or a consequence of matter interactions, in which case it is usually referred to as dark energy. In fact, Einstein himself inserted a cosmological constant into his equations when he first formulated the theory of general relativity. At the time there was no evidence for the expansion of the Universe and his motivation was to produce a static Universe as a solution to the equations. After the discovery of the cosmological expansion by Hubble, Einstein is reported to have referred to this insertion as his "biggest blunder". The origin and true nature of the cosmological constant is still hotly debated. Theoretical predictions of its value have over time deviated significantly from its currently accepted value of approximately 10^{-26} m^{-2}.

Since the density of other forms of matter (including the visible galaxies and dark matter) invariably decreases as the Universe expands while the cosmological constant by definition stays the same, the cosmic acceleration can only increase with time. In the end it may be so large that the Universe will practically tear itself apart (the *Big Rip*). Luckily for us this is not expected to happen for billions of years.

- **The Beginning of Time**

By extrapolating Hubble's expansion law as far as possible backwards in time, one concludes that the Universe started 14 billion years ago with zero scale factor $a(t) = 0$. At this precise time, the volume of the Universe was zero, while the density of matter and the curvature of the Universe were infinite. Just as in the case of black holes, the prediction of infinite physical quantities constitutes a singularity that signals the breakdown of general relativity. The singularity, this time at zero scale factor, again implies that the theory must be modified or completely replaced near the Big Bang. This provides added motivation for the search for the successful formulation of a theory of quantum gravity unifying quantum mechanics with general relativity. It is even possible that such a theory will unify gravity with the other three forces in nature as well. Although there exist promising candidate theories such as String Theory and Loop Quantum Gravity, none has been tested in the laboratory or via astrophysical observations.

- **Why is the Universe so boring at large scales?**

The high degree of homogeneity, isotropy and flatness of space also remains to be understood. One can show that at no time since the Big Bang could the distant portions of the Universe in different directions in the sky have interacted with each other,[22] because they were never within each others light cones. Why then do they look so much alike?

In addition, as we have previously discussed, the observed homogeneity and isotropy of the Universe does not require spatial flatness. These symmetries are also consistent with a Universe in which space is equivalent geometrically to a three-dimensional sphere (i.e. with positive spatial curvature) or even to a three-

---

[22] A startling piece of evidence for the homogeneity and isotropy of the early Universe is the Cosmic Microwave Background Radiation (CMBR). This will be discussed in detail in Sect. 8.3.5.

dimensional saddle (i.e. with negative spatial curvature). In the former case the Universe, like the surface of the Earth, is finite. In the latter case it is necessarily infinite in extent, as it is if the Universe is flat. Given that there are three possibilities for the geometry consistent with the observed homogeneity and isotropy, why is the Universe flat? Until a satisfactory scientific explanation is found, these observations will be merely a startling coincidence. One popular theory that attempts an explanation is known as *inflation*. It postulates that soon after the Big Bang, a scalar field that produces negative pressure caused the Universe to expand exponentially fast for a short period of time, but long enough for thermal equilibrium to be established throughout the Universe and for it to become spatially flat (just as a rapid expansion of a gigantic balloon will make small portions of its surface look flat). Although inflation is promising in many respects it remains unproven, but there is hope that observational evidence may follow in the form of gravitational waves that originated during this rapid expansion.

Reference

1. G. Lemaitre, Un Univers homogne de masse constante et de rayon croissant rendant compte de la vitesse radiale des nébuleuses extra-galactiques. Annales de la Socit Scientifique de Bruxelles **47**, 49 (1927). (in French)

Chapter 8
Introduction to the Quantum

8.1 Learning Outcomes

Conceptual

- Understand how the three experimental clues (photoelectric effect, Compton scattering and ultraviolet catastrophe) suggested the particle nature of light.
- Understand the experimental clues that suggested electrons have wave-like properties.
- Understand the physical meaning of the uncertainty principle and its implications.

Acquired Skills

- Problem solving using formulae associated with photo-electric effect, Compton scattering and blackbody radiation intensity curve, including derivation of Wien's law.
- Quantitative use of the relations between wave number, momentum and energy for light and massive particles.
- Calculation of resolving power.

8.2 Light as Particles

8.2.1 Review: Light as Waves

In the late 1800s James Clerk Maxwell formulated a unified theory in terms of partial differential equations that described electricity and magnetism (or electromagnetism for short) as waves whose speed of propagation was determined by fundamental constants in the theory. This speed turned out to be roughly the same as the speed of

© The Author(s), under exclusive license to Springer Nature Switzerland AG 2022
G. Kunstatter and S. Das, *A First Course on Symmetry, Special Relativity and Quantum Mechanics*, Undergraduate Lecture Notes in Physics,
https://doi.org/10.1007/978-3-030-92346-4_8

light as measured at the time, which led Maxwell to speculate that light was a form of electromagnetic disturbance.

Since then, the wavelike properties of electromagnetic phenomena have been verified using interference and diffraction experiments. It has also been confirmed that visible light is indeed a particular form of electromagnetic radiation. The colours we perceive are determined by the frequency of the radiation. The wavelength λ of electromagnetic radiation can in principle be anything between zero and infinity, with a corresponding frequency

$$f = c/\lambda = \frac{3 \times 10^8}{\lambda} \frac{m}{s} \tag{8.1}$$

ranging from infinity to zero. Electromagnetic waves of a wide range of frequencies and wavelengths occur in our everyday lives, including low frequency radio waves, infrared radiation that manifests itself as heat, potentially dangerous ultraviolet radiation from the sun and x-radiation that is used for medical imaging.

For reasons we will see in Sect. 8.3.1 each frequency has a temperature associated with it. For example the Sun emits radiation at a temperature of about 6000K, which appears as yellow to our eyes, and has a frequency and wavelength of 5×10^{24} Hz (1 Hz corresponds to one cycle per second, or sec^{-1}) and 600 nm (600×10^{-9} m), respectively. The full range of electromagnetic waves is referred to as the *electromagnetic spectrum*. Table 8.1 lists a sampling of relevant frequencies and wavelengths. Visible light ranges from about 700 nm (red) to 500 nm (blue). This is a very narrow part of the electromagnetic spectrum with which we are familiar, including radio waves (3000 m) to gamma rays (3×10^{-12} m).

In the late 19th century and early 20th century Maxwell's elegant description of electromagnetism as waves was brought into question by a series of very surprising experimental results.

Table 8.1 Electromagnetic spectrum. Associated temperatures assume approximate blackbody radiation as described in Sect. 8.3.1, specifically using Wien's law Eq. (8.49)

Type of radiation	Frequency (Hz)	Wavelength (m)	Temperature (K)
Radio wave	10^4	3×10^4	10^{-7}
Microwave	10^{10}	3×10^{-2}	10
Infrared	10^{12}	3×10^{-5}	100
Visible (red)	0.4×10^{15}	7×10^{-7}	4000
Visible (yellow)	0.5×10^{15}	6×10^{-7}	5000
Visible (blue)	0.6×10^{15}	5×10^{-7}	6000
Ultraviolet	10^{16}	3×10^{-8}	10^5
X-rays	10^{18}	3×10^{-10}	10^7
Gamma rays	10^{20}	3×10^{-12}	10^9

8.2.2   Photoelectric Effect

In 1887 the German physicist Heinrich Hertz observed that shining ultraviolet light on two metal electrodes with a voltage applied across them affects the voltage required to send an electric current through the air (i.e. spark) between them. The light was apparently transferring energy to the electrons in the metal, making them easier to eject from the surface using an applied voltage. A quantitative version of this *photoelectric effect* has been performed numerous times since Hertz's original discovery, with results that were difficult to interpret in terms of Maxwell's wave theory of light.

- **The Experiment**:
  The experiment consists of shining light of some fixed frequency and intensity (brightness) on the surface of a metal target and then measuring the kinetic energy with which the electrons are ejected. The wave theory of light led physicists of the day to predict that the electrons at the surface would "heat up" gradually as they absorbed the radiation, until they had too much kinetic energy to stay bound to the metal. At this point they would be ejected from the metal with some characteristic velocity v. The time scale for this process and the velocity were expected to depend on how much radiation energy per unit area hit the surface per unit time (i.e. the *intensity of the radiation*). The frequency of the radiation was not expected to play a role.
  How is the kinetic energy with which the electrons are ejected measured? One starts by shining light on a target metal plate T. This causes electrons to be ejected from the surface of the metal and produce a current between the plate T and a second nearby plate C. Next, a reverse electrostatic potential V is set up between T and C in order to decrease the kinetic energy of the electrons that reach C. The potential V is then increased until no more electrons make it to C. At this point, the work done on the electrons by the potential V as they cross the gap between T and C is just enough to reduce the kinetic energy of all the ejected electrons to zero. The minimum value of the potential required to stop the electrons from reaching C is called the *stopping potential* V_{stop}. The interpretation is that V_{stop} measures the maximum kinetic energy K_{MAX} with which the electrons are ejected from T:

$$K_{MAX} = eV_{stop} \qquad (8.2)$$

  It is the maximum kinetic energy because all the electrons are stopped, even the most energetic.
  The above procedure is then repeated for a variety of frequencies and intensities of radiation and a selection of different target metals to see how the maximum kinetic energy is affected.
- **Results**:
  Surprisingly, the experiments revealed that electrons were ejected as soon as the radiation hit the surface of the metal. In contrast to the prediction based on Maxwell's theory, they didn't seem to need any time to gradually absorb the electromagnetic energy. Even more startling was the fact that the maximum kinetic

energy of the ejected electrons did not depend on the intensity of the radiation as one might expect from normal wave phenomena. Instead, K_{MAX} depended strongly on the frequency of the incident radiation, as depicted in Fig. 8.1. The graphs plot K_{MAX} in electron volts vs incident radiation frequency f for three metals. In all cases, the maximum kinetic energy, which must be positive, depends linearly on the frequency:

$$K_{MAX} = hf - \phi_T \tag{8.3}$$

where ϕ_T is different for the different metals. The slope h is universal. That is, it has the same numerical value for all metals tested. The three lines in the graph represent experiments with silver, platinum and sodium, which have stopping potential $\Phi_{silver} = 4.72$ eV, $\Phi_{platinum} = 6.35$ eV and $\Phi_{sodium} = 2.28$ eV, respectively. These determine the intercepts $-\Phi_T$ of the lines extended below the x-axis. The slopes of all three lines are the same to within experimental error, with a value of:

$$h = 4.1 \times 10^{-15} \, \text{eV} \cdot \text{s}$$
$$= 6.6 \times 10^{-34} \, \text{J} \cdot \text{s} \tag{8.4}$$

These results were a complete mystery in the context of the wave theory of light! Einstein explained them in 1905[1] in a Nobel prize winning paper.[2]

- **The Explanation**:
Einstein realized that the experimental results summarized in Eq. (8.3) could most simply be explained by assuming that the incident radiation was behaving like a stream of particles, each with energy and momentum given, respectively, by:

$$\text{Energy} \quad E = hf = \frac{hc}{\lambda} \tag{8.5}$$

$$\text{Momentum} \quad p = \frac{E}{c} = \frac{hf}{c} = \frac{h}{\lambda} \tag{8.6}$$

where h is the universal constant corresponding to the slope of the maximum energy versus frequency graphs in Fig. 8.1. When such a particle of light, called a *photon*, collides with a surface electron in the metal, it transfers energy and momentum to it. Since the collisions are essentially instantaneous, there is no time delay between the first photons hitting the metal and the first electrons being ejected.

If the energy E_γ of each photon is not large enough to overcome the potential energy, Φ_T, with which the electrons are bound to the metal, then no electrons

---

[1] 1905 is called Einstein's miraculous year because in that year he published three papers that each had a profound effect on a different field of physics. In these papers he announced his special theory of relativity, revealed the cause of Browning motion (the random motion of tiny particles through a fluid) and explained the photoelectric effect.

[2] Einstein [1]. It is interesting that Einstein never received a Nobel Prize for either special or general relativity. These theories were thought to be too radical and of little practical use.

Fig. 8.1 Photoelectric Effect

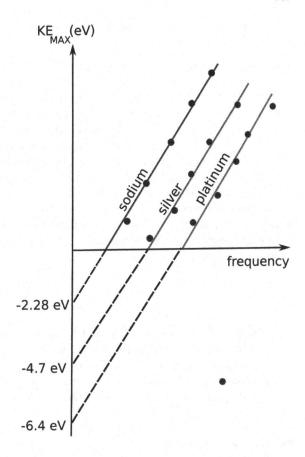

can be ejected. The maximum kinetic energy, K_{MAX}, with which electrons emerge equals the difference between the energy of each photon, which is the maximum energy that could be transferred to an electron in a collision, and the binding potential Φ_T. The binding potential is then the value of the y-intercept in the measured graphs, and the slope of the graphs is h as given in Eq. (8.3).[3] Thus, the assumption that light is made up of particles with energy proportional to the frequency explains the results in a particularly simple and intuitive way even though light also exhibits wave phenomena such as interference. How can light consist of both particles and waves? Einstein's solution to the problem presented physicists of the day with a new deep conundrum.

---

[3] It is remarkable that Einstein's intuitive leap along with this simple argument was sufficient to earn Einstein the Nobel prize for physics.

8.2.3 Compton Scattering

- **The experiment**: In 1932 A. H. Compton fired x-rays onto a graphite target. The energy of the x-rays was very high compared to the binding energy of the electrons in the graphite, so the electrons could be treated as effectively free. A schematic of the experiment is shown in Fig. 8.2. According to Maxwell's wave theory of light, the electron absorbs the radiation and accelerates. This acceleration in turn causes the electron to emit radiation and slow down again. The radiation absorbed and emitted by the target electrons have different wavelengths/frequencies and the precise relationship between them can be calculated from Maxwell's wave theory. This calculation is beyond the scope of the present text, so we will just focus on the results of Compton's experiment.
- **Results**: The experiments gave a very different relationship between incident and emitted radiation than what was expected from the wave theory of light. The observed relationship is:

$$\lambda_f - \lambda_i = \frac{h}{m_e c}(1 - \cos(\theta)) \tag{8.7}$$

where m_e is the electron mass, λ_f and λ_i are the wavelengths of the emitted and absorbed radiation, respectively, θ is the angle between the incident and emitted radiation. The constant h in Eq. (8.7) quite remarkably (or so it seemed at the time) has the same numerical value as the constant h that appears as the slope of K_{MAX} versus frequency in the photoelectric effect (Eq. 8.3).

Fig. 8.2 Compton scattering of a photon and electron

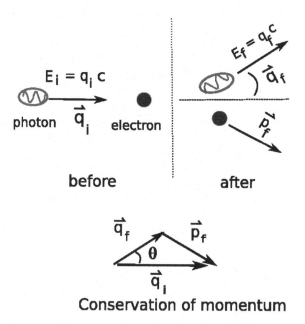

You are probably getting the sense that the constant h plays a significant role in quantum mechanics. It is called *Planck's constant* after German physicist Max Planck, one of the founders of quantum theory. His proposal of a fundamental unit of energy in the context of blackbody radiation (see Sect. 8.3.1) won Planck the Nobel Prize in 1918.

The coefficient $\frac{h}{m_e c}$ on the right hand side of Eq. (8.7) occurs often in quantum physics. It is called the *Compton wavelength*, λ_c, of the electron:

$$
\begin{aligned}
\lambda_c &:= \frac{h}{m_e c} \\
&= \frac{hc}{m_e c^2} \\
&= 2.4 \times 10^{-3}\,\text{nm}
\end{aligned}
\tag{8.8}
$$

Exercise 1 Verify that λ_c as given has units of length.

The second line in Eq. (8.8) above is particularly useful because by choosing suitable units, it yields an expression with manageable numbers. Specifically, one has:

$$
hc = 1241\,\text{eV} \cdot \text{nm} \tag{8.9}
$$
$$
m_e c^2 = 511\,\text{keV} \tag{8.10}
$$

The numerical value for hc in terms of eV $\cdot$ nm is important to remember because it simplifies the exercises in the following discussions.

- **The explanation**: The physical explanation and mathematical derivation of the observed experimental results are incredibly straightforward compared to the complexity of the equations that must be solved in order to obtain the (incorrect) prediction from Maxwell's wave theory.

  As in the photoelectric effect, we assume that the radiation consists of a stream of particles (photons) that undergo elastic collisions with the electrons. The energy of each photon is related to the frequency of the light by:

$$
E = pc = hf = \frac{hc}{\lambda} \tag{8.11}
$$

where f is the frequency and

$$
h = 6.626 \times 10^{-34}\,\text{J} \cdot \text{s} \tag{8.12}
$$

is again Planck's constant.[4]

---

[4] We have added a few significant digits to the measured value for good measure.

The photons are of course relativistic, and it turns out that the electrons acquire relativistic velocities in the collisions. We will now use the relativistic scattering techniques that we learned in Sect. 6.7.1 to show that the relation Eq. (8.7) follows directly from special relativity and the assumption that the light incident on the electron is in the form of particles (photons) with zero rest mass and energy given by Eq. (8.11).

Consider the collision between a photon and an initially stationary electron, as illustrated in Fig. 8.2.

The initial four-momentum $\mathbf{Q_I}$ and $\mathbf{P_I}$ for the photon and electron, respectively, are:

$$\mathbf{Q_I} = \begin{pmatrix} E_I/c \\ \mathbf{q}_I \end{pmatrix} \tag{8.13}$$

$$\mathbf{P_I} = \begin{pmatrix} m_e c \\ 0 \end{pmatrix} \tag{8.14}$$

where $E_I = hf_I = hc/\lambda_I$ and $\mathbf{q}_I$ is the initial three-momentum of the electron. After the collision the respective four-momenta are:

$$\mathbf{Q_F} = \begin{pmatrix} E_F/c \\ \mathbf{q}_F \end{pmatrix} \tag{8.15}$$

$$\mathbf{P_F} = \begin{pmatrix} \sqrt{m_e^2 c^4 + p_F^2 c^2} \\ \mathbf{p}_F \end{pmatrix} \tag{8.16}$$

where now $\mathbf{q}_F$ and $\mathbf{p}_F$ are the final three-momenta of the electron and photon and $q_F = E_F/c$. We get from energy conservation:

$$E_I + m_e c^2 = E_F + \sqrt{m_e^2 c^4 + p_F^2 c^2} \tag{8.17}$$

Solving Eq. (8.17) for p_F^2:

$$p_F^2 = \frac{E_F^2}{c^2} + \frac{E_I^2}{c^2} - 2\frac{E_F E_I}{c^2} + 2mc\frac{(E_I - E_F)}{c} \tag{8.18}$$

From the vector form of momentum conservation

$$\mathbf{p}_f + \mathbf{q}_f = \mathbf{p}_i + \mathbf{q}_i \tag{8.19}$$

one can use the triangle inequality to get:

$$\begin{aligned} p_F^2 &= q_F^2 + q_I^2 - 2q_F q_I \cos(\theta) \\ &= \frac{E_F^2}{c^2} + \frac{E_I^2}{c^2} - 2\frac{E_F}{c}\frac{E_I}{c}\cos(\theta) \end{aligned} \tag{8.20}$$

where θ is the angle between $\mathbf{q}_F$ and $\mathbf{q}_I$. Eliminating p_F by using Eqs. (8.18) and (8.20) yields:

$$E_I - E_F = \frac{E_I E_F}{mc^2}(1 - \cos(\theta)) \qquad (8.21)$$

Using the relationship Eq. (8.11) to express the photon energy in terms of the wavelength gives the desired formula given in Eq. (8.7).

Exercise 2 Verify that Eq. (8.7) follows from Eq. (8.21).

As in the photoelectric effect, treating light as particles when they interact with electrons solved a very important observational problem.

8.3 Blackbody Radiation and the Ultraviolet Catastrophe

8.3.1 Blackbody Radiation

A blackbody is an idealized object that absorbs all the radiation incident on it. As it gains energy by absorbing radiation, it heats up and starts emitting radiation. The emitted radiation is independent of the nature of the incident radiation. A simple model often used for an ideal blackbody consists of a closed cavity containing radiation in thermal equilibrium within the walls of the cavity, as illustrated in Fig. 8.3. In this model, the cavity walls contain a single tiny hole that can absorb radiation shined directly on it. If the hole is small enough, the incident radiation will disappear into the blackbody and not re-emerge for a very long time because the chance of it hitting exactly the right spot after bouncing off the inside wall is very small.

One can think of the absorbed radiation as trapped, getting absorbed and re-emitted by the inside walls of the cavity over and over again until the system reaches

Fig. 8.3 Cavity model for ideal blackbody. Any radiation that is absorbed through the hole is trapped in the cavity and interacts within the walls many times before it has any chance of escaping again

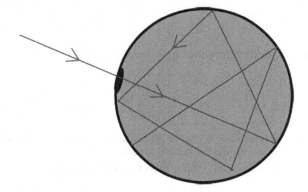

thermal equilibrium. The assumption of thermal equilibrium implies that on average the inside walls absorb and emit equal amounts of radiation energy so that they do not cool down or heat up. The radiation inside the cavity is "thermal radiation" that depends only on the temperature of the walls and not their structure or composition. All details are washed out. If you have ever looked through the window of a pottery kiln as it heats up, you would have noticed that after a while, you can no longer make out the details of the pottery inside. You see only a featureless glow. The colour of that glow depends on the temperature of the kiln. In complete analogy with the kiln, one can observe the radiation inside our idealized blackbody cavity by looking through the hole. The light that emerges when the hole is arbitrarily small is ideal blackbody radiation. Like the radiation in the kiln, it depends only on temperature.

The defining characteristic of blackbody radiation is its *intensity* $I(\lambda)$, which is defined as follows: $I(\lambda)d\lambda$ is the energy of radiation with wavelength between λ and $\lambda + d\lambda$ emitted per unit time by the blackbody per square meter of the emitting surface per steradian.[5] The most convenient units of $I(\lambda)$ are Watts per m^2 per μm. The unit micrometer (μm) is used because in most cases of physical interest, the relevant wavelengths are in the micrometer range. The intensity, $I(\lambda)$, tells us how much radiation is emitted and also how the energy is distributed among all possible wavelengths. The key fact to remember about the intensity is that it depends only on the temperature of the blackbody. It is independent of the composition or other attributes of the blackbody itself.

We emphasize that we are talking about an ideal blackbody. Virtually all physical systems have temperature and can to a close approximation be treated as blackbodies. This includes the Sun, lightbulbs, the Earth as a whole and human beings. Since these are not ideal blackbodies, the radiation intensity does depend slightly on the nature of the physical system.

The intensity of radiation can also be expressed in terms of frequency, but one has to take into account the relationship between wavelength and frequency. Specifically:

$$d\lambda = d\left(\frac{c}{f}\right)$$
$$= -\frac{c}{\lambda^2}df$$
$$= -\frac{f^2}{c}df \tag{8.22}$$

Ignoring the minus sign[6] because intensity should be a positive quantity, we find:

$$I(\lambda)d\lambda = I(\lambda(f))\frac{f^2}{c}df \tag{8.23}$$

---

[5] The radiation comes out of the blackbody at all angles. A steradian is a unit solid angle. There are, by definition 4π steradians on a unit sphere.

[6] The minus sign is simply telling us that as wavelength increases frequency decreases, and vice versa.

where $I(\lambda(f))$ means that you replace λ by its expression in terms of f in $I(\lambda)$. Finally, the expression for intensity in terms of frequency is:

$$\tilde{I}(f) = I(\lambda(f))\frac{f^2}{c} \tag{8.24}$$

The formula for $I(\lambda)$ was derived by British physicists Lord Rayleigh and Sir James Jeans in the early 1900s using classical electromagnetic wave theory and the laws of thermodynamics. It is called the Rayleigh-Jeans law and takes the form:

$$I_{classical} = 2c\frac{k_B T}{\lambda^4} \tag{8.25}$$

where k_B is a fundamental constant that appears in thermodynamics, whose numerical value is:

$$k_B = 1.38064852 \times 10^{-23}\,\text{J} \cdot \text{K}^{-1} \tag{8.26}$$

k_B is call the Boltzmann constant.[7] In the above, K denotes degrees Kelvin

$$K = {}^\circ\text{C} + 273.15. \tag{8.27}$$

Exercise 3 Check that the units on the left and right hand side of Eq. (8.25) agree.

The Kelvin temperature scale has physical significance as a fundamental thermodynamic quantity. $K = 0$ (or $^\circ\text{C} = -273.15$) is the lowest temperature achievable by any physical system. $K = 0$ is therefore called *absolute zero*. It signifies that the entire system, however large or complex, is in a unique state that undergoes no thermal fluctuations.[8] In both classical thermodynamics and its quantum version statistical mechanics, there is a law stating that absolute zero cannot be achieved in a finite number of steps by a system originally in a state of non-zero temperature. It is called the third law of thermodynamics.

8.3.2 Derivation of the Rayleigh-Jeans Law

8.3.2.1 Normal Modes of Radiation in a Cavity

The general form of the Rayleigh-Jeans Law Eq. (8.25) can be understood in the context of Maxwell's wave theory of light and classical thermodynamics. We will

---

[7] It is named after Austrian physicist Ludwig Boltzmann who played a major role in the development of *statistical mechanics*, the microscopic theory that underlies classical thermodynamics.

[8] In quantum mechanics, this unique state is called the ground state, a terminology that will be explained in Chaps. 10 and 11.

now go through this derivation in detail because it is a wonderful example of the use of symmetry to simplify the analysis of a complicated system thereby clarifying the underlying physics.

First, we note from Maxwell's theory that the electromagnetic field (light) is a wave which in the case of our ideal blackbody is confined to the cavity. We assume the electric and magnetic fields vanish at the walls of the cavity. This makes sense given that the fields are assumed to be zero everywhere outside and we do not wish the magnitude of the fields to jump suddenly from the outside of the walls to the inside. In more technical terms, we are assuming that the walls of the cavity are electrical conductors so that any abrupt change in the fields across the walls would be immediately corrected by an induced charge.

Before analyzing the case of three dimensional electromagnetic waves, we will consider the simple analogy of waves on a string. As with the cavity, we must fix boundary conditions at the ends of the string. We assume that the string is held fixed at both ends so that the amplitudes of the waves must vanish there.[9] Recall that a wave on a string fixed at both ends can be decomposed into a sum of normal modes, each labelled by an integer $n = 1, 2, 3, \ldots \infty$, called the *mode number*. The normal modes are *standing waves* whose wavelength λ_n is such that the waves vanish at the ends of the string.

A standing wave is one whose *nodes* (points where the wave vanishes) do not move while the points on the string between the modes oscillate up and down. Normal modes are represented mathematically as sine waves. It is possible for the sine wave to vanish at both ends if exactly half a wave fits on the string, or a full wave fits on a string, or three halves of a wave fits on a string, etc. The *mode number* is defined as the number of half waves in the given sine wave. As the mode number goes up, so does the number of *nodes* along the string. The number of nodes not counting end points is given by $n - 1$, where n is the mode number. The corresponding wavelengths are given by:

$$\lambda_n = \frac{2L}{n}, \; n = 1, 2, 3, \ldots, \infty \tag{8.28}$$

For $n = 1$, half a wave fits on the string, for $n = 2$ an entire wave fits on the string, etc. The parts of the string between the nodes vibrate with a frequency determined by the wavelength and speed v_s of the waves along the string:

$$f_n = \frac{v_s}{\lambda_n}$$
$$= \frac{n v_s}{2L} \tag{8.29}$$

---

[9] Other boundary conditions are possible. For example, if the string is attached via a frictionless hoop to a post so that it is free to move up and down at the ends, then the boundary condition is that the slope of the wave must vanish at the post.

The speed v_s depends only on the properties of the string and is the same for every normal mode. From Eqs. (8.28) and (8.29) we can see that there are an infinite number of normal modes with arbitrarily large frequencies and arbitrarily small wavelengths.

We are interested in electromagnetic waves confined to a cavity. The *normal modes* in this case are solutions to the electromagnetic wave equations that vanish at the walls. This is analogous to the normal modes of a wave on a string with both ends fixed.[10] Not surprisingly, in a three dimensional cavity, the solutions are labelled by three independent mode numbers, one for each possible direction in which the wave can propagate. In terms of Cartesian coordinates, these mode numbers are n_x, n_y and n_z, each of which can be any integer between 0 and ∞. If all three mode numbers are zero, then there is no wave. Each set of integers, such as $(n_x, n_y, n_z) = (4, 6, 1)$, corresponds to a particular normal mode of the waves in the cavity. Each of these normal modes has associated with it a wavelength $\lambda(n_x, n_y, n_z)$ and corresponding frequency $f(n_x, n_y, n_z) = \frac{c}{\lambda(n_x, n_y, n_z)}$, where c is the speed of light. $\lambda(n_x, n_y, n_z)$ and $f(n_x, n_y, n_z)$ are as yet unknown functions of the mode numbers. Our first task is to derive the dependence of the wavelength $\lambda(n_x, n_y, n_z)$ on the mode numbers (n_x, n_y, n_z) by using a simple symmetry argument.

We are interested in the behaviour of the intensity at short wavelengths since this is part of the spectrum most relevant to the ultraviolet catastrophe. For short wavelengths, which correspond to large mode numbers, the cavity is effectively infinite so one expects the shape of the walls of the cavity to not play a significant role. In this case, there is no physical way to distinguish the x-axis from the y-axis, etc. The problem is essentially rotationally invariant. This implies that the wavelength can only depend on the rotationally invariant quantity:

$$\mathcal{N} = \sqrt{n_x^2 + n_y^2 + n_z^2}. \tag{8.30}$$

For a wave that propagates only along the x-axis, we know that $\lambda(n_x, 0, 0) = 2L/n_x$. By symmetry, the wavelength for arbitrary (n_x, n_y, n_z) must take the form:

$$\lambda(\mathcal{N}) = \frac{2L}{\mathcal{N}} \tag{8.31}$$

One can verify this result rigorously by solving the wave equation in a three dimensional cavity, but symmetry has made this rather difficult calculation unnecessary.

We can now count the states with $\mathcal{N}$ in the range $\mathcal{N} \to \mathcal{N} + d\mathcal{N}$ for short wavelengths when $\mathcal{N}$ is very large compared to $d\mathcal{N}$. In essence we need to know how many ways we can choose three integers (n_x, n_y, n_z) and still keep $\mathcal{N}$ in Eq. (8.30) in the range $\mathcal{N} \to \mathcal{N} + d\mathcal{N}$. To this end, imagine plotting all the allowed mode

---

[10] Despite the similarities between electromagnetic radiation and waves on a string, there is one big difference. Light propagates in a vacuum, which of course is impossible for so-called mechanical waves such as waves on a string and sound.

numbers on a three dimensional graph, with n_x along one axis, n_y along the second and n_z along the third. This is a three dimensional infinite array of equally spaced dots.

We now have to count the dots with values of $\mathcal{N}$ in the required range. An important observation is that for very large $\mathcal{N}$, we can treat the mode numbers as continuous: even though the smallest dn_x is unity, this can be considered infinitesimal compared to $\mathcal{N} \gg 1$. This simplifies the counting immensely because the dots are so close together that they essentially fill in the diagram. The only exception is near the origin $n_x = n_y = n_z = 0$, where there are very few states, so that they can safely be ignored.

We now consider all the points that fall inside a spherical shell with inside radius $\mathcal{N}$ and outside radius $\mathcal{N} + d\mathcal{N}$. Since we are assuming that the dots are infinitesimally close we can use calculus to deduce that the total number N_{tot} of dots within a volume of this radius is approximately[11]:

$$N_{tot}(\mathcal{N}) \approx \frac{4\pi}{3}\mathcal{N}^3 \tag{8.32}$$

so that the number of dots in the shell is approximately:

$$dN_{tot}(\mathcal{N}) \approx 4\pi\mathcal{N}^2 d\mathcal{N} \tag{8.33}$$

Since the mode numbers have to be positive only 1/8th of the shell is physically relevant. In addition, the electric and magnetic fields are perpendicular to each other and to their direction of motion. As the wave propagates, they rotate together about the direction of motion in either a clockwise or counter-clockwise direction. These two choices are called polarization states of the photon, so that for each fundamental radiation mode in the cavity there are two possible states. We therefore have:

$$dN_{tot}(\mathcal{N}) \approx 2\frac{1}{8}4\pi\mathcal{N}^2 d\mathcal{N}$$
$$= \pi\mathcal{N}^2 d\mathcal{N} \tag{8.34}$$

We now use Eq. (8.31) to obtain:

$$d\lambda(\mathcal{N}) = -\frac{2L}{\mathcal{N}^2}d\mathcal{N} \tag{8.35}$$

The minus sign signifies that an increase in wavelength requires a decrease in wavenumber. Replacing $d\mathcal{N}$ and $\mathcal{N}$ in Eq. (8.34) by their expressions in terms of λ from Eqs. (8.31) and (8.35), the number of states dN_{tot} is[12]:

$$dN_{tot}(\lambda) \approx \pi(2L)^3\frac{d\lambda}{\lambda^4} \tag{8.36}$$

---

[11] We use $\approx$ here and henceforth to denote "approximately equal to".

[12] It must be a positive number.

We now call on a result from statistical mechanics[13] called the *equipartition theorem*, which we will prove directly below.

Consider a system with a very large number, N, of independent degrees of freedom in thermal equilibrium at fixed temperature T. One example is a system of N independent simple harmonic oscillators. The equipartition theorem states that the kinetic energy associated with each degree of freedom contributes on average $\frac{1}{2}k_B T$ to the energy of the system, where k_B is the Boltzmann constant. The total average kinetic energy contributed by the N degrees of freedom is therefore:

$$KE(N) = \frac{1}{2}Nk_B T \tag{8.37}$$

The equipartition theorem also states that if the potential energy of the particle is quadratic in position $V = \frac{1}{2}kx^2$ like that of a simple harmonic oscillator, then each particle contributes another $\frac{1}{2}k_B T$ to the average energy. For N harmonic oscillators:

$$PE(N) = \frac{1}{2}Nk_B T \tag{8.38}$$

We are now able to state the complete *equipartition theorem*: a large number N of simple harmonic oscillators in thermal equilibrium at temperature T has a total average energy (kinetic plus potential) of

$$E(N) = KE(N) + PE(N) = 2 \times N \times \frac{1}{2}k_B T$$
$$= Nk_B T \tag{8.39}$$

This is a very important result because simple harmonic oscillators play a major role in many areas of physics. They are, in fact, everywhere. The reason is that all objects that are displaced slightly from a position of stable equilibrium (pendulums, cars stuck in a ditch, molecules in a solid...) undergo simple harmonic motion when released. You can convince yourself that this is true either by doing experiments or by examining a Taylor expansion of the potential around the stable equilibrium position, x_0: $V(x) = V(x_0) + V'(x_0)(x - x_0) + \frac{1}{2}V''(x_0)(x - x_0)^2 + \ldots$. The constant term is irrelevant since it produces no forces, and the linear term must vanish or it would push the object away from equilibrium. The first, and dominant term, is therefore the SHO potential with $k = V''(x_0)$. This is yet another situation in which Nature has been kind to physicists, since the simple harmonic oscillator is one of the few systems that we can quantize exactly!

---

[13] Statistical mechanics is the microscopic theory from which the laws of classical thermodynamics are derived.

8.3.2.2   Proof of Equipartition Theorem

In a system consisting of a large number of particles in thermal equilibrium at fixed temperature, the energy of each individual particle is not fixed. The energy of the particle can undergo "thermal fluctuations" as it bounces off other particles in the system, or the walls of the container. As one might imagine, higher energies are less likely than low energies. On the other hand, the higher the temperature of the system, the greater the energy on average of its individual constituents.[14] Thus, one expects the probability for an individual particle to pick up a great deal of energy in a collision to be greater at higher temperature. The normalized probability $P(E)$ for an individual particle in a large system at fixed temperature, T to have energy E turns out to be:

$$P(E) = \frac{e^{-\frac{E}{k_B T}}}{\int dE\, e^{-\frac{E}{k_B T}}} \tag{8.40}$$

You can verify that $P(E)$ in Eq. (8.40) has the required properties. It is called the Boltzmann distribution.

For free particles:

$$E = \frac{1}{2}mv^2 \tag{8.41}$$

To calculate the average or mean energy,[15] $\langle E \rangle$, that this particle contributes to the total energy of the system, we must take the average over all possible speeds that the particle can possess:

$$\begin{aligned}
\langle E \rangle &= \int_0^\infty dv\, E\, P(E) \\
&= \frac{\int_0^\infty dv\left(\frac{1}{2}mv^2 e^{-\frac{mv^2}{2k_B T}}\right)}{\int_0^\infty dv\left(e^{-\frac{mv^2}{2k_B T}}\right)} \\
&= \frac{1}{2}k_B T
\end{aligned} \tag{8.42}$$

The denominator in the above expression is the normalization constant required on the right hand side of Eq. (8.40) in order for the probability distribution to be normalized.

Exercise 4 Perform the integrals to obtain the last line in Eq. (8.42). You will need the following generic integrals:

---

[14] Temperature is in fact a direct measure of the average kinetic energy of its constituents.

[15] see Appendix 15.2 for a general discussion of continuous probabilities.

$$\int_0^\infty dy e^{-by^2} = \frac{1}{2}\sqrt{\frac{\pi}{b}}$$

$$\int_0^\infty dy y^2 e^{-by^2} = -\frac{d}{db}\left(\int_0^\infty dy e^{-by^2}\right)$$

$$= -\frac{d}{db}\left(\frac{1}{2}\sqrt{\frac{\pi}{b}}\right)$$

$$= \frac{1}{4}\sqrt{\frac{\pi}{b^3}} \tag{8.43}$$

We now consider a large number of simple harmonic oscillators in thermal equilibrium with fixed temperature T. In this case the energy of each harmonic oscillator is:

$$E = \frac{1}{2}mv^2 + \frac{1}{2}kx^2 \tag{8.44}$$

To calculate the average energy we need to sum over all possible speeds v and positions x:

$$\langle E\rangle = \int_0^\infty dv \int_{-\infty}^\infty dx\, E\, P(E)$$

$$= \frac{\int_0^\infty dv \int_{-\infty}^\infty dx\,\left(\frac{1}{2}mv^2 + \frac{1}{2}kx^2\right) e^{-\frac{mv^2+kx^2}{2k_BT}}}{\int_0^\infty \int_{-\infty}^\infty dv\, dx\,\left(e^{-\frac{mv^2+kx^2}{2k_BT}}\right)}$$

$$= \frac{\int_0^\infty dv\,\left(\frac{1}{2}mv^2\right) e^{-\frac{mv^2}{2k_BT}} \int_{-\infty}^\infty dx\, e^{-\frac{kx^2}{2k_BT}}}{\int_0^\infty dv\,\left(e^{-\frac{mv^2}{2k_BT}}\right) \int_{-\infty}^\infty dx\,\left(e^{-\frac{kx^2}{2k_BT}}\right)}$$

$$+ \frac{\int_0^\infty dv\, e^{-\frac{mv^2}{2k_BT}} \int_{-\infty}^\infty dx\,\left(\frac{1}{2}kx^2\right) e^{-\frac{kx^2}{2k_BT}}}{\int_0^\infty dv\,\left(e^{-\frac{mv^2}{2k_BT}}\right) \int_{-\infty}^\infty dx\,\left(e^{-\frac{kx^2}{2k_BT}}\right)}$$

$$= \frac{1}{2}k_BT + \frac{1}{2}k_BT$$

$$= k_BT \tag{8.45}$$

In a box containing N one dimensional harmonic oscillators, the total kinetic energy is Nk_BT, as stated in the equipartition theorem Eq. (8.39).

8.3.2.3 Rayleigh-Jeans Law

With the equipartition theorem in our tool belt, we now re-enter the cavity filled with radiation in thermal equilibrium with the walls at temperature T. Each normal mode

of the radiation behaves like an independent degree of freedom: a detailed analysis
reveals that the electric field provides the analogue of kinetic energy for the mode and
the magnetic field provides the corresponding potential energy. Each mode therefore
provides $2 \times \frac{1}{2} k_B T = k_B T$ to the average energy in the cavity. The number of modes
in a given range of wavelengths is given by Eq. (8.36), so the total average energy
(kinetic plus potential) of the modes in this range is equal to:

$$
\begin{aligned}
dE_{EM} &= k_B T |dN_{TOT}| \\
&= \pi k_B T (2L)^3 \frac{d\lambda}{\lambda^4} \\
&= 8\pi L^3 \frac{d\lambda}{\lambda^4}
\end{aligned}
\tag{8.46}
$$

To reiterate, Eq. (8.46) gives the total average electromagnetic energy in the cavity
in the normal modes with wavelength between λ and $\lambda + d\lambda$. It is accurate for large
$\mathcal{N}$ or small λ.

We can now complete the derivation of the Rayleigh-Jeans law for the intensity
$I(\lambda)d\lambda$ emitted by an ideal blackbody at temperature T. Recall that this is the
radiation energy in the required range of wavelengths emitted per unit time per unit
area from the surface of the blackbody. For simplicity, we assume that the blackbody
is in the shape of a cube of length L. We will again use symmetry to conclude that
the radiation is approximately equally distributed among modes moving along the
x-axis, y-axis and z-axis. Given this assumption, at any given time roughly one sixth
of the total radiation in the cavity is moving towards each of the six sides of the box.
Since it takes light $dt = L/c$ to get from the left hand side of the box to the right
hand side, the amount of radiation hitting the right hand wall per unit time is roughly
equal to one sixth of the energy in the required range divided by the time it takes
each mode to get from the left to the right hand side of the box[16]:

$$
\frac{1}{6} \frac{dE_{EM}^{right}}{dt} \sim \frac{1}{6} \frac{dE_{EM}}{L/c}
\tag{8.47}
$$

The area of the right hand wall is L^2, so we must divide by L^2 to get the radiation
per unit area hitting the right hand wall. Finally, we get

$$
\begin{aligned}
I_{right}(\lambda)d\lambda &= \frac{1}{L^2} \frac{1}{6} \frac{dE_{EM}}{L/c} \\
&= \frac{c}{6L^3} (2L)^3 \frac{d\lambda}{\lambda^4} \\
&= \frac{4c\pi}{3} \frac{k_B T}{\lambda^4} d\lambda
\end{aligned}
\tag{8.48}
$$

[16] More accurately, the radiation travels equally in all 4π directions of the unit sphere, so to get the
amount of radiation that hits each wall we must divide by 4π instead of 6. This gives the energy per
unit time per steradian hitting that wall. This slight error in our calculation comes from assuming
cubic symmetry to simplify the calculation of the modes, instead of the correct spherical symmetry.

Note that the length L cancels out of the final expression.

This is a truly remarkable result. Our relatively simple (but somewhat long) argument produced the Rayleigh-Jeans law, including the numerical factor up to a factor of two![17] Now that we understand the origin and apparent inevitability of the Rayleigh-Jeans expression for the intensity of blackbody radiation in Maxwell's wave theory, we are in a better position to appreciate the dilemma that was faced by physicists at the turn of the twentieth century.

8.3.3 The Ultraviolet Catastrophe

The expression for radiation intensity in Eq. (8.48) goes to infinity as the wavelength approaches zero. This means that it predicts that the energy per unit time per unit area emitted at small wavelengths is infinite, which is clearly nonsense. The experimental data at the time quite sensibly showed that this was not the case. The observed intensity versus wavelength curve vanished at small wavelengths. There was therefore a huge mismatch between the apparently unavoidable predictions of thermodynamics (the Rayleigh-Jeans law Eq. (8.48)) and the actual properties of blackbody radiation. The experimental results for various temperatures are represented in Fig. 8.4 and compared to the classical prediction Eq. (8.48).

The curves in Fig. 8.4 depend only on the temperature T and carry no information about the composition or physical properties of the black body. Moreover, there is a maximum of the radiation intensity at a wavelength λ whose value depends inversely on temperature, as follows:

$$\lambda_{max} = \frac{2.9 \times 10^{-3}}{T} \quad \text{m} \cdot \text{K} \tag{8.49}$$

Equation (8.49) is called *Wien's law*. For objects at room temperature, $T = 20° = 293$ K, $\lambda_{max} \sim 10^{-5}$ m $\sim 10\,\mu$m which is well into the infrared. On the other hand, the temperature of the Sun is about 6000 K, so the corresponding wavelength is 5×10^{-7} m, or about 500 nm, which is the wavelength of yellow light.

We will now show how the observed properties of the blackbody radiation intensity curve at small wavelengths (illustrated in Fig. 8.4) follow rather directly from another quantum conjecture.

---

[17] As mentioned above, the error in numerical factor is that we assumed that the radiation in the cavity was moving in one of six directions. In fact it moves in all directions of the unit sphere, i.e. 4π steradians, so we should have divided by 4π steradians in Eq. (8.47) instead of a factor of 6, which would have given us precisely the right formula for the intensity.

Fig. 8.4 Observed blackbody intensity versus wavelength for three temperatures, 3000 K (infrared), 4000 K (red) and 6000 K (yellow), which is the temperature (colour) of our Sun. The Rayleigh-Jeans prediction for 6000 K is also illustrated for comparison. (Color figure online)

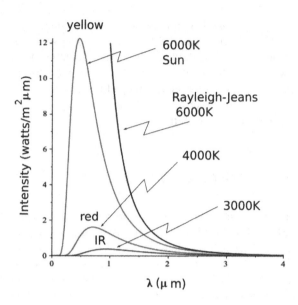

8.3.4 Quantum Resolution

Partly out of desperation to solve the ultraviolet catastrophe, Planck made the bold hypothesis that the energy of a simple harmonic oscillator and hence that of each mode of radiation inside the cavity can only come in lumps that are proportional to the frequency of the oscillator. That is[18]

$$E_m = mhf = mhc/\lambda, \quad \text{where} \quad m = 1, 2, 3, \ldots \quad (8.50)$$

where h was at the time yet another new constant of nature that, when measured, turned out to be the same as Planck's constant encountered in the context of the photoelectric effect and Compton scattering.

Recall that the key to the derivation of the Rayleigh-Jeans law Eq. (8.48) was the Boltzmann distribution Eq. (8.40), which when integrated over all possible positions and speeds of the simple harmonic oscillator lead to the equipartition theorem. The Rayleigh-Jeans law then followed directly from the counting of normal modes in the cavity. Let us repeat the derivation of the equipartition theorem under the assumption that the possible energies of the modes of the simple harmonic oscillators are given by Eq. (8.50). In order to calculate the average value of energy contributed by each harmonic oscillator, we now need to do an infinite sum instead of two integrals:

---

[18] As we will see in Sect. 10.2, the energies predicted by quantum mechanics are $E_m = (m + \frac{1}{2})hf$. The shift of $\frac{1}{2}$ is called the "zero point energy" and has very interesting consequences. It can be ignored in the present discussion.

$$\langle E \rangle = \frac{\sum_{m=1}^{\infty} mhf e^{-m\frac{hf}{k_B T}}}{\sum_{m=1}^{\infty} e^{-m\frac{hf}{k_B T}}}$$

$$= \frac{hf}{e^{\frac{hf}{k_B T}} - 1}. \tag{8.51}$$

This is the famed Bose-Einstein distribution for massless bosons (particles with integer spin) as described in Sect. 13.4.

Exercise 5 1. Do a Taylor expansion of the bottom line in Eq. (8.51) assuming $\frac{hf}{k_B T} \ll 1$ and show that to first order it reproduces the equipartition theorem result Eq. (8.45) for the contribution of a single simple harmonic oscillator to a large system in thermal equilibrium at temperature T.
2. Show that the sum in the first line of Eq. (8.51) yields the expression in the second line.

We can consider each mode of radiation in the cavity to behave like a three dimensional simple harmonic oscillator. As in Eq. (8.46), we use Eq. (8.51) to calculate the total average energy contributed by the modes in the range $N_{tot} \rightarrow N_{tot} + dN_{tot}$[19]:

$$dE_{EM} = \langle E \rangle$$

$$= |\frac{hf}{e^{\frac{hf}{k_B T}} - 1} dN_{tot}| \tag{8.52}$$

Replacing f by hc/λ, using Eq. (8.36) for $dN_{tot}(\lambda)$, dividing by the light transit time of the cavity L/c, and finally dividing by the number 4π of steradians in the unit sphere and by the area L^2 of each side, we get:

$$I(\lambda)d\lambda = \frac{2}{\lambda^5} \frac{hc^2}{e^{\frac{hc}{k_B T\lambda}} - 1} d\lambda. \tag{8.53}$$

Equation (8.53) is the experimentally observed intensity for blackbodies at any temperature T. $I(\lambda)$ reduces to Eq. (8.25) for large wavelengths (low frequencies) and high temperatures, while yielding a drastically different behaviour for small wavelengths (high frequencies) and low temperatures, as required by the blackbody intensity curves shown in Fig. 8.4. As in the case of the photoelectric effect and Compton scattering, one simple, albeit strange, assumption, namely Eq. (8.50), was sufficient to explain one of the most important conundrums of the early twentieth century.

By putting numerical values for the fundamental constants in MKS units into the formula for intensity $I(\lambda)$ one obtains:

---

[19] Recall that dN_{tot} is in reality 1, but very small compared to N_{tot}, so it can be considered infinitesimal.

$$I(\lambda) = \frac{40}{\lambda(\mu m)} \frac{1}{\exp(\frac{14000}{\lambda(\mu m)T(K)} - 1)} \frac{W}{m^2 \, \mu m} \qquad (8.54)$$

where the notation $\lambda(\mu m)$ indicates that the wavelength is to be expressed in micrometers. The temperature T must be given in Kelvin (K).

Exercise 6   Verify that the equation:

$$\left.\frac{dI(\lambda)}{d\lambda}\right|_{\lambda_{max}} = 0 \qquad (8.55)$$

yields Eq. (8.49) for λ_{max}. *Hint: you may need the fact that the solution to* $5(e^x - 1) = xe^x$ *is* $x \approx 5$.

Exercise 7

1. Verify that Eq. (8.54) follows from Eq. (8.53).
2. What is $I(\lambda)$ for $\lambda = 0.6 \, \mu m$ at $T = 6000 \, K$ (the temperature of the Sun)? What is $I(\lambda)$ for $\lambda = 0.6 \, \mu m$ at $T = 310 \, K$ (the temperature of the human body)?

Exercise 8

1. Verify by explicit calculation that Eq. (8.53) follows from Eq. (8.52).
2. Verify that for $hf/(k_B T) \ll 1$ Eq. (8.53) corresponds precisely to Eq. (8.25).

Using the analytic expression for the intensity Eq. (8.53), we can calculate the total energy emitted by a blackbody per unit area per unit time at all wavelengths. This is the *power per unit area P/A*, and is obtained by integrating the intensity over λ:

$$\frac{P}{A} = \int_0^\infty d\lambda I(\lambda)$$
$$= \frac{2\pi^5}{15h^3 c^2} k_B^4 T^4$$
$$= \sigma T^4 \qquad (8.56)$$

where

$$\sigma := \frac{2\pi^5}{15h^3 c^2} k_B^4 = 5.67 \times 10^{-8} \, W/(m^2 \cdot K^4) \qquad (8.57)$$

Exercise 9   1. What is the thermal power output (heat due to blackbody radiation) of the Sun, assuming that it is a perfect blackbody at temperature 6000 K and that the radius is 700,000 km?

2. What is the power output of a healthy human, assuming that they are a perfect cylinder of height 3 m and diameter and circumference 0.3 m, and that they are a perfect blackbody at temperature 37 °C = 310 K? Is this an over-estimate or an under-estimate? Explain.

8.3.5   The Early Universe: The Ultimate Blackbody

The vacuum of space surrounding the Earth is filled with electromagnetic radiation that is almost perfectly blackbody, or thermal, in nature (see Fig. 8.5). This radiation originates in very deep space, and is *isotropic*: it looks the same in all directions in the sky. The Nobel Prize was awarded to Arno Penzias and Robert Woodrow Wilson in 1978 for the discovery of this radiation. While it was detected earlier by Canadian astronomer Andrew McKellar in 1941, who thought it came from inter-stellar molecules, Penzias and Wilson were the first to verify that it came from all directions in space, thereby supporting its interpretation as having originated shortly after the big bang.[20] Using a giant radio telescope built by Bell Laboratories, Penzias and Wilson were able to measure the temperature of the radiation to be around 3 K. Their radio telescope was able to measure the intensity of the radiation only at radio wavelengths, so the temperature was deduced by assuming that the radiation was blackbody at all wavelengths and then fitting their few data points to a suitable curve as in Fig. 8.5. The radiation they discovered was later named the *Cosmic Microwave Background Radiation (CMBR)*.

Balloon and rocket-borne measurements of the radiation intensity for different wavelengths made in the 1960–1980s continued to fit the curve in Fig. 8.5 very well, but experimental challenges limited the data to wavelengths below the 3 K blackbody peak of about 0.3 cm. The Cosmic Background Explorer (COBE) satellite, launched in 1989, was the first to verify the excellent agreement with the blackbody spectrum well beyond the 0.3 cm peak, firmly establishing the Planckian nature of CMBR. When the data was presented at the American Astronomical Society meeting in 1990, it received a standing ovation.[21] More recently, the blackbody nature of the spectrum at all wavelengths has been confirmed with even more accuracy by the WMAP and Planck satellite missions. The CMBR temperature has now been measured to four

---

[20] The existence of this radiation was first predicted in the late 1940s by Ralph A. Alpher, Robert Hermann and George Gamow, but they were largely ignored. Just prior to the Penzias and Wilson discovery, a group of theorists at Princeton University, led by Robert H. Dicke and including Jim Peebles revived the prediction. Jim Peebles, a Canadian by birth and undergraduate education, was awarded a Nobel prize for his contributions to cosmology in 2019.

[21] Weinberg [2].

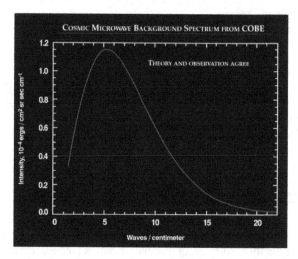

Fig. 8.5 Intensity plot obtained by the COBE satellite. The data points cannot be distinguished from the theoretical curve. *Credit* https://lambda.gsfc.nasa.gov/product/cobe/

digit accuracy to be 2.7255 ± 0.0006 K.[22] It is nothing short of amazing that we are able to measure a fundamental property of the Universe with such great precision!

One of the many remarkable properties of the CMBR is that it is highly isotropic. Satellite measurements are able to detect the very small differences in temperature in the CMBR at different locations in the sky. As shown in Fig. 8.6, the variations are one part in 100,000. These minute *anisotropies*, as they are called, provide a wealth of information about the history of our observable Universe.

The explanation for the CMBR comes from cosmology (see Sect. 7.8). Four hundred thousand years after the Big Bang 14 billion years ago, the Universe was filled with a very hot, dense hydrogen gas. The temperature, about 4000 K, was enough to ionize the hydrogen, breaking it up into a plasma of free electrons and protons. These hot charged particles scattered light back and forth, so that the hot ionized plasma was filled with a cloud of radiation that was in thermal equilibrium with it.

As the Universe expanded, it cooled and the temperature of the plasma fell below the value needed to keep it ionized. The electrons and protons then re-combined to form neutral hydrogen. At this point the thermal radiation had no charged particles with which to interact and was "set free". To a close approximation the radiation has been travelling unchanged throughout the Universe since that moment. The CMBR is this residual thermal radiation and therefore provides an accurate snapshot of what the ionized plasma looked like at the time of "last scattering" 14 billion years ago. As the Universe expanded, the wavelengths of all the radiation underwent a cosmological redshift and the temperature of the cosmic background radiation has cooled to its current 3 K.[23]

---

[22] Fixsen [3].

[23] The CMBR was described in more detail in Sect. 7.8.

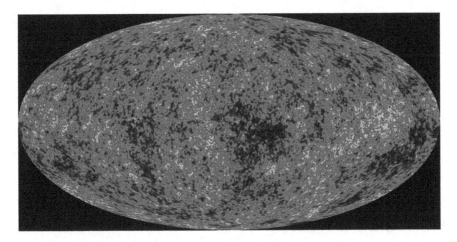

Fig. 8.6 This figure shows a map of the temperature distribution of the Cosmic Microwave Background as a function of direction in the sky. The hot (blue) and cold (red) regions denote fluctuations of about 1 part in 100,000 with respect to the overall temperature of 2.725 K. It is to great accuracy the same in all directions (isotropic). *Credit* https://map.gsfc.nasa.gov/media/121238/index.html. (Color figure online)

Example 1 1. Use Wien's law in Eq. (8.49) to determine the wavelength λ_{max} of peak intensity at the time of last scattering.

$$\lambda_{then} = \frac{3 \times 10^{-3} \text{m} \cdot \text{K}}{T_{then}} = \frac{3 \times 10^{-3}}{3 \times 10^{3}} m = 10^{-6} m \qquad (8.58)$$

2. What is the wavelength of peak intensity now?

$$\lambda_{now} = \frac{3 \times 10^{-3} \text{m} \cdot \text{K}}{T_{now}} = \frac{3 \times 10^{-3}}{3} m = 10^{-3} m \qquad (8.59)$$

3. What is the corresponding red-shift factor

$$z := \frac{\lambda_{now} - \lambda_{then}}{\lambda_{then}} \qquad (8.60)$$

Since $\lambda_{then} \ll \lambda_{now}$ we can use the approximation:

$$z \sim \frac{\lambda_{now}}{\lambda_{then}} = 10^{3} \qquad (8.61)$$

4. If this were a relativistic red-shift instead of a cosmological red-shift, what would v/c be for the hydrogen gas that emitted the CMB radiation?
Using Eq. (6.52) in Sect. 6.7

$$\frac{v}{c} = \frac{(1+z)^2 - 1}{(1+z)^2 + 1}$$

$$= \frac{1 - (1+z)^{-2}}{1 + (1+z)^{-2}}$$

$$\sim 1 - 2\frac{1}{(1+z)^{-2}}$$

$$\sim 1 - 2 \times 10^{-6} = 0.999998 \tag{8.62}$$

where we have used the Taylor approximation, given that $(1+z)^{-2} \sim 10^{-6} \ll 1$.

8.4   Particles as Waves

8.4.1   Electron Waves

We are used to the idea that light is a wave that exhibits interference and diffraction. Although we may not realize it, we are also familiar with the particle-like behaviour of light such as the dots of light that appear on a photographic plate as well as in digital images obtained using photoelectric detectors and Charge Coupled Devices (CCD's). It is more difficult to think that electrons can behave like waves, but there is by now substantial experimental evidence that they do so. A particularly striking example of this is presented in Fig. 8.7. It is an image taken with a *scanning tunnelling electron microscopic (STEM)* of a ring of 48 iron atoms that were arranged in a circle also using a STEM.[24] The waves represent the distribution of surface electrons that are trapped inside the ring of atoms. Such a pattern would not be present if electrons were particles as described by Newtonian mechanics.

Another example of the wave-like properties of electrons is provided by the double slit interference experiment normally associated with light (see Sect. 8.4.3.1). In the electron version, one directs a beam of electrons, all with the same momentum and energy, through two slits onto a screen on the other side. The resulting interference pattern of bright and dark bands shown in Fig. 8.8 suggests the electrons behave like plane waves that pass through both slits! More will be said about this in Sect. 8.4.3.1. We will see in Chap. 9 that it is not only electrons that exhibit wave-like properties. In fact all particles (electrons, protons, billiard balls...) are described by quantum mechanics as probability waves that provide them with both particle-like and wave-like attributes.

---

[24] Crommie et al. [4].

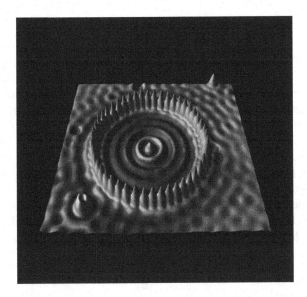

Fig. 8.7 A ring of iron atoms arranged in a circle using a scanning tunnelling electron microscope. This "quantum corral" beautifully illustrates the probability waves that provide the quantum description of the electrons trapped inside the ring of atoms. *Credit* G. Binnig and H. Rohrer, https://doi.org/10.1103/RevModPhys.71.S324, American Physical Society

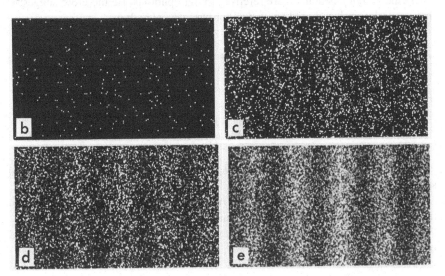

Fig. 8.8 Electron double slit experiment. As individual electrons pass through the slits, an interference pattern gradually appears on the screen, suggesting that each individual electron behaves like an extended wave as it propagates from the source to the screen (see Sect. 8.4.3.1). *Credit* WikiCommons, with permission from Dr. Tonomura

8.4.2 de Broglie Wavelength

The ultraviolet catastrophe, photo-electric effect and Compton scattering all point to the conclusion that electromagnetic waves carry energy in quanta, or lumps. They therefore exhibit both wave-like and particle-like attributes. The relationship between the wavelength and energy of light involves a new constant of nature, namely the Planck constant, h, which is universal in the sense that it is the same for all distinct phenomena.

In 1942, the French physicist Louis de Broglie suggested that if light exhibits wave-like and particle-like properties, then it is reasonable to assume that particles such as electrons should also exhibit both attributes. One can think of this as a type of symmetry argument since it postulates similar behaviour for all forms of matter/energy. In order to deduce the wavelength associated with particles, de Broglie noted that for light one has:

$$\lambda_\gamma = \frac{hc}{E_\gamma}$$

$$= \frac{h}{p_\gamma} \tag{8.63}$$

where the γ signifies that we are referring to light/photons. He therefore suggested the same relationship should be obeyed for particles, namely:

$$\lambda = \frac{h}{p} \tag{8.64}$$

Since particles have non-zero mass m[25] and momentum $p = mv$, we can write:

$$\lambda = \frac{h}{mv} \tag{8.65}$$

Equation (8.65) is called the *de Broglie wavelength* of a particle with mass m and speed v.

Equation (8.63) relates the wavelength of light to its energy, $E_\gamma = hc/\lambda_\gamma$, which in turn provides a relationship between the energy of light and its frequency: $E_\gamma = hf_\gamma$. Pushing the analogy further, de Broglie figured that a similar relationship must hold for particles such as electrons:

$$E = hf \tag{8.66}$$

But what is the frequency, f, in this case? For any classical wave, the frequency and wavelength are related by

$$f = v_w/\lambda, \tag{8.67}$$

---

[25] This discussion is non-relativistic so all mass is essentially rest mass.

where v_w is the speed of the wave.[26] In the case of a massive particle, v_w cannot be the actual speed of the particle as determined by the momentum that appears in Eq. (8.64). Keeping in mind that we are assuming the particle to be non-relativistic and free, its energy is just the kinetic energy:

$$E = K = \frac{1}{2}mv^2 \tag{8.68}$$

Using Eqs. (8.66) and (8.67) gives:

$$\begin{aligned} \frac{1}{2}mv^2 &= h\frac{v_w}{\lambda} \\ &= h\frac{v_w}{h/(mv)} \\ &= mvv_w \end{aligned} \tag{8.69}$$

Thus we find that the relationship between the speed of the particle and the speed of its quantum wave is:

$$v_w = \frac{1}{2}v \tag{8.70}$$

At this stage, the physical origin of this quantum velocity is far from clear. We will see how it comes about when we study probability waves for electrons in Chap. 9 and more rigorously in the Appendix (Sect. 15.4.1).

8.4.3 Consequences

8.4.3.1 Electron Double Slit Experiment

Recall that a double slit interference pattern is produced when a beam of monochromatic light (plane waves) passes through two slits onto a screen on the other side. If the slit widths are less than the wavelength of the light, they act as point sources as shown in Fig. 8.9 producing circular wave fronts that interfere with each other constructively wherever a peak meets a peak, and destructively wherever a peak meets a trough. In the limit $\lambda \ll d$ the interference fringes on the screen are separated by:

$$\Delta y = \frac{\lambda L}{d} \tag{8.71}$$

where L is the distance from slits to screen and d is the distance between the slits.

---

[26] We have labeled v_w with a subscript w to denote that it is a velocity associated with the quantum wave and distinct from the classical velocity v.

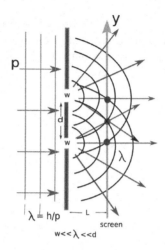

Fig. 8.9 Double slit experiment

   Interference is a property of plane waves that literally pass through two slits at the same time. There is no puzzle here since waves are extended objects. Now consider an experiment in which the electron beam is made weak enough so that on average only one electron passes through the slits at a time and leaves a dot on the screen. Gradually, as more and more dots appear, one sees that there are regions on the screen that are preferred destinations for the electrons and regions that appear to be avoided. Eventually an interference pattern forms consisting of "bright" bands, where many electrons hit, and "dark" bands, where virtually no electrons end up. It is very difficult to imagine at first glance how the electrons in this experiment can exhibit interference. Nonetheless, as illustrated in Fig. 8.8, the famous experiment performed by Dr. Akira Tonomura and collaborators in 1989[27] showed that electrons which pass through two slits one at a time do form an interference pattern on the screen on the other side. This can only happen if each individual electron is described by a wave. As we will see in Chap. 9, quantum mechanics provides an explanation for this strange phenomena by describing electrons, in fact all particles, in terms of waves.

8.4.3.2 The Electron Microscope

In our discussion of the double slit experiment, we relied on the observation that when the size of a slit through which a plane wave passes is smaller than the wavelength, the wave emerges on the other side of the slit as a circular wave (spherical if the slit is a small hole in the screen). If one thinks of light rays, instead of plane waves, the rays are bent in all directions as they pass through the slit. This effect is called diffraction.

---

[27] Tonomura et al. [5].

On the other hand if the slit is much larger than the wave length, the plane wave passes through the opening more or less unimpeded so the individual rays move in a straight line. This means that if one wants to measure the size of an opening using light of wavelength λ_γ, then the opening must be larger than the wavelength of the light. This is true also when one tries to view objects by examining their shadow or using reflected light. We say that the resolving power of optical instruments is *diffraction limited*: it is impossible to clearly image anything smaller than the wavelength of the light being used.

Given that electrons behave quantum mechanically as waves, it is not surprising to learn that they can be used instead of light to image objects. This is useful because there is an inherent limit to the size of microscopic objects that can be imaged via electromagnetic waves. You may wonder why one cannot just use higher and higher frequencies, hence shorter and shorter wavelengths, of light to view arbitrarily small structures. The reason is that the energy of the observing wave must not be large enough to destroy what you are looking at. For light the relationship between resolving power and the energy of the corresponding photons is given by

$$E_\gamma = \frac{hc}{\lambda_\gamma} \tag{8.72}$$

Visible light has a wavelength between 400 and 700 nm, which is the lower limit on the size of the things that one can resolve using an optical microscope. X-rays on the other hand can resolve images down to 0.1 nm, but the associated energy is:

$$E_\gamma = \frac{1241\,\text{eV} \cdot \text{nm}}{0.1\,\text{nm}} = 12\,\text{keV} \tag{8.73}$$

This is high enough to damage whatever atomic size object you are trying to observe. For example the ionizing energy for hydrogen (the amount of energy it takes to knock the electron out of orbit), is only 13.6 eV, one thousand times smaller than the energy of light required to resolve it.

Electrons with the same wavelength and frequency as X-rays have 100 times less energy. An electron with de Broglie wavelength λ has momentum $p = h/\lambda$ and non-relativistic energy:

$$
\begin{aligned}
E_e &= \frac{1}{2} m_e v^2 = \frac{1}{2} m_e \left(\frac{p}{m_e}\right)^2 \\
&= \frac{1}{2} \cdot \frac{1}{m_e c^2} \cdot \left(\frac{hc}{\lambda}\right)^2 \\
&= \frac{1}{2} \cdot \frac{E_\gamma^2}{m_e c^2}
\end{aligned}
\tag{8.74}
$$

For an atomic scale wavelength of $\lambda \approx 0.1$ nm,

$$E_e = \frac{1}{2} \cdot \frac{1}{511\,\text{keV}} \cdot \left(\frac{1241\,\text{eV} \cdot \text{nm}}{0.1\,\text{nm}}\right)^2$$
$$\sim \frac{12.41}{1022} \cdot 12\,\text{keV} = 120\,\text{eV} \tag{8.75}$$

We have been using non-relativistic formulae for the electron. We can verify that such an electron is non-relativistic by comparing this kinetic energy to the invariant mass of the electron:

$$\frac{E_e}{m_e c^2} \sim \frac{0.12\,\text{keV}}{500\,\text{keV}} \ll 1 \tag{8.76}$$

8.4.3.3 Neutron Interferometry

The wave nature of matter is exhibited not only by electrons, but also for much heavier particles. Neutrons are about 2000 times more massive than electrons and have a de Broglie wavelength 2000 times smaller than that of electrons with the same speed. Neutron diffraction is difficult to achieve, but under the right circumstances it is possible to observe neutron interference. The key is to split a coherent stream of neutrons into two beams, such that each of the beams passes through a region of space with different potential energy. One then recombines the beams in order to see interference effects that depend on the strength and variation of the potentials through which they pass. Since neutrons interact via all the forces of nature, the scope of neutron interferometry experiments is enormous.[28]

One particularly interesting experiment performed in 1974[29] consists of two neutron beams that pass through regions of differing gravitational fields, that is at different heights above the Earth's surface. Calculations show that the difference in arrival time between the higher and lower beams is

$$\delta t = s H m^2 g \lambda / \hbar^2 \tag{8.77}$$

where $g = 9.8\,\text{m/s}^2$ is the gravitational acceleration, H is the height difference between the two beams, s the length of the horizontal interference arms and m the neutron mass. This phase difference gives rise to an interference pattern that is measured in a neutron interferometer.

Since gravity affects the 'gravitational mass' (m_g) of the neutron, while the quantum equations depend on its 'inertial mass', (m_i) this experiment provides a test of the equivalence principle described in Sect. 7.2.4. Specifically, if one assumes that these two quantities are not identical, then m^2 in Eq. (8.77) gets replaced by $m_i m_g = m_i^2 \frac{m_g}{m_i}$. An independent measurement of the inertial mass of the neutron then allows the ratio m_g / m_i to be determined. The result is that the inertial and grav-

---

[28] Rauch and Werner [6].

[29] Overhauser and Colella [7].

itational mass are equal to 1 part in 10^4. Any deviation between m_i and m_g would require a fundamental modification to the general theory of relativity.

8.5 The Heisenberg Uncertainty Principle

Particle-wave duality has another well known consequence, namely the *Heisenberg uncertainty principle*, which is stated in many different forms. Here we adopt the following:

No physical quantum states exist for a particle in which both the position and momentum have arbitrarily precise values. The less the uncertainty in position, the greater the uncertainty in momentum and vice versa.

The Heisenberg Uncertainty principle is sometimes stated as follows:

One cannot know with arbitrary accuracy at any instant in time both the position and momentum of a given particle.

This second statement of the uncertainty principle is somewhat misleading, since it focuses on what we are able to "know" or measure. In fact, the uncertainty in position and momentum may not come from a practical limitation on our ability to measure. It may be hinting at something fundamental about the nature of microscopic reality. More will be said about this in Chap. 13.

The Heisenberg uncertainty principle is embodied in the equation:

$$\Delta x \, \Delta p \geq \frac{\hbar}{2} = \frac{h}{4\pi}, \tag{8.78}$$

where $\hbar := h/(2\pi)$ is the *reduced Planck constant* and Δx and Δp are the uncertainties in the position and momentum, respectively, of the particle.[30] Equation (8.78) implies that if the uncertainty in position Δx is small, then the uncertainty in momentum Δp must be large.

Heuristic derivation: Suppose you wish to use light of some wavelength λ to measure the position of a particle. The accuracy is diffraction limited:

$$\Delta x \sim \lambda. \tag{8.79}$$

You can decrease the uncertainty by decreasing the wavelength. However, the light you are using has momentum $p = hf = h/\lambda$ and part or all of this momentum can be transferred to the particle during the observation process. The measurement process therefore introduces an uncertainty in the particle's momentum equal to:

$$\Delta p \sim \frac{h}{\lambda}. \tag{8.80}$$

---

[30] The mathematical and physical meaning of these uncertainties will be discussed in Sect. 9.5.2.

The uncertainty in momentum given in Eq. (8.80) depends on the inverse of the wavelength. When you multiply the two uncertainties you get:

$$\Delta p \Delta x \sim h \qquad (8.81)$$

Apart from the factor of 4π, this is consistent with the exact quantum mechanical bound Eq. (8.81).

Exercise 10 Heisenberg was stopped by a police officer one day while driving his sports car. The policeman asked Heisenberg: "Do you know how fast you were going?", to which Heisenberg replied: "No, but I know where I am". The police officer then pointed out "you were going 146.5 km/h". "Great", said Heisenberg, "now I am lost".

1. Assuming that Heisenberg was able to use GPS to locate his position to an accuracy of 10 m, what was the minimum uncertainty in his momentum?
2. Assuming that the uncertainty in the police officer's measurement of Heisenberg's speed was in the last digit, i.e. ±0.1 km/h, what was the uncertainty in Heisenberg's position after the measurement?

References

1. A. Einstein, Physik **17**, 132 (1905)
2. S. Weinberg, *Cosmology* (Oxford University Press, 2008), p. 105
3. D.J. Fixsen, The temperature of the cosmic microwave background. Astrophys. J. **707**, 916–920 (2009)
4. M.F. Crommie, C.P. Lutz, D.M. Eigler, Science **262**(5131), 218–220 (1993)
5. A. Tonomura, J. Endo, T. Matsuda, T. Kawasaki, H. Ezawa, Am. J. Phys. **57**, 117–120 (1989)
6. H. Rauch, S.A. Werner, *Neutron Interferometry*, 2nd edn. (Oxford, 2015)
7. A.W. Overhauser, R. Colella, Phys. Rev. Lett. **33**, 1237 (1974)

Chapter 9
The Wave Function

9.1 Learning Outcomes

Conceptual

- Understand the probabilistic interpretation of the wave function, including the physical interpretation of expectation values and standard deviation.
- Understand the mathematical and physical origins of the Heisenberg uncertainty principle.
- Understand the physical significance of the energy-time uncertainty principle and how it differs from the Heisenberg uncertainty principle.

Acquired skills

- Given a wave function describing the quantum state of a particle:
  - Normalize the wave function
  - Calculate expectation value of its position and the standard deviation of its position.

9.2 Quantum Versus Newtonian Mechanics

Our next task is to understand how quantum mechanics leads to particle-wave duality, the uncertainty principle and many other interesting consequences.

To keep things simple, we will consider a single particle moving along the x-axis under the influence of a conservative force $F(x) = -\frac{dV}{dx}$. We will assume that the particle speed is always much less than the speed of light so that we do not have to worry about special relativity. A relativistic formulation of quantum mechanics exists (see Sect. 10.10), but it is significantly more complicated.

An important thing to keep in mind is that quantum mechanics differs from classical mechanics in a very profound way. The basic constructs of the theory, the quantities from which the equations are built, are distinct from the physical quantities such as position and momentum that we measure in experiments. This is in

© The Author(s), under exclusive license to Springer Nature Switzerland AG 2022
G. Kunstatter and S. Das, *A First Course on Symmetry, Special Relativity and Quantum Mechanics*, Undergraduate Lecture Notes in Physics,
https://doi.org/10.1007/978-3-030-92346-4_9

effect what requires us to abandon much of our physical intuition when studying the microscopic world.

9.2.1   Newtonian Description of the State of a Particle

In Newtonian mechanics, the state of a particle at some instant in time t_0 is determined by two quantities: the position $x(t_0)$ and the velocity $v(t_0)$. It will prove convenient for the subsequent discussion to consider the momentum $p(t) = mv(t)$ since it plays a more fundamental role in quantum mechanics. If you know the position and momentum of a particle at some specific $t = t_0$, then you know everything there is to know about the state of the particle. You can calculate its energy, both kinetic and potential (assuming you know the potential energy function). In addition, if you know all the forces acting on the particle you can use Newton's laws to uniquely determine the motion and hence the future values of the position, momentum and any other associated quantities.

We emphasize that Newtonian mechanics assumes that all observable quantities can in principle be measured simultaneously with arbitrary accuracy. It makes sense that Nature should behave this way, namely that our ability to know physical quantities is limited only by our powers of observation. We will see in Chap. 13, however, that it is precisely this statement that fails to be true in quantum mechanics.

9.2.2   Quantum Description of the State of a Particle

The quantum description of the state of a particle is drastically different from that of the Newtonian description. Instead of specifying the precise position and momentum of the particle, quantum mechanics describes the state of a particle by a function $\psi(x, t)$ that assigns a *complex* number to each point x at time t. $\psi(x, t)$ is called the **wave function** of the particle. The fact that it is complex is crucial to understanding some of the more bizarre and useful predictions of quantum mechanics. It also provides yet another example of how physics and mathematics are inseparable. One might naively think that the "square root of -1" is a purely mathematical construct that has no correspondence in the real world. In fact one cannot account for the behaviour of the microscopic world without it.[1]

Whereas Newtonian physics assigns a specific position to a particle, the quantum wave function in general extends throughout space, just like an electromagnetic wave, for example. It is in this sense that a particle is a wave in quantum mechanics.

---

[1] Or something mathematically equivalent, such as a matrix whose square is minus the unit matrix.

It cannot in general be said to have a unique position any more than a wave on a string or an electromagnetic wave can.[2]

Physically measurable quantities such as the position and momentum of a particle are mathematically and physically distinct from the wave function describing the quantum state of the particle. They are called *observables* associated with the system. This dichotomy between the state of a system and its observables is at the root of the strange properties of quantum systems that will be described in Chap. 13.

So what is the physical interpretation of the wave function? The somewhat counter-intuitive answer is that its magnitude[3] at each point in space provides the probability of measuring the position of the particle to be in the immediate vicinity of that point. In terms of equations:

$$\mathcal{P}(x,t)dx := \psi^*(x,t)\psi(x,t)dx \tag{9.1}$$

gives the probability that a measurement of the position of the particle at time t will yield a value between x and $x + dx$. $\mathcal{P}(x,t)$ is called the *probability density* associated with the state $\psi(x,t)$. The probability of finding the particle somewhere in a specific region of space, say between x_1 and x_2 at a given time t is then :

$$\begin{aligned} P(x_1 \to x_2) &= \int_{x_1}^{x_2} dx \mathcal{P}(x,t) \\ &= \int_{x_1}^{x_2} dx \psi^*(x,t)\psi(x,t) \end{aligned} \tag{9.2}$$

The wave function $\psi(x,t)$ is often called the *probability amplitude*.

Since the probability of finding the particle somewhere on the real line must be one (assuming that the particle exists):

$$P(-\infty \to \infty) = \int_{-\infty}^{\infty} \psi^*(x,t)\psi(x,t)dx = 1 \tag{9.3}$$

This is called the *normalization condition* that must be satisfied at all times by any quantum wave function describing the physical state of a particle. Thus the time dependence of any physical state must be such that the integral in Eq. (9.3) stays constant. Note that the normalization condition requires that $\mathcal{P}$ vanish at infinity in order for the integral to remain finite.

The definition Eq. (9.1) implies that the probability density does not change if one multiplies a wave function $\psi(x,t)$ by an arbitrary complex number of unit norm, $e^{i\alpha}$ for any real number α. Specifically, if $\psi(x,t) \to e^{i\alpha}\psi(x,t)$ then

---

[2] One important difference is that electromagnetic waves correspond to physical quantities that can also be observed classically by measuring the force the electromagnetic fields exert on charged particles. The quantum wave function cannot be observed directly.

[3] Recall that the magnitude of a complex number z is given by $|z| = \sqrt{z^*z}$. See Sect. 15.1 in the Appendix.

$$\psi^*(x, t)\psi(x, t) \rightarrow e^{-i\alpha}\psi^*(x, t)e^{i\alpha}\psi(x, t)$$
$$= e^{-i\alpha}e^{i\alpha}\psi^*(x, t)\psi(x, t)$$
$$= \psi^*(x, t)\psi(x, t) \tag{9.4}$$

Such an overall *phase factor* is physically irrelevant as long as α is independent of x but when we define the quantum versions of momentum and energy in Sect. 9.5.2, we will see that the phase factor does have physical relevance when it is not constant.

The ability to shift the wave function by a constant phase factor corresponds to a continuous global symmetry of quantum mechanics as described in Sect. 9.2.2. There is consequently a conserved charge associated with this symmetry as required by Noether's theorem. This conserved charge is in fact the total probability of finding the particle anywhere, as given in the integral in Eq. (9.3). If this were not constant, total probability would not be conserved. Such unphysical behaviour is dubbed *non-unitary* evolution.

9.3 Measurements of Position

If the wave function predicts only probabilities, how does one use it to make predictions for the outcome of experiments? Instead of thinking about experiments involving individual particles, it is necessary to consider a large number of identically prepared particles, called an *ensemble*, each in a state described by the same wave function $\psi(x)$.[4] One can then use the wave function to predict the statistics of the outcomes when a large number of measurements of position, or in fact any function $f(x)$ of the position, are done on the particles in this ensemble. An important key to understanding quantum mechanics is to realize that only one measurement can be done on each particle since after this measurement the particle will be in a different state, one that depends on the outcome of the measurement. In other words measurement changes the wavefunction of each individual particle in the ensemble so that subsequent measurements will in effect constitute a different experiment.

We now look in more detail at the statistical predictions one can extract from the quantum wavefunction. Using the *probability density* $\mathcal{P}(x, t) = \psi^*(x, t)\psi(x, t)$, one can calculate the averages, or *expectation* values, that result from a large number of measurements of position, and functions of position (see Sect. 15.2.2 in the Appendix):

$$\langle x \rangle = \int_{-\infty}^{\infty} dx \, x \psi^*(x)\psi(x).$$

$$\langle x^2 \rangle = \int_{-\infty}^{\infty} dx \, x^2 \psi^*(x)\psi(x).$$

---

[4] For the time being we restrict attention to states of particles at fixed time, so that we drop the argument t of the wave function. We will look more carefully at the time dependence in Sect. 10.7.

$$\langle f(x) \rangle = \int_{-\infty}^{\infty} dx \ f(x) \psi^*(x) \psi(x).$$

$$\Delta x_{sd} = \sqrt{|\langle x^2 \rangle - \langle x \rangle^2|}. \tag{9.5}$$

Δx_{sd} is the *standard deviation* of the probability distribution $\mathcal{P}$. If it is large, then measurements will yield many different values for x that are not close to the average value.

9.4 Example: Gaussian Wave Function

Suppose a large number of particles are prepared in a state with wave function:

$$\psi(x) = A e^{-\frac{(x-x_0)^2}{2b^2}} \tag{9.6}$$

where A is a complex number. The wave function in Eq. (9.6) is called *Gaussian*.

1. The probability distribution $\mathcal{P}(x)$ for $b = 1/2$ and $x_0 = 2$ is plotted in Fig. 9.1.
2. The value of the normalization constant A is determined by the condition:

Fig. 9.1 Gaussian probability distribution

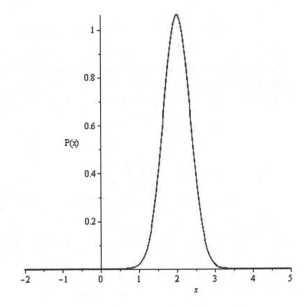

$$1 = \int_{-\infty}^{\infty} dx \; \psi^*(x)\psi(x)$$

$$= A^*A \int_{-\infty}^{\infty} dx \; \exp(-(x - x_0)^2/(2b^2)) \exp(-(x - x_0)^2/(2b^2))$$

$$= |A|^2 \int_{-\infty}^{\infty} dx \; \exp(-(x - x_0)^2/(b^2))$$

$$= |A|^2 \int_{-\infty}^{\infty} b \, dy \exp(-y^2) \quad \text{where } y := (x - x_0)/b$$

$$= |A|^2 b \sqrt{\pi}, \tag{9.7}$$

where we have used the basic integral:

$$\int_{-\infty}^{\infty} dy \exp\left(-\frac{y^2}{2}\right) = \sqrt{2\pi}. \tag{9.8}$$

The normalization condition requires:

$$|A|^2 = \frac{1}{b\sqrt{\pi}}, \tag{9.9}$$

The normalized wave function is therefore:

$$\psi(x) = \frac{1}{\sqrt{b\sqrt{\pi}}} \exp(-(x - x_0)^2/(2b^2)). \tag{9.10}$$

Note that the normalization condition does not determine the phase of the complex number A, only its magnitude.

3. The most probable value occurs at the value of x where:

$$\frac{d\mathcal{P}(x)}{dx} = 0$$

$$= \frac{d(\psi^*(x)\psi(x))}{dx}$$

$$= \frac{d(|A|^2 \exp(-(x - x_0)^2/b^2))}{dx}$$

$$= -\frac{2|A|^2}{b^2}(x - x_0) \exp(-(x - x_0)^2/b^2). \tag{9.11}$$

The most probable value is therefore $x = x_0$.

4. The average value, or expectation value, of the particles' measured position is calculated as follows:

$$\langle x \rangle = \int_{-\infty}^{\infty} dx \, x \psi^*(x)\psi(x)$$

$$= \frac{1}{b\sqrt{\pi}} \int_{-\infty}^{\infty} dx \, x \exp(-(x-x_0)^2/(b^2))$$

$$= \frac{1}{b\sqrt{\pi}} \int_{-\infty}^{\infty} dx \, [(x-x_0)+x_0] \exp(-(x-x_0)^2/(b^2))$$

$$= 0 + x_0. \tag{9.12}$$

The first term is zero by symmetry of the integral, the second term gives x_0 because the wave function is normalized.

5. $\langle x^2 \rangle$ for this state is

$$\langle x^2 \rangle = \int_{-\infty}^{\infty} dx \, x^2 \psi^*(x)\psi(x)$$

$$= \frac{1}{b\sqrt{\pi}} \int_{-\infty}^{\infty} dx \, (x-x_0)^2 \exp(-(x-x_0)^2/(b^2))$$

$$+ \frac{1}{b\sqrt{\pi}} \int_{-\infty}^{\infty} dx \, (2xx_0 - x_0^2) \exp(-(x-x_0)^2/(b^2))$$

$$= \frac{1}{b\sqrt{\pi}} \int_{-\infty}^{\infty} dy \, y^2 \exp(-y^2/(b^2)) + x_0^2$$

$$= \frac{b^2}{2} + x_0^2, \tag{9.13}$$

where we have changed variables to $y := x - x_0$.

6. The *standard deviation* is:

$$\Delta x_{sd} = \sqrt{|\langle x^2 \rangle - (\langle x \rangle)^2|}. \tag{9.14}$$

Consequently, for the Gaussian probability distribution above:

$$\Delta x_{sd} = \sqrt{\frac{b^2}{2} + x_0^2 - x_0^2} = \frac{b}{\sqrt{2}}. \tag{9.15}$$

9.5 Momentum in Quantum Mechanics

9.5.1 Pure Waves

De Broglie told us that the momentum p of a particle is proportional to the inverse of its wavelength. A wave function that describes a particle with fixed momentum must therefore be a *pure wave*, by which we mean a periodic cosine or sine function with fixed wavelength λ. The general form of such a wave is:

$$\psi_p(x) = A \cos\left(\frac{2\pi x}{\lambda}\right) + B \sin\left(\frac{2\pi x}{\lambda}\right)$$
$$= A \cos(kx) + B \sin(kx). \tag{9.16}$$

In the last line of Eq. (9.16) we have defined the *wave number*, k, by:

$$k := \frac{2\pi}{\lambda}. \tag{9.17}$$

In terms of wave number, the momentum has the form:

$$p = \hbar k, \tag{9.18}$$

where $\hbar$ is the reduced Planck's constant:

$$\hbar := \frac{h}{2\pi} = 1.0545718 \times 10^{-34} \text{ J} \cdot \text{s}. \tag{9.19}$$

A quantum wave function that is a pure wave can therefore be written in terms of the momentum:

$$\psi_p(x) = A \cos(px/\hbar) + B \sin(px/\hbar). \tag{9.20}$$

Recall that in quantum mechanics the wave function, and hence the constants A and B in Eq. (9.16) are complex numbers. Thus Eq. (9.16) can be rearranged algebraically as follows:

$$\psi_p(x) = \tilde{A} \left(\cos(kx) + i \sin(kx)\right) + \tilde{B} \left(\cos(kx) - i \sin(kx)\right), \tag{9.21}$$

with $A = \tilde{A} + \tilde{B}$ and $B = i(\tilde{A} - \tilde{B})$. Using *Euler's formula*:

$$e^{\pm ikx} = \cos(kx) \pm i \sin(kx), \tag{9.22}$$

we find that a complex pure wave can be written in the form:

$$\psi_p(x) = \tilde{A} e^{ikx} + \tilde{B} e^{-ikx}. \tag{9.23}$$

We henceforth also use the term pure waves to denote e^{ikx} and/or e^{-ikx}. When looking at time dependent solutions to the full Schrödinger equation in Sect. 10.3.1 (see Eq. (10.17)), we will see that the functions e^{ikx} and e^{-ikx} describe pure waves with momentum $p = \hbar k$ moving in the positive (negative) x-directions, respectively. By allowing the wave number k and hence momentum p to be positive or negative, we can account for waves moving in both directions.

The mathematics of Fourier transforms described in the Appendix 15.3 informs us that **any** complex function $\psi(x)$ can be written as a sum or superposition of pure waves with wave number k, where k can be positive or negative, depending on which way the momentum of the pure wave is pointing. This leads to the expression:

$$\psi(x) = \frac{1}{\sqrt{2\pi}} \int_{-\infty}^{\infty} dk\phi(k) \exp(ikx). \tag{9.24}$$

In Eq. (9.24), $\phi(k)$ is a complex function of the wave number k. It specifies the amplitude of each pure wave that is required for the linear superposition of pure waves to add up to the desired function $\psi(x)$. This is accomplished mathematically by having just the right amount of constructive and destructive interference between all the pure waves to produce cancellation where required, and enhancement where required. $\phi(k)$ contains all the information required to reproduce $\psi(x)$ and *vice versa*. It therefore represents a complementary mathematical description of the wave function $\psi(x)$.

If the function $\psi(x)$ is very narrow, or highly localized in space, then one needs to sum many different pure waves with different wavenumbers in order to produce the required cancellation over a large part of the real line. Its complementary function (i.e Fourier transform) $\phi(k)$ will then be a very wide (spread out) function of wavenumber. Conversely, if $\phi(k)$ is narrow, then the function $\psi(x)$ will be close to a pure wave and therefore very wide. As shown in Appendix 15.3, the width Δx of $\psi(x)$ in x and the width Δk of its complementary description $\phi(k)$ in k satisfy the following inequality:

$$\Delta x \Delta k \geq \frac{1}{2}. \tag{9.25}$$

Equation (9.25) is a result in pure mathematics. The interesting physics associated with Eq. (9.25) emerges from de Broglie's observation that in quantum mechanics momentum is proportional to wave number. If one replaces k by the de Broglie wave relation $p/\hbar$ in Eq. (9.25) and then multiplies both sides by $\hbar$, one obtains the Heisenberg uncertainty principle:

$$\Delta x \Delta p \geq \frac{\hbar}{2}. \tag{9.26}$$

The take away message is that the uncertainty principle is a consequence of two equally important ingredients. The first is the physical connection postulated by de Broglie between the momentum of a particle and its wavelength/wavenumber. The second is a mathematical property of the complementary descriptions of quantum particle states in terms of the wave function on one hand and its Fourier transform on the other.

9.5.2   The Momentum Operator

We know how to calculate the expectation value of x for any given physical wave function $\psi(x)$, but what about the momentum? Using the de Broglie relation, this requires calculating $\langle p \rangle = \langle \hbar k \rangle$, which can most easily be done using the complementary description (Fourier transform) $\phi(k)$ of $\psi(x)$. For reasons explained in more detail in Appendix 15.3 the probability of measuring the wave number in a given quantum state to be between k and $k + dk$ is:

$$\Phi(k)dk = \phi^*(k)\phi(k)dk, \tag{9.27}$$

so that the probability density for wavenumber is:

$$\Phi(k) = \phi^*(k)\phi(k). \tag{9.28}$$

This is at least plausible because $\Phi(k)$ as defined in Eq. (9.27) is non-negative. Moreover, as proven in Appendix 15.3.5, if the wave function is normalized, then so is its Fourier transform:

$$\int_{-\infty}^{\infty} dk\ \Phi(k) := \int_{-\infty}^{\infty} dk\ \phi^*(k)\phi(k) = 1. \tag{9.29}$$

In terms of the probability density, $\Phi(k)$, the average value of momentum is therefore:

$$\langle p \rangle = \langle \hbar k \rangle$$
$$= \hbar \int_{-\infty}^{\infty} dk \phi^*(k) k \phi(k). \tag{9.30}$$

It is also proven in Appendix 15.4.5 that one can get the same expectation value directly from the wave function $\psi(x)$ by using the following expression:

$$\langle p \rangle = \int dx \psi^*(x)(-i\hbar)\frac{\partial \psi(x)}{\partial x} \tag{9.31}$$

While the expression looks a bit odd, it can readily be understood by writing $\psi(x)$ and $\psi^*(x)$ in Eq. (9.31) as an integral over their Fourier transforms $\phi(k)$ and $\phi^*(k)$, in which case all the x dependence is in the pure complex exponentials e^{ikx}. The derivative with respect to x in Eq. (9.31) therefore pulls down a factor of ik in front of each exponential in $\psi(x)$. The exponentials in the Fourier transform of $\psi(x)$ effectively cancel with their corresponding complex conjugate in $\psi^*(x)$ and one is left with Eq. (9.30). We leave the detailed proof as an exercise.

Exercise 1 Prove that if $\psi(x)$ is defined in terms of $\phi(k)$ by Eq. (9.24), then

1.

$$\langle\psi|\psi\rangle = \int_{-\infty}^{\infty} dk\phi^*(k)\phi(k). \tag{9.32}$$

2.

$$\int_{-\infty}^{\infty} dx\psi^*(x)(-i\hbar)\frac{\partial\psi(x)}{\partial x} = \int_{-\infty}^{\infty} dk\phi^*(k)\hbar k\phi(k) \tag{9.33}$$

You will need to use Eq. (15.36) in Sect. 15.2.3.

The left hand side Eq. (9.31) formally looks like one is calculating the average value of a derivative, or to use a fancier term *differential operator*:

$$\hat{p} = -i\hbar\partial/\partial x. \tag{9.34}$$

In fact this is precisely what one is doing. For this reason, $\hat{p}$ in Eq. (9.31) is called *the momentum operator*.

The calculation of $\langle\hat{p}^2\rangle$ is obtained by inserting the momentum operator twice into the integral for the expectation value, as follows:

$$\langle\hat{p}^2\rangle = \frac{1}{b\sqrt{\pi}}\int dx\psi^*(x)\left((-i\hbar)\frac{\partial}{\partial x}(-i\hbar)\frac{\partial}{\partial x}\right)\psi(x)$$

$$= \frac{1}{b\sqrt{\pi}}\int dx\psi^*(x)(-i\hbar)^2\frac{\partial^2\psi(x)}{\partial x^2}, \tag{9.35}$$

where we have made use of the fact that differential operators in this notation always act on whatever is found on their right.

Example 1 For a particle in a state described by a Gaussian Eq. (9.6) calculate Δp_{sd} by calculating the expectation value of the operators $\hat{p}$ and $(\hat{p})^2$ using $\hat{p} = -i\hbar\partial/\partial x$.

Solution:

$$\langle\hat{p}\rangle = \frac{1}{b\sqrt{\pi}}\int dx\psi^*(x)(-i\hbar)\frac{\partial\psi(x)}{\partial x}$$

$$= \frac{1}{b\sqrt{\pi}}\int dx\exp(-(x-x_0)^2/(2b^2))(-i\hbar)\frac{-2(x-x_0)}{2b^2}\exp(-(x-x_0)^2/(2b^2))$$

$$= \frac{1}{b\sqrt{\pi}}\frac{i\hbar}{b^2}\int dx(x-x_0)\exp(-(x-x_0)^2/(b^2))$$

$$= 0. \tag{9.36}$$

The vanishing of $\langle \hat{p} \rangle$ in Eq. (9.36) is to be expected because the Gaussian in Eq. (9.6) is symmetric under $x \to -x$. A non-zero average momentum would pick out a preferred direction, namely along the positive x-axis if $p > 0$ or along the negative x-axis if $p < 0$, which is inconsistent with the symmetry of the wave function.

This symmetry argument does not apply to $\langle \hat{p}^2 \rangle$, since $\hat{p}^2$ is also symmetric under a change of direction $x \to -x$ in the double derivative. We therefore must do the calculation to find the answer:

$$
\begin{aligned}
\langle \hat{p}^2 \rangle &= \frac{1}{b\sqrt{\pi}} \int dx \, \psi^*(x)(-i\hbar)^2 \frac{\partial^2 \psi(x)}{\partial x^2} \\
&= \frac{1}{b\sqrt{\pi}} \int dx \exp(-(x-x_0)^2/(2b^2))(-i\hbar)^2 \frac{\partial}{\partial x} \left(\frac{-(x-x_0)}{b^2} e^{-(x-x_0)^2/(2b^2)} \right) \\
&= \frac{\hbar^2}{b\sqrt{\pi}} \int dx \left[\frac{1}{b^2} + \frac{(x-x_0)^2}{b^4} \right] \exp(-(x-x_0)^2/(b^2)) \\
&= \frac{\hbar^2}{b^4} \left(b^2 - \langle (x-x_0)^2 \rangle \right) \\
&= \frac{\hbar^2}{b^4} \left(b^2 - \frac{b^2}{2} \right) \\
&= \frac{\hbar^2}{2b^2}.
\end{aligned}
\tag{9.37}
$$

The standard deviation is:

$$
\Delta p_{sd} = \frac{\hbar}{\sqrt{2}b}.
\tag{9.38}
$$

Combining the result in Eq. (9.38) with Eq. (9.15), for a Gaussian wave function one finds:

$$
\Delta x_{sd} \Delta p_{sd} = \frac{\hbar}{2}.
\tag{9.39}
$$

This is the best one can do with regard to balancing uncertainty of position and uncertainty of momentum. Gaussian wave functions are said to *saturate* the uncertainty principle.

9.6   Energy in Quantum Mechanics

The total Newtonian energy for a particle moving in a potential $V(x)$ is:

$$
E = \frac{p^2}{2m} + V(x)
\tag{9.40}
$$

In quantum mechanics we need to calculate the average energy of a particle of mass m in a given quantum state $\psi(x)$. This will tell us the average value for energy that would be obtained if one measured the energy of a large number of particles each prepared in the same state $\psi(x)$. In Sect. 9.5.2 we learned how to construct a differential operator $\hat{p}$ that allowed us to calculate the average momentum of a particle in a given state as well as the average value p^2, or more generally p^n. This suggests the following operator definition for kinetic energy in quantum mechanics:

$$
\widehat{K} := \frac{\hat{p}^2}{2m}
$$

$$
= \frac{1}{2m}(-i\hbar)^2\frac{\partial^2}{\partial x^2} \tag{9.41}
$$

The resulting operator for the total energy $\hat{H}$, kinetic plus potential, is therefore:

$$
\hat{H} = \frac{\hat{p}^2}{2m} + V(x)
$$

$$
= \frac{1}{2m}(-i\hbar)^2\frac{\partial^2}{\partial x^2} + V(x) \tag{9.42}
$$

The expression for the total energy in quantum mechanics is given the letter H because it is the quantum version of the *Hamiltonian* that plays such an important role in classical mechanics.

The average energy of a particle in a given state described by the wave function $\psi(x)$ is:

$$
\langle\hat{H}\rangle = \int dx\, \psi^*(x)\left(\frac{-\hbar^2}{2m}\frac{\partial^2}{\partial x^2} + V(x)\right)\psi(x)
$$

$$
= \int dx\left(\psi^*(x)\frac{-\hbar^2}{2m}\frac{\partial^2\psi(x)}{\partial x^2} + \psi^*(x)V(x)\psi(x)\right) \tag{9.43}
$$

We are also interested in the uncertainty, or standard deviation, ΔE of the energy. It is defined as usual by:

$$
\Delta E = \sqrt{\left|\langle\hat{H}^2\rangle - \langle\hat{H}\rangle^2\right|} \tag{9.44}
$$

This requires calculating $\langle\hat{H}^2\rangle$ by applying $\hat{H}$ twice underneath the expectation value integral:

$$
\langle\hat{H}^2\rangle = \int dx\left(\psi^*(x)\left(\frac{-\hbar^2}{2m}\frac{\partial^2}{\partial x^2} + V(x)\right)\left(\frac{-\hbar^2}{2m}\frac{\partial^2}{\partial x^2} + V(x)\right)\psi(x)\right)
$$

$$
\tag{9.45}
$$

Note that in the above expression the differential operators act on everything to their right. In particular, the differential operator $(\partial^2/\partial x^2)$ in the first factor in brackets in the integrand differentiates the factor $\left(\frac{-\hbar^2}{2m}\frac{\partial^2\psi(x)}{\partial x^2} + V(x)\psi(x)\right)$ to yield:

$$\langle \hat{H}^2 \rangle = \int dx \left(\psi^*(x) \left(\frac{-\hbar^2}{2m} \right)^2 \frac{\partial^4 \psi(x)}{\partial x^4} + \psi^*(x) V(x) \left(\frac{-\hbar^2}{2m} \frac{\partial^2 \psi(x)}{\partial x^2} \right) \right.$$

$$\left. + \psi^*(x) \left(\frac{-\hbar^2}{2m} \frac{\partial^2 V(x) \psi(x)}{\partial x^2} \right) + \psi^*(x) V(x) V(x) \psi(x) \right) \qquad (9.46)$$

9.6.1 Energy-Time Uncertainty Relation

There is another uncertainty relation involving the standard deviation of the energy ΔE in any time-dependant state and the minimum time Δt_O required for the expectation value of any operator $\hat{O}$ to change significantly.[5]

In mathematical terms the *energy-time uncertainty relation* states:

$$\Delta E \Delta t_O \geq \frac{\hbar}{2} \qquad (9.47)$$

The time interval Δt_O must be defined rather carefully. In particular, we need to specify what constitutes a *significant* change in the expectation value of the operator $\hat{O}$. A significant change in this context is one that is approximately equal to or greater than the uncertainty in the expectation value of that observable in the same state.

While Eq. (9.47) looks the same as the Heisenberg uncertainty principle, and is often used in similar ways, it is in fact very different in nature. Heisenberg's uncertainty principle implies that if the uncertainty in x in any state is small then the uncertainty in p is necessarily large, and *vice-versa*. The best one can do is optimize the uncertainties in complementary observables.

Equation (9.47) on the other hand does not say anything about uncertainty in time measurements, and for good reason. Time is absolute in non-relativistic quantum mechanics, just as in Newtonian mechanics.[6] Instead, Eq. (9.47) gives us a relationship between energy uncertainty in a time dependent state, and the rate at which observables can evolve in that state. If the energy of a given state is nearly certain, so that it is approximately stationary, then Δt_O must be large because the expectation values of all observables are roughly constant. On the other hand, if the energy of a given state is highly uncertain, then it is possible for expectation values of operators to evolve very rapidly and Δt_O is correspondingly small.

We stress that although the time interval Δt_O is defined with respect to a particular observable $\hat{O}$, the energy-time uncertainty relation Eq. (9.47) applies to Δt_O for **all** operators. It is a general relation between the rate at which expectation values can change significantly and how close a given state is to being stationary.

---

[5] We will study the time dependence of quantum states further in the next Chapter. In particular Sect. 10.2.1 will show that the expectation value of all time independent operators in a given state will be constant if and only if the state has definite energy, i.e. if $\Delta E = 0$. Such a state is called *stationary*.

[6] The relativistic version of quantum mechanics will be discussed in Sect. 10.10.

The energy-time uncertainty relationship is often described as stating that in quantum mechanics energy conservation can be violated for short periods of time: the greater the violation, the shorter the allowed time. This is not the case. ΔE refers not to a violation of energy conservation but to the uncertainty in the energy in a given state. It is a measure of the spread in the probability distribution of energy for an ensemble of identical systems that are prepared in the same state.

Example 2 For a free particle of mass m, derive the energy-time uncertainty principle for the position observable from the Heisenberg uncertainty principle.[7]

Solution:

The operator $\hat{O}$ in this case is the position operator $\hat{x}$. Δt_x is the time it takes for the average position of the particle whose state is described by a Gaussian wave packet to change by an amount greater than the uncertainty Δx in its position. Thus:

$$\Delta t_x = \frac{\Delta x}{\frac{d\langle \hat{x}\rangle}{dt}} = m\frac{\Delta x}{\langle \hat{p}\rangle} \tag{9.48}$$

where we have used

$$\langle \hat{p}\rangle = m\frac{d\langle \hat{x}\rangle}{dt} \tag{9.49}$$

To calculate the change in the expectation value of the energy for a free particle we will use that:

$$\Delta E = \Delta\left(\frac{p^2}{2m}\right)$$
$$= \frac{1}{2m}\Delta p^2$$
$$\sim \frac{1}{m}\langle p\rangle \Delta p \tag{9.50}$$

The step from the second line to the third in Eq. (9.50) seems reasonable but it needs to be checked explicitly that (see Exercise 2 directly below):

$$\Delta(p^2) \sim 2\langle p\rangle \Delta p \tag{9.51}$$

Using Eqs. (9.51) and (9.50) we find:

---

[7] This is a variation of an example given in *Quantum Mechanics*, D. J. Griffiths and D. F. Schroeter, 3rd edition, Cambridge University Press (2018).

$$\Delta E \Delta t_x = m \frac{\Delta x}{\langle p \rangle} \frac{1}{2m} \langle p \rangle \Delta p$$

$$= \Delta x \Delta p \geq \frac{\hbar}{2} \tag{9.52}$$

as expected from the energy-time uncertainty relationship.

Exercise 2 Verify Eq. (9.51) explicitly for a Gaussian of width $b\sqrt{2}$ and mean position $x_0 = 0$. You can use the generalization of the result in Eq. (9.33) of Example 1, namely:

$$\langle \psi(x) | \hat{p}^n | \psi(x) \rangle = \hbar^n \int_{-\infty}^{\infty} \phi^*(k) k^n \phi(k) \tag{9.53}$$

and the Fourier transform of the Gaussian given in Eq. (15.54) in the Appendix.

Exercise 3 In March, 2013, it was announced that the data obtained previously by two independent measurements using the Large Hadron Collider at CERN, Switzerland, had detected the Higgs boson. This illusive particle was the last missing piece to the standard model that unified the electromagnetic and weak interactions. The Higgs is an unstable particle and its mass was measured to be 125 ± 0.2 GeV. Based on the uncertainty in energy, use the energy-time uncertainty principle to deduce the minimum time it would take for the Higgs to undergo a major change in state, i.e. to decay to other elementary particles. Note that this exercise will yield the relationship well known in particle physics between the lifetime of an unstable elementary particle created in the laboratory, and the measured spread (uncertainty) in the energy with which it was created.

Chapter 10
The Schrödinger Equation

10.1   Learning Outcomes

Conceptual

- Definition and interpretation of stationary states.
- Qualitative justification (not proof) of how to calculate expectation values of functions of momentum using both position and momentum wave functions.
- Qualitative understanding of quantum observables as linear operators.
- Qualitative understanding of eigenvalues and eigenfunctions and their role in measurement.
- Symmetries as operators in quantum mechanics.
- Proof of the quantum version of Noether's theorem.
- Stationary states as standing wave solutions to the time-dependent Schrödinger equation.
- Qualitative understanding of quantum tunnelling and some of its physical applications.

Acquired Skills

- Given a general wave function:

  1. Normalize the wave function.
  2. Calculate average momentum and standard deviation.
  3. Check whether it satisfies the time independent Schrödinger equation and if so determine the energy.

- Given the stationary state for a free particle in a box or simple harmonic oscillator, calculate physical consequences such as values of energy, expectation values and standard deviations of operators, sketch the wave functions and probability densities.
- Calculation of tunnelling amplitudes.

© The Author(s), under exclusive license to Springer Nature Switzerland AG 2022   213
G. Kunstatter and S. Das, *A First Course on Symmetry, Special Relativity and Quantum Mechanics*, Undergraduate Lecture Notes in Physics,
https://doi.org/10.1007/978-3-030-92346-4_10

10.2   The Time Independent Schrödinger Equation

10.2.1   Stationary States

So far we have dealt with wave functions at some instant in time, without worrying about time dependence. In quantum mechanics, like in classical mechanics, waves do not generally stand still. To illustrate the properties of moving waves, we will first consider a wave propagating along a string.

We will take x to be the coordinate that locates a particular point on the string, and $y(x, t)$ to be the displacement from the equilibrium position of the point on the string at location x at time t. We define the *node* $x_n(t)$ of a wave to be the location on the string where the displacement vanishes at time t: $y(x_n, t) = 0$. In general, the nodes move as the wave moves. An exception is the *standing wave* which is created by taking a *linear superposition* (i.e sum) of two separate waves of the same amplitude and wavelength moving in opposite directions with the same velocity. The result is a wave with *nodes* that stand still, with all points between the nodes oscillating up and down. As an example, consider a standing cosine wave by setting $k_1 = k_2 = k$ and $\omega_1 = \omega_2 = \omega$ in Eq. (15.73) of Appendix 15.4.2. The resultant wave is:

$$y_{tot} = 2A \cos(kx) \cos(\omega t) \tag{10.1}$$

Equation (10.1) describes a *standing wave*. Each point on the string at location x oscillates up and down with amplitude $2A \cos(\omega t)$ and frequency $f = 2\pi\omega$. At fixed t it is a pure wave with wavelength $\lambda = 2\pi/k$ and amplitude $2A \cos(\omega t)$.

The most general standing wave takes the form:

$$y(x, t) = f(x)g(t) \tag{10.2}$$

At fixed time t, the shape of the wave is $f(x)$. This shape stays the same for all time but each point on the string oscillates with an amplitude $g(t)$ that changes with time.

Exercise 1 On a single graph sketch the wave in Eq. (10.1) over one wavelength at $t = 0$, $\omega t = \frac{\pi}{2}$ and $\omega t = \pi$. You can take A = 1.

Stationary states in quantum mechanics provide the quantum analogue of Eq. (10.1).

Definition: A *stationary state* in quantum mechanics is defined to have the following properties:

1. It has definite energy, which in quantum terms requires $\Delta E_{sd} = 0$.
2. The average values (expectation values) of all measurable quantities are independent of time. This condition requires the probability density $\mathcal{P}(x) = \Psi^*(x, t) \Psi(x, t)$ to be independent of time.

The above two conditions are satisfied by a time dependent wave function of the form:

$$\Psi(x, t) = \psi_E(x)e^{-iEt/\hbar} \tag{10.3}$$

where $\psi(x)$ is a complex valued function of position that obeys the *time independent Schrödinger equation* (10.4):[1]

$$\hat{H}\psi_E(x) := \frac{-\hbar^2}{2m}\frac{d^2\psi_E(x)}{dx^2} + V(x)\psi_E(x)$$
$$= E\psi_E(x) \tag{10.4}$$

and we have used the definition of the Hamiltonian in Eq. (9.42).

$\Psi(x, t)$ in Eq. (10.3) is essentially a complex standing wave. It is a product of a purely spatial part, and an oscillating time dependent amplitude. We now verify that it has the key properties specified in the definition of a stationary state defined above.

1. The time dependence cancels out of the probability density:

$$\Psi^*(x, t)\Psi(x, t) = \left(\psi_E(x)e^{-iEt/\hbar}\right)^* \psi_E(x)e^{-iEt/\hbar}$$
$$= \psi_E^*(x)e^{+iEt/\hbar}\psi_E(x)e^{-iEt/\hbar}$$
$$= \psi_E^*(x)\psi_E(x) \tag{10.5}$$

2. The state has energy E with $\Delta E = 0$:

$$\langle\hat{H}\rangle = \int dx\,\psi_E^*(x)\left(\frac{-\hbar^2}{2m}\frac{d^2\psi_E(x)}{dx^2} + V(x)\psi_E(x)\right)$$
$$= \int dx\,\psi_E^*(x)E\psi_E(x)$$
$$= E \tag{10.6}$$

Where we have used the time independent Schrödinger equation (10.4) to get the second line, and normalization of the wave function to get the third line.

$$\langle\hat{H}^2\rangle = \int dx \left(\psi_E^*(x)\hat{H}\hat{H}\psi(x)_E\right)$$
$$= \int dx \left(\psi_E^*(x)\hat{H}E\psi(x)_E\right)$$
$$= \int dx \left(\psi_E^*(x)E^2\psi(x)_E\right)$$
$$= E^2 \tag{10.7}$$

---

[1] We now change to ordinary derivatives to emphasize that $\psi_E(x)$ depends only on one variable.

Thus,

$$\Delta E = \sqrt{\langle \hat{H}^2 \rangle - \langle \hat{H} \rangle^2}$$
$$= \sqrt{E^2 - E^2}$$
$$= 0 \tag{10.8}$$

3. The frequency of oscillation of the complex exponential on the right hand side of Eq. (10.4) is:

$$f = \frac{2\pi}{\hbar/E}$$
$$= \frac{E}{h}$$
$$\rightarrow \quad E = hf \tag{10.9}$$

10.3   Examples of Stationary States

10.3.1   Free Particle in One Dimension

The potential $V(x) = 0$ in this case, so the energy is

$$E = \frac{p^2}{2m} \tag{10.10}$$

The time independent Schrödinger equation is then:

$$\frac{-\hbar^2}{2m}\frac{d^2\psi_E(x)}{dx^2} = E\psi_E(x) \tag{10.11}$$

In order to solve this, we first multiply both sides of the equation by $\frac{-2m}{\hbar^2}$ to get

$$\frac{d^2\psi_E(x)}{dx^2} = -\frac{2m}{\hbar^2}E\psi_E(x)$$
$$= -k^2\psi_E(x) \tag{10.12}$$

where we have defined

$$k := \sqrt{\frac{2mE}{\hbar^2}} \tag{10.13}$$

We note that Eq. (10.12) is the same form as the simple harmonic equation for a particle of mass M moving under the influence of an ideal spring with spring constant K and fundamental frequency $\Omega = \sqrt{\frac{K}{M}}$:

$$\frac{d^2 x(t)}{dt^2} = -\frac{K}{M} x(t)$$
$$= -\Omega^2 \, x(t) \tag{10.14}$$

One difference is that instead of being a function of time, $\psi(x)$ is a function of position. Another is that the fundamental frequency $\Omega = \sqrt{K/M}$ in the simple harmonic oscillator equation is replaced by the energy dependent frequency $\omega := \sqrt{2mE/\hbar^2}$ in the Schrödinger equation. The solutions are identical in form. They can be written in terms of sine and cosine functions, or in complex form:

$$\psi_E(x) = A \cos kx + B \sin kx$$
$$= \tilde{A} e^{ikx} + \tilde{B} e^{-ikx} \tag{10.15}$$

where

$$\hbar^2 k^2 = 2mE = p^2 \tag{10.16}$$

and the constants A, B, $\tilde{A}$, $\tilde{B}$ are complex in general.

The full time-dependent form of this stationary state is:

$$\Psi(x, t) = \left(A e^{ikx} + B e^{-ikx} \right) e^{-i\omega t}$$
$$= A e^{i(kx - \omega t)} + B e^{-i(kx + \omega t)} \tag{10.17}$$

$$\text{where} \quad \hbar\omega = E = \frac{\hbar^2 k^2}{2m} \tag{10.18}$$

The solution is an arbitrary linear combination of pure waves with wave number k and $-k$, and angular frequency $\omega = \hbar k^2/2m$. The exponential function $e^{i(kx - \omega t)}$ describes a pure wave moving to the right with speed

$$v_w = \frac{\omega}{k}$$
$$= \frac{v}{2} \tag{10.19}$$

where we have used Eq. (10.13) and $E = mv^2/2$ to get the last line. $e^{-i(kx + \omega t)}$ is a wave moving to the left with the same speed (see Sect. 15.4.1.). The wave functions in Eq. (10.17) represent states with fixed magnitude of momentum $|p|$ and energy $E(p) = p^2/(2m)$. Note that the speed of the wave is one half the speed v that would be associated with a free particle with classical energy $E = mv^2/2$, as discussed in Sect. 8.4.2 (see Eq. (8.70)).

Such pure waves can NOT be normalized. The probability density $\Psi^*(x,t)\Psi(x,t)$ is constant so that the probability of finding the particle is the same everywhere. One way to get around this problem is to construct a wave packet, as in Eq. (15.88) of Sect. 15.4.4, to describe the quantum state for a particle that is localized and hence normalizable. In this case, however, the state does not have a precise momentum or energy since it will necessarily be a linear combination of pure waves. Another remedy is to assume that the particle is free in the sense of not having any forces acting on it, but confined to a box, as in the example below.

10.3.2   Particle in a Box with Impenetrable Walls

Consider a one-dimensional free particle (i.e. no forces acting on it), of mass m confined to a box of length L with impenetrable walls, so that the particle is restricted to the line segment $0 \leq x \leq L$. In practical terms this means that the probability of finding the particle outside the box is zero, so the wave function must vanish for $x \leq 0$ and $x \geq L$. Inside the box, the wave function $\psi_E(x)$ is a solution to Eq. (10.4) with $V(x) = 0$. Just as in the case of electromagnetic waves inside a blackbody cavity (See Sect. 8.3.1), the wave function must vanish at the walls of the box on the inside as well as on the outside. This insures the continuity of the wave function across the wall. The boundary conditions on the solutions to the Schrödinger equation inside the box are therefore[2]:

$$\psi_E(0) = \psi_E(L) = 0 \tag{10.20}$$

As mentioned above, $\psi_E(x)$ obeys the free particle Schrödinger equation inside the box:

$$\frac{d^2\psi_E(x)}{dx^2} = -\frac{2mE}{\hbar^2}\psi_E(x) \tag{10.21}$$

Since this is again the same form as the simple harmonic oscillator equation $\ddot{x}(t) = -\omega^2 x(t)$, we know that the solution is a linear combination of cosine and sine functions. In order to satisfy the boundary conditions in Eq. (10.20) we must restrict to solutions of the form:

$$\psi_{E_n}(x) = A_n \sin\left(\frac{\pi n x}{L}\right), \qquad n = 1, 2, 3, \dots \tag{10.22}$$

---

[2] Strictly speaking this is only true because we are considering the impenetrability of the walls to be an approximation. Completely impenetrable walls would require an infinite repulsive potential, something that is not physically attainable. In the case of truly impenetrable walls more general boundary conditions are possible.

The integer n is the *quantum number* that labels all physically allowed states. The constant A is again determined up to an arbitrary phase by the normalization condition. The solutions depend on an arbitrary positive integer. Since $\sin(a) = -\sin(-a)$ for all a, choosing $n = -4$, say, would give precisely the same physical wave function as $n = +4$. Only the normalization constant A would change by a physically irrelevant sign. In Eq. (10.22) we have labelled the energy E and the normalization constant A by a subscript n to emphasize that both depend on the value of the integer.

Exercise 2 Show that the normalization constant A_n for the wave function Eq. (10.22) is:

$$|A_n|^2 = \frac{2}{L} \tag{10.23}$$

and is therefore independent of n. Note that this is special to this particular example. In most examples of stationary states, the normalization constant does depend on the energy quantum number.

The wave number and momentum associated with Eq. (10.22) are given in terms of the quantum number n by:

$$k_n = \frac{\pi n}{L} \tag{10.24}$$

$$p_n = \hbar k_n = \frac{\hbar \pi n}{L} \tag{10.25}$$

Recall that the p_n are to be thought of as magnitudes of momentum because Eq. (10.22) describes standing waves that move neither to the left nor to the right. The probability of measuring the momentum to be $+p_n$ and $-p_n$ are equal despite the fact that a classical particle with the same energy moves either to the left or right.[3]

The energy E_n of the state $\psi_{E_n}(x)$ given in Eq. (10.22) is:

$$E_n = \frac{p_n^2}{2m}$$
$$= \hbar^2 \frac{\pi^2 n^2}{2mL^2}, \quad n = 1, 2, 3, ... \tag{10.26}$$

The energy in Eq. (10.26) cannot take on any positive real value. Instead there is an infinite, but discrete set of possible energies labelled by the positive integers. We say that the energy is *quantized*.

As discussed further in Sect. 10.6.1 the small value of $\hbar$ in MKS units guarantees that the discrete behaviour is only observable for microscopic particles with tiny energies. For very large energies, namely those labelled by very large integers, the

---

[3] Technically, we say that Eq. (10.22) is not an eigenstate of the momentum operator $p = -i\hbar\partial/\partial x$ because the derivative operator changes the sine function to a cosine function. Such a quantum state therefore does not have a definite momentum.

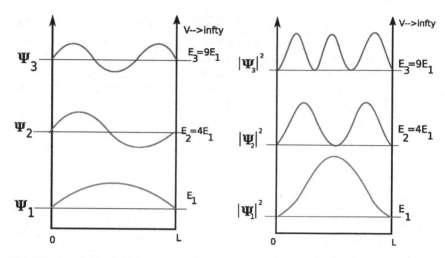

Fig. 10.1 Amplitudes (left) and probability densities (right) of the first three energy eigenstates of a particle in a box with impenetrable walls

classical continuum of allowed energies is approximately recovered in the sense that changing n by 1 produces a very small fractional change in the energy. We can verify this by taking the expression Eq. (10.26) and changing n to $n + \Delta n$ for $\Delta n \ll n$:

$$
\begin{aligned}
\Delta E_n &= E_{n+\Delta n} - E_n \\
&= \hbar^2 \frac{\pi^2 (n + \Delta n)^2}{2mL^2} - \hbar^2 \frac{\pi^2 n^2}{2mL^2} \\
&= \hbar^2 \frac{\pi^2 n^2}{2mL^2} \left[\left(1 + \frac{\Delta n}{n} \right)^2 - 1 \right]^2 \\
&= \hbar^2 \frac{\pi^2 n^2}{2mL^2} \left(2\frac{\Delta n}{n} + \frac{(\Delta n)^2}{n^2} \right) \\
&\to 0 \quad \text{as} \quad \frac{\Delta n}{n} \to 0
\end{aligned}
\tag{10.27}
$$

The wave functions and probability densities are illustrated in Fig. 10.1.

As indicated earlier each allowed state represents a complex standing wave that vanishes at the sides of the box.

10.3.3 Simple Harmonic Oscillator

The potential energy for a simple harmonic oscillator is:

$$V(x) = \frac{1}{2}Kx^2 \tag{10.28}$$

where K is the spring constant. For a mass m, the angular frequency in radians per second of the oscillator is:

$$\Omega = \sqrt{\frac{K}{m}} \tag{10.29}$$

The corresponding Schrödinger equation is

$$-\frac{\hbar^2}{2m}\frac{d\psi_E(x)}{dx^2} + \frac{1}{2}Kx^2\psi_E(x) = E\psi_E(x) \tag{10.30}$$

Since the wave function ψ must be normalizable, it must go to zero as $x \rightarrow \pm\infty$. The simplest solution and, as it turns out, the one with the lowest energy is:

$$\psi_0(x) = Ae^{-Cx^2} \tag{10.31}$$

The lowest energy state of any quantum system is called the *ground state*.

Exercise 3 Verify that Eq. (10.31) satisfies the Schrödinger equation (10.30) if and only if the constant C is related to the parameters of the harmonic oscillator by:

$$C = \frac{m\Omega}{2\hbar} \tag{10.32}$$

and that the corresponding energy E of the state is

$$E_0 = \frac{1}{2}\hbar\Omega \tag{10.33}$$

As shown in Exercise 3 above, the lowest energy that the stationary state of a quantum simple harmonic oscillator can possess is $E_0 > 0$. E_0 is called the *ground state energy*, or "zero point energy", and plays a very important role in physics.
The calculation of the normalization constant proceeds as follows:

$$\int_{-\infty}^{\infty} dx\, \psi_0^*(x)\psi_0(x) = A^2 \int_{-\infty}^{\infty} dx e^{-2Cx^2}$$

$$= A^2\sqrt{\frac{\pi}{2C}} \tag{10.34}$$

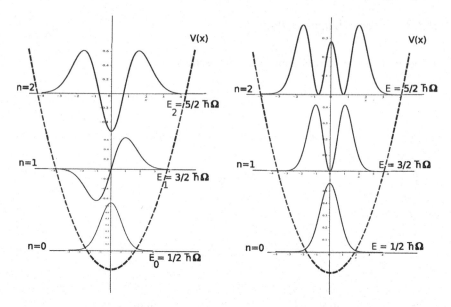

Fig. 10.2 First few energy levels of the simple harmonic oscillator

This gives:

$$|A| = \left(\frac{2C}{\pi}\right)^{1/4} \tag{10.35}$$

As in the case of the particle in a box, the quantum simple harmonic oscillator turns out to have an infinite number of states, each labelled by a non-negative integer n, with quantized energies given by:

$$E_n = (n + \frac{1}{2})\hbar\Omega, \quad n = 0, 1, 2, 3... \tag{10.36}$$

Note that for the simple harmonic oscillator, $n = 0$ is allowed and corresponds to the ground state discussed above. Figure 10.2 illustrates the wave functions of the lowest energy states of the simple harmonic oscillator as well as their probability densities and energies.

The classical continuum is recovered for $n \gg 1$ as required because the difference between the energies in adjacent states ($\Delta n = 1$) is much smaller than the energies themselves. This is explained more generally in Sect. 10.6.1:

Exercise 4 Consider a simple harmonic oscillator of mass 1 kg, with initial amplitude 1 m and spring constant 2 Newtons per meter. If its energy is quantized according to Eq. (10.36) above, what is the value of the integer n?

10.4 Absorption and Emission

We saw in Sect. 10.3 that there is a discrete spectrum of allowed energies for a particle in an impenetrable box and for a simple harmonic oscillator. A discrete spectrum is a general feature of quantum particles that are confined to a finite region of space either in a box or by a potential well and therefore are not free.

The spectrum necessarily has a lowest energy state, called the *ground state*.[4] For the particle in an impenetrable box, for example, the energy of the ground state is:

$$E_1 = \hbar^2 \frac{\pi^2}{2mL^2}, \qquad p_1 = \sqrt{2mE_1} \qquad (10.37)$$

Such a minimum energy implies that such a quantum particle trapped in a box cannot "sit still", in the sense that the probability of measuring it to have zero momentum is zero. In addition, a particle in its ground state cannot emit radiation because it has no lower energy state to which to go. It can however absorb energy and jump to a higher energy state.

A particle in a state other than the ground state can either emit or absorb energy. Energy conservation requires this energy to be precisely the amount needed to take it from its initial stationary state i, with energy E_{n_i}, to the final state f, with energy E_{n_f}. If the energy is absorbed or emitted in the form of a photon, then the photon energy and corresponding frequency must be:

$$E_\gamma = hf_\gamma = E_{n_f} - E_{n_i} \qquad (10.38)$$

For the particle in an impenetrable box:

$$E_\gamma = hf_\gamma = \hbar^2 \frac{\pi^2}{2mL^2} \left(n_f^2 - n_i^2 \right) \qquad (10.39)$$

If $n_f > n_i$, the energy is positive and a photon is absorbed. If $n_f < n_i$, E_γ is negative which means that a photon is emitted, not absorbed

For a simple harmonic oscillator,

$$E_\gamma = hf_\gamma = \left(n_f - n_i \right) hf \qquad (10.40)$$

where f is the frequency of the oscillator. Note that in this case the energy of the emitted photons must be an integer times hf. This provides the explanation for Planck's intuitive leap, described in Sect. 8.3.4, which resolved the ultraviolet catastrophe.

---

[4] There is an exception that proves the rule, namely a particle in a potential of the form $V(x) \propto 1/x^2$, but its quantization requires special care and is outside the scope of this text.

Exercise 5 An electron is in an electron trap of length $L = 0.6\,\mu$.

1. What is the energy of the electron in its ground state?
2. What are the energy and frequency of the photon it would have to absorb in order to go from the ground state to the $n = 4$ state?
3. What frequency photon would it subsequently emit if it dropped into the $n = 2$ state?

10.5   Tunnelling

10.5.1   Particle in a Box with Penetrable Walls

In Sect. 10.3.2 we derived the quantum states of a particle in an impenetrable box. The probability of finding the particle outside the box was, by assumption, zero.

If the wall of the box is penetrable, the potential energy barrier that the particle classically has to overcome in order to escape is finite. Consider a potential energy function $V(x)$ of the following form:

$$
\begin{aligned}
V(x) &= 0 & 0 < x < L, & \quad \text{in the box} \\
V(x) &= V_0 > 0 & x < 0;\, x > L, & \quad \text{outside the box}
\end{aligned}
\tag{10.41}
$$

As shown in Sect. 10.5.2, things get interesting for particles with $E < V_0$ that are forbidden by Newtonian mechanics to escape from the box. In quantum mechanics the wave function "leaks" into the forbidden region on both sides of the box, namely $x > L$ and $x < 0$. The solution to Schrödinger's equation inside the box, where $V = 0$, is:

$$
\psi_{in} = A\cos(k_n x) + B\sin(k_n x)
\tag{10.42}
$$

where

$$
k_n = \frac{\sqrt{2m E_n}}{\hbar}
\tag{10.43}
$$

We have labelled the solutions by an integer, anticipating that the energy will be quantized leading to a discrete set of solutions.

Outside the box, the Schrödinger equation is:

$$
-\frac{\hbar^2}{2m}\frac{d\psi_n(x)}{dx^2} + V_0\psi_n(x) = E_n\psi_n(x)
\tag{10.44}
$$

Taking the potential energy term to the right hand side, and remembering that $E_n < V_0$, we can write this as:

$$\frac{d^2\psi_n(x)}{dx^2} = \kappa_n^2 \psi_n(x) \tag{10.45}$$

where

$$\kappa_n = \frac{\sqrt{2m(V_0 - E_n)}}{\hbar} \tag{10.46}$$

This looks like the free particle Schrödinger equation, but with a positive sign on the right instead of negative and its solution is almost the same. The positive sign leads to solutions consisting of real exponentials rather than complex exponentials, or if you wish, hyperbolic cosines and sines rather than the usual cosine and sine functions. The general solution is:

$$\psi_n^{out}(x) = Ce^{\kappa_n x} + De^{-\kappa_n x} \tag{10.47}$$

Exercise 6 Verify explicitly that Eq. (10.47) solves Eq. (10.45).

The first term on the right hand side of Eq. (10.47) is unphysical. When we square the wave function to get the probability density, it will yield a contribution that grows exponentially with distance from the box. This implies that the probability of finding the particle outside the box grows rapidly with distance, something that makes no sense. A more technical objection to the first term in the solution is that it would make the wave function non-normalizable. We must therefore set $C = 0$ in order to obtain a physical solution. This leaves:

$$\psi_n^{out}(x) = \tilde{D}e^{-\kappa(x-L)} \tag{10.48}$$

where we have absorbed a factor of $e^{\kappa L}$ into the arbitrary constant D in order to make it clear that we expect the wave function to drop off exponentially from $x = L$. One can repeat this analysis on the left hand side of the box ($x < 0$).

In order to completely determine the energies and corresponding wave functions one must determine the constants multiplying the wave functions in each of the three regions. By symmetry the constants are the same on the left and the right. This leaves three complex constants, two inside and one outside. Two of them are determined by requiring the wave function to be continuous and smooth at the walls of the box, the third is determined by normalizing the final wave function and dropping the overall phase.

Finally, we see that the probability of finding a particle outside a penetrable box is non-zero, even if it does not possess enough energy classically to get over the barrier. The probability density:

$$\mathcal{P}(x) = \tilde{D}^2 e^{-2\kappa(x-L)} \tag{10.49}$$

drops off exponentially as the distance from the box increases. As with the finite single barrier in Sect. 10.5.2, Eq. (10.49) gives the relative probability of finding the particle in the forbidden region at $x > L$ compared to finding it at the wall of the box at $x = L$:

$$
\begin{aligned}
T &:= \frac{\mathcal{P}(x > L)}{\mathcal{P}(x = L)} \\
&= \frac{\tilde{D}^2 e^{-2\kappa(x-L)}}{\tilde{D}^2} \\
&= e^{-2\kappa(x-L)}, \quad x > L
\end{aligned}
\tag{10.50}
$$

where T is called the *transmission coefficient*.

10.5.2 Tunnelling Through a Potential Barrier of Finite Width

Consider now a potential barrier of finite height V_0 and finite width L, as shown in Fig. 10.3. In this case one must solve Schrödinger's equation in three distinct regions, region I to the left of the barrier, region II inside the barrier and region III to the right of the barrier. The wave functions in all three regions have the same energy E. In regions I and III the potential is zero so the particle wave function is that of a pure wave with fixed energy E and corresponding wave number $\sqrt{\frac{2mE}{\hbar}}$.

If the energy of the particle is greater than the height of the potential energy barrier ($E > V_0$), then the particle has enough energy to go through the barrier. In this case the solution is a pure wave everywhere, but the wave numbers in different regions are different. On the left and right hand side of the barrier, regions I and III, the wave number is that of a free particle with energy E, namely $k_{III} = k_I = \frac{\sqrt{2mE}}{\hbar}$. In region II however, the total energy of the particle is the same as in the other regions, but the kinetic energy is decreased as the particle gains V_0 in potential energy. The corresponding wave number in region II is therefore $k_{II} = \frac{\sqrt{2m(E-V_0)}}{\hbar}$. This type of solution is called a *scattering solution*. Because it is a pure wave that is not normalizable, there is no quantization condition on the energy E.

If the energy $E < V_0$, then classically the particle would remain either on the right or on the left of the barrier. Quantum mechanically the situation is quite different. Solving the Schrödinger equation in region II results in a wave number $k_{II} = \sqrt{2m(E - V_0)}\hbar$ that is an imaginary number. This tells us that inside the barrier the wave function is not an oscillating wave. The complex exponentials of the free particle solution become real exponentials, as in Sect. 10.5.1 above. That is, in Region II the solution is of the form:

$$
\begin{aligned}
e^{\pm i \times i \frac{\sqrt{2m(V_0-E)}}{\hbar}} &= e^{\mp \frac{\sqrt{2m(V_0-E)}x}{\hbar}} \\
&= e^{\mp \frac{\kappa_{II} x}{\hbar}}
\end{aligned}
\tag{10.51}
$$

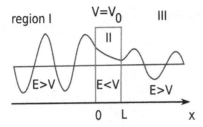

Fig. 10.3 Quantum wave function for free particle in the presence of a potential energy barrier V_0 with energy $E < V_0$. Note that the wave function is smooth and continuous on both sides of the barrier

where

$$\kappa_{II} = \frac{\sqrt{2m(V_0 - E)}}{\hbar} \tag{10.52}$$

This is either a growing or a decreasing exponential. Assuming without loss of generality that the particle is most likely be on the left of the wall, one must choose the decaying exponential in region II with corresponding exponentially decreasing probability density. The wave function must be continuous and smooth at the boundaries between the regions. This ensures that both the probability density and the momentum are smooth.[5] The ratio of the wave function amplitudes just to the left of the wall and just to the right, is given by:

$$\frac{\psi_{III}(L)}{\psi_I(0)} := e^{-\kappa_{II}L} \tag{10.53}$$

where L is the width of the wall in this case, and κ_{II} is given in (10.52). The relative probability of finding the particle on the left as opposed to the right is:

$$T := \left| \frac{\psi_{III}(L)}{\psi_I(0)} \right|^2 := e^{-2\kappa_{II}L} \tag{10.54}$$

T is called the transmission coefficient.

Since the solution is a pure wave on either side of the barrier, it is again non-normalizable, so there is no quantization condition on the energy E.

Example 1 A 30 keV electron is confined by a potential barrier that is 40 eV high and 1.0 nm wide.

---

[5] Smoothness is no longer required if the potential is infinite, since even classically a perfectly elastic collision with a perfectly reflecting wall causes the momentum to change direction discontinuously.

1. What is the transmission coefficient?
   Solution:

$$
\begin{aligned}
\kappa &= \frac{\sqrt{2m(V - E)}}{\hbar} \\
&= \frac{\sqrt{2mc^2(V - E)}}{\hbar c} \\
&= \frac{\sqrt{2 \times 500 \text{ keV} \times (40 - 30) \text{ eV}}}{1241 \text{ eV} \cdot \text{nm}} \\
&= 2.6 \text{ nm}^{-1} \\
\rightarrow T &= \exp(-2\kappa L) = e^{-5.2} = 5 \times 10^{-3}
\end{aligned}
\tag{10.55}
$$

2. What if the barrier width was increased by a factor of 2:
   Answer:

$$
T = e^{-2\kappa 2L} = e^{-1.2 \times 2} = 2 \times 10^{-5}
\tag{10.56}
$$

The moral of this calculation is that the transmission coefficient is VERY sensitive to the width of the barrier because it appears in the exponential.

Example 2 An enthusiastic 70 kg physics professor paces back and forth from wall to wall at 0.5 m/s. It requires 20 Joules of energy to break through the classroom wall which is 0.25 m thick. What is the transmission coefficient for finding the professor in the corridor on the other side of the wall?
Solution:
The energy of the professor is

$$
\begin{aligned}
E &= \frac{1}{2}mv^2 \\
&= \frac{1}{2}70 \text{ kg}(0.5 \text{ m/s})^2 \\
&= 8.6 \text{ Joules}
\end{aligned}
\tag{10.57}
$$

In addition:

$$
\begin{aligned}
\kappa &= \frac{\sqrt{2m(V - E)}}{\hbar} \\
&= \frac{\sqrt{2 \times 70 \times (20 - 8.6) \text{ kg Joules}}}{1.05 \times 10^{-34} \text{ Joule sec}} \\
&\approx 4 \times 10^{35} \text{ m}^{-1}
\end{aligned}
\tag{10.58}
$$

The transmission coefficient is:

$$T = e^{-2\kappa L}$$
$$= e^{-2\times(0.25\ \text{m})\times1\times10^{35}\ \text{m}^{-1}}$$
$$\approx e^{-2.5\times10^{33}} \tag{10.59}$$

While tunnelling through the wall is not impossible, it is HIGHLY unlikely. The professor would need to pace back and forth $e^{10^{33}}$ times before appearing on the other side of the wall. At the rate of one bounce off the wall per minute, this would far exceed the age of the Universe!

10.5.3 Applications of Tunnelling

Fission via Alpha Decay

Alpha decay occurs when an unstable nucleus emits a Helium nucleus, otherwise known as an alpha particle. The Helium nucleus has two protons and two neutrons, so that Z, the *atomic number*, which is equal to the charge of the original nucleus, decreases by two. The *mass number* A, which is defined as the total number of nucleons, (i.e. protons and neutrons) decreases by four. We use the notation $_Z^A X$ to specify the mass number A and charge Z of a nucleus, X. Using this notation, alpha decay takes the form:

$$_Z^A X \rightarrow {}_{Z-2}^{A-4}Y + {}_2^4He \tag{10.60}$$

X is called the "parent" nucleus, Y is called the "daughter".

Exercise 7 In one form of nuclear fission, Uranium nuclei decay into Thorium nuclei by emitting an alpha particle, as follows:

$$_{88}^{226}U \rightarrow {}_{86}^{222}Th + {}_2^4He \tag{10.61}$$

The kinetic energy of the alpha particle in this decay is between about 4 and 5 MeV. Does the alpha particle emerge with relativistic velocity? Justify your answer.

There exists a relatively simple model that describes alpha decay in terms of an alpha particle that is initially trapped in the nuclear potential of the Uranium nucleus but escapes via quantum tunnelling. This model is called *Gamow's Theory*, after George Gamow, the Ukrainian-American physicist who invented the model in 1928. The effective potential barrier that traps the alpha particle is illustrated in Fig. 10.4. Remarkably, the Gamow theory correctly produces the lifetime for alpha decay, despite its conceptual and mathematical simplicity.

Fig. 10.4 Quantum wave function for the stationary state of an alpha particle classically trapped in the nuclear potential of a Uranium nucleus

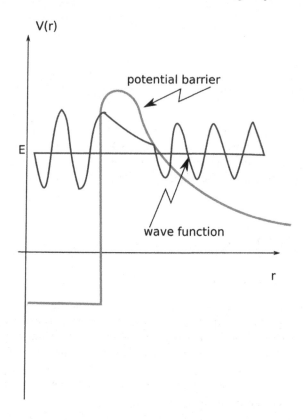

Fusion

It is energetically favourable for two nuclei with mass number greater than 62 to join together to form a nucleus with an even larger mass number. The resulting nucleus has greater binding energy per nucleon than the original two nuclei.[6] This process, called *fusion*, releases huge amounts of energy per unit mass of fusion material. Fusion requires the two nuclei to get close enough together for the strong nuclear forces to dominate their interaction. This is difficult because the nuclei, being positively charged, repel each other via electrostatic forces. The only way to get the nuclei to overcome the electrostatic potential barrier is to heat them to very high temperatures and/or subject them to large pressures.

The requisite conditions occur inside the Sun due to the strong gravitational forces pulling on the Sun's plasma. The energy that sustains us on Earth is powered by these fusion reactions in the Sun. Scientists have been trying for many years to produce such high temperatures and pressures in laboratories, since the resulting fusion reaction would provide us with energy more efficiently, safely and cleanly than either carbon based sources or fission. The key technological problem is to

---

[6] For nuclei with mass number less than 62 fission is energetically favoured.

Fig. 10.5 Potential energy of two $^2_1 H$ nuclei as a function of their separation. The red line is their energy, while the green lines indicate the energy levels of the stationary states for the potential (Color figure online)

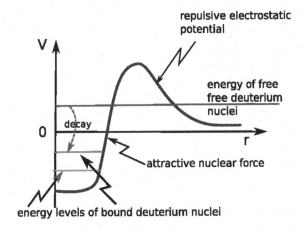

keep the plasma confined in the laboratory long enough for the fusion reaction to get going.

Tunnelling plays an important role in fusion. Just as in the case of alpha decay, there exists a simple model for the basic fusion reaction that converts two deuterium nuclei ($^2_1 H$) into a single Helium 3 ($^3_2$He) nucleus plus one neutron, with a release of energy $\Delta E = 3.27 \, \text{MeV}$:

$$^2_1 H + {}^2_1 H \rightarrow {}^3_2 He + N \tag{10.62}$$

Long range electromagnetic forces keep the two charged nuclei apart. If they get close enough, then the short range strong interaction takes over, and makes it energetically favourable for the $^2_1 H$ nuclei to be bound together. This is represented by the potential energy function in Fig. 10.5.

The initially free $^2_1$H nuclei outside the electrostatic potential barrier are able to tunnel through the potential and settle into one of the stationary states that have lower energy, with a corresponding release of the energy.

Scanning Tunnelling Electron Microscope

Scanning tunnelling electron microscopes (STEMs) are able to image very small objects, such as the circular ring of iron atoms in Fig. 8.7 in Sect. 8.4.1. Remarkably, they can also distinguish the electron density waves trapped within the corral.

The basic mechanism underlying the scanning tunnelling electron microscope is shown in Fig. 10.6. A potential barrier is set up between the material being scanned and the point of the needle of the STEM. Electrons are able to create a current by tunnelling from the material surface to the needle. This current is therefore very sensitive to the distance between the material and the needle (exponentially sensitive in fact) so that changes in the current provide a high resolution image of the surface.

By moving the tip of the needle to be within a few atomic diameters of the surface, the STEM is able to pick up individual atoms and move them around. This is how the Quantum Corral in Fig. 8.7 was constructed.

Fig. 10.6 Schematic of a scanning tunnelling electron microscope. The current between the solid and the needle is created by quantum tunnelling against the induced potential, and is therefore very sensitive to the distance between the atoms in the material and the needle

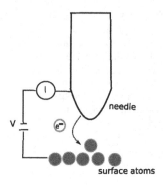

Resonant Tunnelling Diode

A resonant tunnelling diode is a tiny switching device with no moving parts that uses fundamental properties of quantum mechanics to operate. At its core is a quantum dot, which is a microscopic box designed to trap electrons via electrostatic forces. As with the particle in a box discussed in Sect. 10.5.1, only a discrete set of energy levels are available to electrons in the box. Electrons outside the quantum dot (see Fig. 10.7) cannot tunnel into the dots if their energies do not match one of the stationary states inside. By subjecting the quantum dot in the centre to an external potential, it is possible to raise and lower the energy levels of the stationary states inside. When one of these energy levels matches the energy of the external electrons a current flows.[7] Otherwise, the switch is off. The resonant tunnelling diode therefore provides a tiny rapid switching device without moving parts.

10.6   The Quantum Correspondence Principle

10.6.1   Recovering the Everyday World

In our every day lives, Newtonian physics and classical electrodynamics (collectively called "classical" physics) describe things very well. We do not notice the effects of quantum mechanics because Planck's constant is a very small number when expressed in terms of kilograms, meters and seconds. It stands to reason that the quantum states of microscopic systems should also be well described by Newtonian physics when their energies get sufficiently large, i.e. for highly excited quantum states for which the quantum number, n, is large.

For macroscopic, i.e. large, systems we do not notice that there are discrete energy levels. We think systems can absorb or emit arbitrarily small quantities of energy. How can this be understood if everything has quantized energy labelled by a discrete

---

[7] Such matching of energies is called resonance, hence the name.

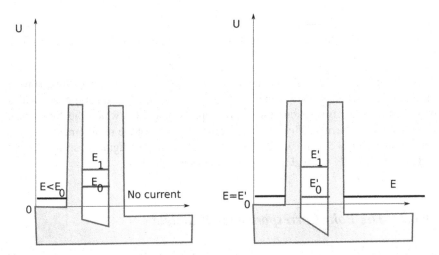

Fig. 10.7 Schematic of a resonant tunnelling diode, consisting of a potential well that traps electrons with specific quantized energies, i.e. the stationary states. On the left, the switch is open: electrons have energy E that does not match the energies of the stationary states of the quantum well so that tunnelling cannot occur and no current flows. On the right, the switch is closed: the potential in the quantum dot has been adjusted so that the ground state corresponds to the energy E of the electrons on the left. Now tunnelling is possible and the current flows

quantum number n? The answer is that when a quantum system has very large energy, the quantum number also gets large so that the energy spacing between adjacent ($\Delta n = 1$) levels is very small compared to the energies of the levels themselves:

$$\frac{\Delta E_n}{E_n} \to_{n \to \infty} 0 \tag{10.63}$$

where

$$\Delta E_n = E_{n+1} - E_n \tag{10.64}$$

We can verify that this works for a free particle inside a box with impenetrable walls:

$$E_n = \frac{\pi^2 \hbar^2 n^2}{2mL^2}$$

$$\Delta E_n = \frac{\pi^2 \hbar^2 2\Delta n}{2mL^2}$$

$$\to \quad \frac{\Delta E_n}{E_n} = \frac{2}{n} \to_{n \to \infty} 0 \tag{10.65}$$

Exercise 8 Verify that the correspondence principle holds for the Simple Harmonic Oscillator in Sect. 10.3.3.

As discussed in Sect. 1.4 this must work for every system described by quantum mechanics because otherwise there would be instances in which quantum mechanics produced different predictions for macroscopic systems for which the Newtonian predictions have been verified. This is another example of the requirement that any new paradigm must contain within it a vestige of the old paradigm whose predictions had been previously verified as valid in some specific regime.

10.6.2 The Bohr Correspondence Principle

Assume the system absorbs or emits energy via electromagnetic waves, i.e. photons. If classical physics applies, the frequency of the emitted radiation should be determined by the classical oscillation frequency of the system itself. The radiation must still obey the quantum rules, so that the energy of the photon emitted must be:

$$E_\gamma = h f_{class}(E) \tag{10.66}$$

where $f_{class}(E)$ is the classical oscillation frequency of the system emitting/absorbing the radiation. As suggested by the notation, this frequency can depend on the energy of the classical system. It is only independent of the energy for the simple harmonic oscillator.

If the emitted, or absorbed radiation is due to a transition between two adjacent energy levels ($\Delta n = 1$) then

$$E_\gamma = \Delta E_n \tag{10.67}$$

Assuming $\Delta n = 1$ is small enough compared to n, we can approximate ΔE_n in Eq. (10.67) by:

$$\Delta E_n \sim \frac{dE_n}{dn}\Delta n = \frac{dE_n}{dn} \tag{10.68}$$

Using Eqs. (10.67) and (10.68), Eq. (10.66) can now be written as:

$$\frac{dE_n}{f_{class}(E_n)} = h\,dn \tag{10.69}$$

For a one dimensional system, $f_{class}(E_n)$ is a unique function of the classical energy so that Eq. (10.69) can be integrated to obtain the quantum spectrum E_n in the semi-classical, large n, limit.

Example 3 To see how this works for the particle in an impenetrable box, we note that a classical particle with energy $E = \frac{p^2}{2m}$ bounces back and forth between the walls in time:

$$T = \frac{2L}{v} = \frac{2mL}{p}$$ (10.70)

This is the time for one complete circuit, so the frequency of oscillations between the walls is:

$$\begin{aligned} f_{class} &= \frac{1}{T} \\ &= \frac{p_n}{2mL} \\ &= \frac{\sqrt{2mE_n}}{2mL} \\ &= \frac{\sqrt{E_n}}{\sqrt{2m}L} \end{aligned}$$ (10.71)

Putting this into Eq. (10.69) yields:

$$dE_n \frac{\sqrt{2m}L}{\sqrt{E_n}} = hdn$$ (10.72)

which can be integrated to yield:

$$2\sqrt{E_n}\sqrt{2m}L = hn + \text{constant}$$ (10.73)

so that

$$E_n = \frac{h^2 n^2}{8mL^2} = \frac{\pi^2 \hbar^2}{2mL^2} n^2 + \text{constant}$$ (10.74)

as given in Eq. (10.26). We can neglect the integration constant in Eq. (10.74) because we expect it to be irrelevant for large n and large energies. In the case of the particle in a box, the integration constant turns out to be zero, but this cannot be derived in the semi-classical limit.

Exercise 9 Use the Bohr Correspondence Principle to derive the semi-classical energy spectrum for the simple harmonic oscillator.

10.7   The Time Dependent Schrödinger Equation

10.7.1   Heuristic Derivation

We now provide a heuristic argument for the form of the *time dependent Schrödinger equation* that governs the evolution of quantum wave functions in general. The time independent Schrödinger must then emerge as a special case for stationary states.

Recall the time dependence of the wave function in the stationary case:

$$\Psi(x, t) = \psi_E(x)e^{-iEt/\hbar} \tag{10.75}$$

where $\psi_E(x)$ is a solution to the time independent Schrödinger Eq. (10.4) which we rewrite here for convenience:

$$\hat{H}\psi_E(x) = E\psi_E(x) \tag{10.76}$$

We can now write a very suggestive equation that is solved by the stationary state wave function $\Psi(x, t)$ in Eq. (10.75). Using Eq. (10.76), we can write Eq. (10.4) in the form:

$$\frac{\partial \Psi(x, t)}{\partial t} = -\frac{i}{\hbar} E \Psi(x, t) = -\frac{i}{\hbar}\hat{H}\Psi(x, t) \tag{10.77}$$

$$i\hbar \frac{\partial \Psi(x, t)}{\partial t} = \hat{H}\Psi(x, t)$$

$$= -\frac{\hbar^2}{2m}\frac{\partial^2 \Psi(x, t)}{\partial x^2} + V(x)\Psi(x, t) \tag{10.78}$$

Equation (10.78) is the *time dependent Schrödinger equation* for which we are searching. The time dependent Schrödinger equation is a linear partial differential equation that describes the time evolution of a general time dependent state in quantum mechanics. It illustrates the special role played by the Hamiltonian in generating time translations. Other important things to note are that it is linear in the wave function and that it is first order in the time derivative. Finally, the stationary states that we have been considering are by construction solutions to the time dependent Schrödinger equation (TDSE).

In order to find physically relevant solutions to the time dependent Schrödinger equation, one needs to impose appropriate boundary conditions. In addition, one needs to know the state $\Psi(x, t_0)$ at some initial time t_0. If the particle is not confined to a finite region of space, then the boundary conditions are provided by the requirement that the wave function be normalized:

$$\int_{-\infty}^{\infty} dx \ \Psi^*(x, t)\Psi(x, t) = 1 \tag{10.79}$$

This forces the wave function, or at least its norm, to vanish sufficiently fast at $\pm\infty$ and makes the mathematical problem well defined and solvable.[8]

You may have noticed that the left hand side of Eq. (10.79) depends on time, whereas the right hand side does not. The beauty of the Schrödinger equation is that if you normalize the wave function at any one time, say t_0 (and you have constructed the Hamiltonian correctly, boundary conditions included), then the wave function stays normalized. This is important because if the normalization were not conserved then quantum mechanics would be inconsistent. For example a particle could over time disappear from the Universe in the sense that the probability of finding it anywhere could go to zero. As one might expect, the conservation of probability is related to a symmetry of the equation of motion, in this case the TDSE. The total probability of finding the particle somewhere between $-\infty$ and $+\infty$ is in fact the Noether conserved quantity associated with the invariance of the wave function under a change of global phase.

Example 4 Prove that an initial normalized wave function $\Psi(x, t)$ that obeys the time dependent Schrödinger equation will stay normalized in time.

Solution:
Take the time derivative under the integral on the left hand side of Eq. (10.79) and use the product rule underneath the integral sign to get:

$$\frac{d}{dt}\left(\int_{-\infty}^{\infty} dx\ \Psi^*(x, t)\Psi(x, t)\right) = \int_{-\infty}^{\infty} dx \frac{\partial\Psi^*(x, t)}{\partial t}\Psi(x, t)$$
$$+ \int_{-\infty}^{\infty} dx\ \Psi^*(x, t)\frac{\partial\Psi(x, t)}{\partial t}$$

$$(10.80)$$

Note that when the time derivative is moved inside the integral sign it is replaced by a partial derivative with respect to time, because it is differentiating the wave function (and its complex conjugate) with respect to time only and not space. Now apply the TDSE, Eq. (10.78), and its complex conjugate to the appropriate factors in the second and first terms, respectively:

$$\frac{d}{dt}\left(\int_{-\infty}^{\infty} dx\,\Psi^*(x, t)\Psi(x, t)\right) =$$
$$\frac{i}{\hbar}\int_{-\infty}^{\infty} dx \left(-\frac{\hbar^2}{2m}\frac{\partial^2\Psi^*(x, t)}{\partial x^2} + V(x)\Psi^*(x, t)\right)\Psi(x, t)$$
$$-\frac{i}{\hbar}\int_{-\infty}^{\infty} dx\ \Psi^*(x, t)\left(-\frac{\hbar^2}{2m}\frac{\partial^2\Psi(x, t)}{\partial x^2} + V(x)\Psi(x, t)\right) \quad (10.81)$$

The two last terms on each line, containing the potential $V(x)$, cancel with each other. The first two terms on each line combine into a total derivative

---

[8] This assumes that the potential is finite everywhere. If this is not the case there are ambiguities in the choice of boundary conditions at the singular points that must be addressed.

$$\frac{-i}{\hbar} \int_{-\infty}^{\infty} dx \left[\frac{\hbar^2}{2m} \frac{\partial^2 \Psi^*(x,t)}{\partial x^2} \Psi(x,t) - \Psi^*(x,t) \frac{\hbar^2}{2m} \frac{\partial^2 \Psi(x,t)}{\partial x^2} \right]$$

$$= -i \frac{\hbar}{2m} \int_{-\infty}^{\infty} dx \frac{\partial}{\partial x} \left[\frac{\partial \Psi^*(x,t)}{\partial x} \Psi(x,t) - \Psi^*(x,t) \frac{\partial \Psi(x,t)}{\partial x} \right]$$

$$= -i \frac{\hbar}{2m} \left[\frac{\partial \Psi^*(x,t)}{\partial x} \Psi(x,t) - \Psi^*(x,t) \frac{\partial \Psi(x,t)}{\partial x} \right]_{-\infty}^{\infty}$$

$$= 0 \tag{10.82}$$

The second to last line follows from $\int_a^b dx \frac{df(x)}{dx} = f(b) - f(a)$. This vanishes because the normalization of the wave function requires it to vanish at $\pm\infty$.

It is difficult to find exact analytic solutions to the TDSE for arbitrary initial conditions. Most solutions used in physics are either approximations, or found numerically via computer. In general, the TDSE causes the initial width of a wave function to increase with time. It "spreads out". For example, under time evolution determined by the TDSE wave function, a free particle in a box that is initially Gaussian spreads out in time until the probability amplitude essentially fills the box.

10.7.2   Coherent States

There is one notable exception to the spreading of wave functions. This occurs for a particle of mass m in a simple harmonic oscillator potential with spring constant K and corresponding fundamental frequency $\Omega = \sqrt{K/m}$. The time dependent quantum state is described by a Gaussian of the form given in Eq. (9.6):

$$\Psi_\alpha(x,t) = \left(\frac{m\Omega}{\pi\hbar} \right)^{1/4} e^{\frac{i}{\hbar} p_\alpha(t)x - \frac{m\Omega}{2\hbar}(x-x_\alpha(t))^2 + i\theta(t)} \tag{10.83}$$

where $x_\alpha(t) := x_0 \cos(\Omega t)$

$$p_\alpha(t) := m \frac{dx_\alpha(t)}{dt} = -x_0 m\Omega \sin(\Omega t)$$

where $i\theta(t)$ is a phase that needs to be determined. It is a function of $x_\alpha(t)$ and $p_\alpha(t)$. α is an arbitrary real positive parameter.

The wave function in Eq. (9.6) is called a *coherent state*. It is not a stationary state but has very specific time dependence. You will determine the precise form of $\theta(t)$ in the following exercise.

Exercise 10   Consider the coherent state given in Eq. (10.83)

1. Verify that it satisfies the TDSE with $V(x) = \frac{1}{2}Kx^2$ for appropriate choice of the function $\theta(t)$.

2. Show that the expectation value of position and momentum are given by $x_\alpha(t)$ and $p_\alpha(t)$ respectively.

In a coherent state, the position probability amplitude and the momentum probability amplitude are Gaussian functions that maintain their shape as the state evolves. The functions $x_\alpha(t)$ and $p_\alpha(t)$ correspond to the expectation values of position and momentum, respectively. They obey the Newtonian harmonic oscillator equations of motion. The parameter α determines the mean energy of the oscillator, which is a constant of the motion as required by Noether's theorem. Coherent states are important in quantum mechanics for a variety of reasons. One of these is that they provide exact, dynamical quantum states that in many ways mimic the behaviour of classical particles.

Exercise 11 Show that if one tries to solve the TDSE with a time dependent wave function of the form:

$$\Psi(x, t) = \psi(x) f(t) \tag{10.84}$$

then the TDSE Eq. (10.78) implies:

1. The time dependent part of the wave function is of the form:

$$f(t) = e^{\frac{-iEt}{\hbar}} \tag{10.85}$$

and
2. The spatial part of the wave function $\psi(x)$ obeys the time independent Schrödinger equation:

$$\hat{H}\psi(x) = E\psi(x) \tag{10.86}$$

where E is an arbitrary constant, independent of x and t.

This exercise uses separation of variables to derive the equations for stationary states from the TDSE. This procedure turns out to be equivalent to assuming that the stationary state wave function takes the form of a standing wave.

10.8 Observables as Linear Operators

This section gives a qualitative overview of some of the mathematics that underlies quantum mechanics. We establish terminology and concepts that will be useful when we discuss symmetry in the context of quantum mechanics (Sect. 10.9) and also when we elaborate on electron spin (Sect. 13.4). A deeper understanding of these concepts must await a more advanced course on quantum mechanics, such as presented in the textbook by Griffiths and Schroeter.[9]

---

[9] Griffiths and Schroeter [1].

We have alluded to the fact that the quantum description of a system involves two parts: the wave function describing the quantum state of the system and the observables, such as position, momentum and energy, associated with that system. The wave function allows one to predict the statistical outcomes of measurements of the observables. In calculating the expectation value of momentum Eq. (9.31) and energy Eqs. (9.42) and (9.43), we saw that both of these observables could be represented as differential operators that act on the wave function. This is very different from classical mechanics in which measurements of all observables in principle yield precise values. We therefore distinguished quantum momentum and quantum energy from their classical counterparts by putting hats over them. Note that the quantum version of position, while not a differential operator, can be thought of as an operator that multiplies the wave-function by another function, namely x:

$$\hat{x}\psi(x) = x\psi(x) \tag{10.87}$$

In fact any function $f(x)$ of x has a quantum version $\hat{f}(x) = f(\hat{x})$. One example is the potential energy, $\hat{V}(x) := V(\hat{x})$.

The key thing to remember here is that in quantum mechanics all observables are represented by operators that act linearly on quantum states. These operators can contain partial derivatives $\partial/\partial x$ with respect to x as well as functions of x.

Consider a generic quantum observable or operator $\hat{O}$. As mentioned previously, $\hat{O}$ acts on a wave function $\psi(x)$ to produce another transformed wave function $\psi'(x)$ as follows:

$$\hat{O} : \psi(x) \rightarrow \psi'(x) = \hat{O}\psi(x) \tag{10.88}$$

It is important to note that the action of such linear operators does not, in general, preserve the normalization of states and does not generate time evolution. This distinction is left to the Hamiltonian operator $\hat{H}$.

Knowing how observables act on states enables us to calculate the physical predictions of quantum mechanics. Specifically, expectation values of an observable $\hat{O}$ obtained from measurements made on a large number of identical systems that are prepared in the same quantum state $\psi(x)$ is given by:

$$\langle \hat{O} \rangle = \int dx \, \psi^*(x)\hat{O}\psi(x) \tag{10.89}$$

In addition

$$\langle \hat{O}^2 \rangle = \int dx \, \psi^* \hat{O}\hat{O}\psi(x)$$

$$\langle \hat{O}^3 \rangle = \int dx \, \psi^* \hat{O}\hat{O}\hat{O}\psi(x), \text{ etc.} \tag{10.90}$$

so that the standard deviation of the measurements is:

$$\Delta O_{sd} = \sqrt{\langle \hat{O}^2 \rangle - \langle \hat{O} \rangle^2} \tag{10.91}$$

It is worth repeating that these equations describe the statistics obtained from measurements on particles in a large number of identical experiments.[10] They do not tell us specific values for single measurements on an individual particle. In general quantum mechanics will not predict with certainty precise values except under special circumstances. The measurements must, however, produce one of a special set of values associated with a given observable. This set of values is called the *spectrum* of that observable and, as we saw in Sect. 10.3, is in general quantized.

Even if the value of an observable, say energy, is uncertain before a measurement, a subsequent measurement made on the same system immediately after the measurement must yield the same value with a probability of one. After a measurement, the quantum system is therefore in a state of definite value for this observable, which implies that measurements of this observable for an ensemble of particles that have all been previously measured will give the same value with zero standard deviation. This is what is meant by wave function collapse.[11]

The necessary and sufficient condition for a quantum state $\psi_O(x)$ to have a definite value of an observable O is that the state obey the following equation.

$$\hat{O}\psi_o(x) = \lambda_o \psi_o(x) \tag{10.92}$$

where λ_0 must be a real number because it corresponds to the physical, measured value of the observable $\hat{O}$. This is a general version of the time-independent Schrödinger equation (10.4) obeyed by stationary states that, by definition, have definite energy.

Mathematically, Eq. (10.92) is called the eigenvalue equation for the operator $\hat{O}$, with λ_o the corresponding eigenvalue. $\psi_o(x)$ is called the eigenfunction (or eigenstate) of $\hat{O}$ with eigenvalue λ_o. Every measured value of the observable $\hat{O}$ must correspond to one of its eigenvalues, and after the measurement the state of the system must be one of the eigenfunctions that obeys Eq. (10.92). The eigenvalue equation normally has many solutions, each of which yields a physically realizable eigenvalue and eigenstate. To be a measurable quantity, the eigenvalue λ_o must be a real quantity, which imposes a mathematical condition on the form of any physical operator.[12] Each allowed value of λ_0 leads to a different solution, but sometimes two different solutions have the same eigenvalue. When this happens the states with the same eigenvalue are said to be *degenerate*.

---

[10] In practical terms, a single experiment normally does a large number of measurements on individual particles in a beam of identically prepared particles. Each particle in the beam is, however, only measured once.

[11] As will be discussed in Sect. 13.8, wave function collapse as described here cannot be described within the standard framework of quantum mechanics, because it is an irreversible process. You cannot reconstruct the state prior to measurement from knowledge of the state after the measurement. All time evolution in quantum mechanics must be reversible.

[12] The condition is that the operator must be Hermitian.

One can show that for an eigenstate ψ_0 of an operator $\hat{O}$ Eq. (10.92) implies:

$$\hat{O}^n \psi_o(x) = \lambda_o^n \psi_o(x), \quad \text{for all integers } n \geq 0 \qquad (10.93)$$

and consequently

$$\langle \hat{O}^n \rangle = \langle \hat{O} \rangle^n = \lambda_0^n \qquad (10.94)$$

This in turn tells us that the statistical variation ΔO is zero for an eigenstate ψ_o of an operator $\hat{O}$. As desired, one can therefore in this case predict with certainty what value of this observable will be measured for individual particles that are known to be in this eigenstate, which is what we set to achieve in imposing the condition Eq. (10.92).

There is a similarity between the action of a quantum operator on a quantum state and the linear transformation of a point on a plane. In both cases the essential mathematics is described by linear algebra, which is the study of vectors and matrices. The big difference is that the quantum states are in general normalized vectors in an *infinite dimensional* vector space.[13] The operators are therefore infinite dimensional matrices, with an infinite number of eigenvalues and eigenvectors. In the case of finite dimensional linear transformations it is not difficult to show that in order to have real eigenvalues, the matrix representing the transformation, T_{ij}, must be Hermitian:

$$T_{ji}^* = T_{ij} \qquad (10.95)$$

The infinite dimensional form of this is a bit trickier to write down but it is completely analogous.

We now have a method of determining the allowed values, or *spectrum* of any observable. We need to find all the eigenvalues and eigenfunctions of the corresponding operator. In the case of the position operator, we have assumed that the position can take on any real value between $-\infty$ and $+\infty$, unless it is confined to a box with impenetrable walls, in which case x has to be in the corresponding interval on the real line, say between $x = 0$ and $x = L$. When the quantum states are required to be normalizable, i.e. for all but the free particle, we found that the energy could only take on discrete values. Such a discrete spectrum occurs for the normalizable state of any observable described by a second order differential operator, such as the Hamiltonian and the angular momentum (see Sect. 11.6.2).

The eigenvalue equation (10.92) in this case is a complex second order differential equation whose most general solution has two complex parameters, or four real parameters. One of these is fixed by the normalization condition and one contributes an overall phase that is physically irrelevant. This leaves two parameters. In order to be normalizable, the wave function must also satisfy boundary conditions at two boundaries. For impenetrable boxes these boundary conditions are imposed at finite values of the coordinate, but in general, the two boundaries are at $-\infty$ and $+\infty$.

---

[13] One notable exception is the quantum state describing the spin of a particle as discussed in Sect. 13.4.

Each boundary condition places a constraint on the value of the wave function at that point.[14] The boundary conditions therefore eliminate two more parameters.[15]

If you have been counting, you will notice that we have no free parameters left in the general solution, so the problem is over-constrained and we should have either no solutions, or at most one solution. There is however a physical parameter that we have been ignoring, namely the eigenvalue itself. It turns out that in most physical situations, the eigenvalue equation does have solutions for a discrete (countable) set of eigenvalues. This set prescribes the only values obtainable via a measurement of the corresponding observable. In this case, the allowed set of eigenvalues is called a *discrete* spectrum. If there are no solutions, then the operator is unphysical.

One exception to the above is the free particle. As we saw in Sect. 10.3.1 a free particle can have values of momentum $\hbar k$ and energy $(\hbar k)^2/(2m)$ that are not quantized. The difference is that the solution to the eigenvalue equation for a free particle is a plane wave. The probability amplitude extends to plus and minus infinity so that plane waves cannot be normalized. Strictly speaking these are not physical states, but as we have seen in Sect. 10.3.1, they can be used to construct physical states that are linear superpositions of plane waves.

We now show that in the case of two observables that are *complementary*, i.e. that obey an uncertainty principle principle such as Eq. (8.78), it is mathematically impossible to find a state that is simultaneously an eigenstate of both. Recall that the Heisenberg uncertainty principle described in Sect. 8.5 implies that the quantum state of a particle can have either a sharply defined position, or a sharply defined momentum, but not both. Given the above discussion this would require the particle to be in a quantum state that is an eigenstate ψ of both the position operator $\hat{x}$ and momentum operator $\hat{p}$, and Heisenberg tells us this cannot happen. The general condition that forbids this can be determined as follows. Suppose ψ is an eigenstate of two observables, $\hat{O}$ and $\hat{Q}$, with eigenvalues λ_o and λ_Q, respectively. This implies that:

$$
\begin{aligned}
\hat{O}\hat{Q}\psi &= \hat{O}\lambda_Q\psi \\
&= \lambda_Q\hat{O}\psi \\
&= \lambda_Q\lambda_o\psi
\end{aligned}
\tag{10.96}
$$

We can apply the two operators in the opposite order to obtain:

---

[14] We are making the mathematically indefensible simplification that the values $\pm\infty$ correspond to points on the real line. Nothing could be further (so to speak) from the truth, but it makes the discussion somewhat simpler.

[15] You may be wondering why each boundary condition on a complex wave function eliminates one real parameter, and not a complex parameter. The requirement that the wave function vanish restricts only the norm to vanish and places no restriction on the phase.

$$\hat{Q}\hat{O}\psi = \hat{Q}\lambda_o\psi$$
$$= \lambda_o\hat{Q}\psi$$
$$= \lambda_o\lambda_Q\psi \tag{10.97}$$

$\hat{Q}$ and $\hat{O}$ can only have simultaneous eigenstates if the order of operation is irrelevant, that is if, for all wave functions $\psi(x)$:

$$\hat{O}\hat{Q}\psi = \hat{Q}\hat{O}\psi \tag{10.98}$$

Using the terminology of linear algebra, condition Eq. (10.98) implies that the operators must *commute*. A very important construct in quantum mechanics is the commutator of two operators, defined by:

$$\left[\hat{O}, \hat{Q}\right]\psi(x) := \hat{O}\hat{Q}\psi - \hat{Q}\hat{O}\psi \tag{10.99}$$

Equation (10.98) can therefore be written:

$$\left[\hat{O}, \hat{Q}\right]\psi(x) := \hat{O}\hat{Q}\psi(x) - \hat{Q}\hat{O}\psi(x)$$
$$= 0 \tag{10.100}$$

Equation (10.100) must be true for any admissible wave function $\psi(x)$, so that it can be written purely as an operator equation, without reference to what the operators act on[16]:

$$\left[\hat{O}, \hat{Q}\right] = \hat{0} \tag{10.101}$$

What we have shown is that the operator $\left[\hat{O}, \hat{Q}\right]$ must vanish for there to exist wave functions that are simultaneously eigenfunctions of both $\hat{O}$ and $\hat{Q}$. Observables whose operators do not commute are called *complementary*: no quantum state exists in which both are known with arbitrary accuracy, and they obey a generalized Heisenberg uncertainty principle. If a quantum state has a precise value for one of the non-commuting observables (i.e. is an eigenstate of that observable), then the value of the other observable must be highly uncertain, and vice versa.

Exercise 12   Show that for all $\psi(x)$:

$$\left[\hat{x}, \hat{p}\right]\psi(x) = i\hbar\psi(x) \tag{10.102}$$

Note: Since it applies to all states, this implies the operator equation:

---

[16] The somewhat bizarre looking hat over the number zero indicates that strictly speaking $\hat{0}$ is the operator that annihilates any state on which it acts.

$$\left[\hat{x}, \hat{p}\right] = i\hbar\hat{\mathbb{1}} \qquad (10.103)$$

where $\hat{\mathbb{1}}$ is the unit operator that by definition leaves all states unchanged.

10.9 Symmetry in Quantum Mechanics

In Chap. 3 we defined a symmetry transformation as an operation that acts either on an object or a set of equations and leaves it unchanged (invariant). We then saw that spacetime symmetries (time translation, spatial translation, rotations, boosts) are linear transformations that can be represented as matrices that act on (i.e. multiply) vectors in space and spacetime. Moreover, Noether's theorem told us that each independent continuous global symmetry (i.e. one that can be labelled by a continuous parameter) of the equations of motion of a system led to a physical quantity that was conserved by the corresponding time evolution. This is such an incredibly powerful result in classical mechanics that it must (and does) have an analogue in quantum mechanics.

Given the close connection classically between symmetry operations and observables that are conserved in time, it is reasonable to assume that there is also a strong connection between symmetry transformations in quantum mechanics and operators that represent observables. This is easiest to explain using a couple of examples. First, the time dependent Schrödinger equation tells us how the wave function behaves under time translation. That is:

$$\psi(x, t + dt) - \psi(x, t) = \frac{\partial\psi(x, t)}{\partial t}dt$$

$$= \frac{-i}{\hbar}\hat{H}\psi(x, t)dt \qquad (10.104)$$

Thus, the Hamiltonian determines (generates) time evolution and the invariance of the equations of motion under time translations implies, as in classical mechanics, that the energy is a constant of motion. Similarly,

$$\psi(x + dx, t) - \psi(x) = \frac{\partial\psi(x, t)}{\partial x}dx$$

$$= \frac{i}{\hbar}\hat{p}\psi(x, t)dx \qquad (10.105)$$

so that the momentum operator $\hat{p}$ generates translations in space. Again, if the quantum equation of motion (i.e Schrödinger's equation) is invariant under spatial translations, then the momentum is a conserved quantity. We will see in Chap. 11 that there also exists operators corresponding to angular momentum that generate rotations in three dimensional space. The quantum observables associated with the three components of angular momentum are conserved when the time dependent Schrödinger equation is invariant under rotations.

How then do we determine whether or not a given quantum operator corresponds to a symmetry of the dynamical equations? First let us recall that a symmetry of the equations of motion is a transformation that leaves the equations unchanged. In the case of time translation, for example, this requires the absence of any explicit dependence on time in the equations. The equations look the same and have the same set of solutions at time t as they do at time $t + a$. Note that this does not imply that all the solutions themselves are independent of time. Instead, it implies that the transformed equations will produce solutions that are the same functions of time as the untransformed solutions. This has the consequence that if one takes a particular solution $x(t)$ and changes $t \rightarrow t + t_0$, then $x(t + t_0)$ will also be a solution, namely the same solution at a later time. As well, Noether's theorem guarantees that the energy will be a constant of the motion.

In order to clarify this point it may be useful to consider the motion of a particle moving in a potential $V(x)$ that is neither constant nor linear in the position variable x. The equations of motion in this case explicitly contain the position x via the force term, so that they are not invariant under translations in space. For example, in the case of the simple harmonic oscillator there is a preferred location in space, namely the equilibrium position. Consequently, for the harmonic oscillator as well as for more general position dependent potentials, the momentum is not a constant of the motion. Only the total energy is conserved, whereas the kinetic energy (which depends only on momentum) and the potential energy change as the solution evolves in such a way as to keep their sum, the total energy, fixed. This total energy is constant again by virtue of Noether's theorem.

Exercise 13 Suppose $x(t) = A \cos(\omega t + \phi)$ solves the simple harmonic oscillator equation of motion:

$$\frac{d^2 x(t)}{dt^2} = -\omega x^2(t) \tag{10.106}$$

Show that if you do a shift $x(t) \rightarrow \tilde{x}(t) = x(t) + x_0$ then $\tilde{x}(t)$ no longer solves Eq. (10.106)

We now move on to quantum mechanics. Consider any state ψ that satisfies the time dependent Schrödinger equation (10.78). If the state

$$\psi' = \hat{O}\psi \tag{10.107}$$

is also a solution for all ψ, then $\hat{O}$ is a symmetry of the quantum dynamics. That is, we require:

$$i\hbar \frac{\partial \psi'}{\partial t} = i\hbar \frac{\partial (\hat{O}\psi)}{\partial t} = \hat{H}\psi'$$

$$= \hat{H}(\hat{O}\psi) \tag{10.108}$$

We consider the case that the operator $\hat{O}$ does not contain time explicitly as a parameter. In this case we can rewrite the left hand side of Eq. (10.108) as:

$$
\begin{aligned}
i\hbar \frac{\partial(\hat{O}\psi)}{\partial t} &= i\hbar\hat{O}\frac{\partial\psi}{\partial t} \\
&= \hat{O}(\hat{H}\psi) \\
&= \hat{H}\hat{O}\psi + \hat{O}\hat{H}\psi - \hat{H}\hat{O}\psi \\
&= \hat{H}\hat{O}\psi + \left[\hat{O}\hat{H} - \hat{H}\hat{O}\right]\psi \\
&= \hat{H}(\hat{O}\psi) + \left[\hat{O}, \hat{H}\right]\psi \qquad (10.109)
\end{aligned}
$$

where we have again used that ψ obeys the Schrödinger equation to get the second line in Eq. (10.109). Comparing the right hand sides of Eqs. (10.108) and (10.109) we see that $\hat{O}\psi$ obeys the TDSE

$$
i\hbar\frac{\partial(\hat{O}\psi)}{\partial t} = \hat{H}(\hat{O}\psi) \qquad (10.110)
$$

if and only if:

$$
\begin{aligned}
\hat{O}(\hat{H}\psi) &= \hat{H}(\hat{O}\psi) \\
\rightarrow \quad [\hat{O}, \hat{H}] &= 0 \qquad (10.111)
\end{aligned}
$$

Thus, $\hat{O}$ transforms solutions onto solutions if and only if $\hat{O}$ commutes with $\hat{H}$.

Having defined a quantum symmetry we now consider the question of conserved quantities. Since quantum mechanics predicts only statistical outcomes, it is natural to define a conserved observable as one whose expectation value remains constant in time. It turns out that the expectation value of an operator $\hat{O}$ will remain constant in time if and only if it commutes with the Hamiltonian. It is a general result in quantum mechanics that:

$$
\frac{d\langle\hat{O}\rangle}{dt} = \frac{1}{\hbar}\langle\left[\hat{O}, \hat{H}\right]\rangle \qquad (10.112)
$$

where we are assuming that $\hat{O}$ does not depend explicitly on time.[17] We now prove Eq. (10.112) for a particle moving in a potential $V(x)$, i.e with $\hat{H}$ given by:

$$
\hat{H} = \frac{-\hbar^2}{2m}\frac{\partial^2}{\partial x^2} + V(x) \qquad (10.113)
$$

---

[17] If $\hat{O}$ does explicitly contain the time t, then Eq. (10.112) gets replaced by

$$
\hbar\frac{d\langle\hat{O}\rangle}{dt} = \langle\left[\hat{O}, \hat{H}\right]\rangle + \left\langle\frac{\partial\hat{O}}{\partial t}\right\rangle
$$

First we note that if $\hat{O}$ does not contain explicit time dependence then:

$$\frac{d\langle\hat{O}\rangle}{dt} = \frac{d}{dt}\int_{-\infty}^{\infty} dx\,\psi^*(x,t)\hat{O}\psi(x,t)$$
$$= \int_{-\infty}^{\infty} dx\,\frac{\partial\psi^*(x,t)}{\partial t}\hat{O}\psi(x,t) + \int_{-\infty}^{\infty} dx\,\psi^*(x,t)\hat{O}\frac{\partial\psi(x,t)}{\partial t}$$

(10.114)

Proceeding as in Exercise 4, we replace the time derivatives of the wave functions by their expressions on the right hand side of the TDSE:

$$\frac{d\langle\hat{O}\rangle}{dt} = \int_{-\infty}^{\infty} dx\left[-\frac{i^*}{\hbar}\left(-\frac{\hbar^2}{2m}\frac{\partial^2\psi^*}{\partial x^2} + V(x)\psi^*\right)\hat{O}\psi(x,t)\right]$$
$$+ \int_{-\infty}^{\infty} dx\,\psi^*(x,t)\hat{O}\left[-\frac{i}{\hbar}\left(-\frac{\hbar^2}{2m}\frac{\partial^2\psi}{\partial x^2} + V(x)\psi\right)\right]$$

(10.115)

Note that we have to complex conjugate the i in the pre-factor of the first term above. We now integrate by parts twice to move the two partial derivatives from acting on ψ^* to acting on the remaining factors in the integrand, and assume that the boundary terms vanish[18] to get:

$$\frac{d\langle\hat{O}\rangle}{dt} = \int_{-\infty}^{\infty} dx\left[\psi^*\left[\frac{i^*}{\hbar}\left(-\frac{\hbar^2}{2m}\frac{\partial^2}{\partial x^2} + V(x)\right)\right]\hat{O}\psi\right]$$
$$+ \int_{-\infty}^{\infty} dx\,\psi^*(x,t)\hat{O}\left[-\frac{i}{\hbar}\left(-\frac{\hbar^2}{2m}\frac{\partial^2\psi}{\partial x^2} + V(x)\psi\right)\right]$$
$$= -\int_{-\infty}^{\infty} dx\left[\psi^*(\hat{H}\hat{O} - \hat{O}\hat{H})\psi\right]$$
$$= \langle[\hat{O},\hat{H}]\rangle$$

(10.116)

which vanishes if $\hat{O}$ is a symmetry, as required.

This proves the quantum version of Noether's theorem, namely that for every symmetry of a quantum system there exists a conserved quantity whose expectation value is a constant.[19] Thus, the quantum version of Noether's theorem is much easier to prove than the classical version.

---

[18] For general operators $\hat{O}$ and Hamiltonians $\hat{H}$ this is a non-trivial assumption that needs to be resolved by some fairly high level functional analysis, especially when there are interesting boundary conditions as in the box with infinite walls.

[19] You should be wondering what has happened to the condition that it be a continuous symmetry. The operators that we are considering generate infinitesimal transformations, as we saw for energy and momentum so that all the symmetries considered in this section are by construction continuous.

We now turn to the interesting question of what happens when a wave function ψ starts at some time t in an eigenstate of an operator $\hat{O}$. Will it remain in the same eigenstate of $\hat{O}$ under time evolution? For this to be true $\hat{H}\psi$ must also be an eigenstate of $\hat{O}$ with the same eigenvalue. That is we want:

$$\hat{O}(\hat{H}\psi) = \lambda_o(\hat{H}\psi)$$
$$= \hat{H}\lambda_o\psi$$
$$= \hat{H}(\hat{O})\psi \qquad (10.117)$$

where in the second line we used the fact that the eigenvalue, a number not an operator, is not affected by $\hat{H}$ (or any operator for that matter) in order to move it to the right. Then we used the eigenvalue equation again in order to replace $\lambda_o\psi$ by $\hat{O}\psi$. The answer to our question then is that the wave function will remain in the same eigenstate of $\hat{O}$ if it commutes with the Hamiltonian, i.e. is a symmetry.

Finally, we make the important observation that when an operator $\hat{O}$ commutes with the Hamiltonian, it is possible to find stationary states (i.e. with fixed energy, E) that are also eigenstates of $\hat{O}$. This is a specific case of the general result that whenever a quantum system has a set of commuting operators, $\{\hat{O}_1, \hat{O}_2, ..., \hat{O}_n\}$ it is possible to find quantum states that are simultaneously eigenstates of all n operators. If n is the maximum number of independent commuting operators that the system allows, then quantum states can be uniquely labelled by the corresponding n eigenvalues $\{\lambda_1, \lambda_2, ...\lambda_n\}$.

To summarize the key points, we have first shown that the set of all symmetries of a given quantum system corresponds to the set of all independent operators that commute with its Hamiltonian. Secondly, we proved the quantum version of Noether's theorem, namely that each such symmetry yields an independent conserved quantity associated with the corresponding operator. Finally, the important physical property of such a symmetry is that one can find simultaneous eigenstates of the Hamiltonian and symmetry operator. For states that are not eigenstates of either, the expectation value of the operator remains constant in time.

10.10 Supplementary: Quantum Mechanics and Relativity

The quantum mechanics described in the previous chapters is not compatible with special relativity. More precisely, the time dependent Schrödinger equation in Eq. (10.78) is not invariant under Lorentz transformations. What we have studied so far is called *non-relativistic quantum mechanics*. The easiest way to see that the TDSE is not relativistically invariant is to recall that its derivation started from the non-relativistic expression for the Newtonian energy in Eq. (9.40).

As we know, the classical world is well described by special relativity, with its non-relativistic description simply a low-velocity ($v \ll c$) approximation. In Sect. 6.7.5, we found that the relativistic energy-momentum relation for a free particle ($V = 0$) is given by

$$E^2 = H^2 = (pc)^2 + (mc^2)^2 \tag{10.118}$$

$$\text{or,} \quad E = H = \pm\sqrt{(pc)^2 + (mc^2)^2} \tag{10.119}$$

Classically, the negative energy solutions are unphysical, so we simply choose to take the positive sign in Eq. (10.119). The correctness of this choice is reinforced by the fact that the Taylor expansion about $p = 0$ of the right hand side of Eq. (10.119) gives the correct Newtonian kinetic energy to first order in (v^2/c^2). In quantum mechanics, both the Hamiltonian H and the momentum p are differential operators. One cannot choose the allowed range of energies to be $E \geq 0$ arbitrarily and it turns out to be difficult to eliminate the negative energy solutions.

In addition, attempting to define an energy operator as a square root, as in Eq. (10.119), gives rise to potential mathematical inconsistencies when the momentum p is an operator. One might try to define H as a Taylor expansion in p, but this in general would result in an infinite order partial differential equation[20] because of the presence of terms of the form $p^n \sim (-i\hbar\partial/\partial x)^n$ for arbitrarily large n.

One can try to avoid this problem by using the operator version of the quadratic equation (10.118), namely

$$\left[-\hbar^2\frac{\partial^2}{\partial t^2}\right]\psi = \left[-\hbar^2c^2\frac{\partial^2}{\partial x^2} + (mc^2)^2\right]\psi \tag{10.120}$$

Equation (10.120) is the one spatial dimension version of the Klein-Gordon equation, named after Oskar Klein and Walter Gordon who proposed it in 1926 to describe relativistic electrons. This turns out not to work. While the Klein-Gordon equation is invariant under Lorentz transformations, the non-relativistic probability density $\mathcal{P} = \psi^*\psi$ is not conserved in time. A conserved relativistic analogue $\mathcal{P}_R$ of the probability density can nonetheless by defined:

$$\mathcal{P}_R = \frac{-i\hbar}{2mc^2}\left[\psi^*\frac{\partial}{\partial t}\psi - \psi\frac{\partial}{\partial t}\psi^*\right] \tag{10.121}$$

Exercise 14   1. Repeat Example 4 in Sect. 10.7 to show that $\mathcal{P} = \psi^*\psi$ does not result in conservation of probability for the Klein-Gordon equation (10.120). *Hint: It is easier to show that the second time derivative of the integrated probability density is non-zero. You may assume vanishing boundary conditions for the wave function at infinity.*
2. Use the same techniques to prove that $\mathcal{P}_R$ in Eq. (10.121) does yield time independent "probabilities" for Eq. (10.120).

---

[20] One that has partial derivatives $\partial^n/\partial x^n$ for arbitrarily large n.

Although $\mathcal{P}_R$ does allow for the time independent normalization of complex wave functions, the relative minus sign between the two terms in Eq. (10.121) means that it is not positive semi-definite so the all-important probabilistic interpretation of the wave function fails remarkably. It turns out that Eq. (10.120) and its conserved density Eq. (10.121) do nonetheless have a useful physical interpretation as describing the dynamics of an electrically charged complex scalar field.[21] In this case $\mathcal{P}_R$ plays the role not of probability density, but charge density, which can be either negative or positive. Thus a physical interpretation for the Klein-Gordon requires the presence of negatively charged scalar particles. This gives a substantial clue to the meaning of the correct relativistic equation describing the quantum wave function of an uncharged single particle that was discovered by Paul Dirac in 1928.

In order to solve the above problems, Dirac needed an equation for the energy that was relativistically invariant, linear in time derivative and at most a finite polynomial in the momentum. In effect he was looking for the "square root" of the right hand side of Eq. (10.118) so that the left hand side contained only a single time derivative. A suitable expression for such a Hamiltonian should be linear in the momentum. In the absence of a potential, Dirac surmised that it might take the general form:

$$H_R\psi = \left[\boldsymbol{\alpha} \cdot \mathbf{p} + \beta m\right]\psi \tag{10.122}$$

where $\boldsymbol{\alpha}$ is an arbitrary constant spatial vector with three components $\{\alpha_i\}$ and β is an arbitrary constant. Dirac further required that Eq. (10.122) be consistent with the time dependent Schrödinger equation (10.120). In other words solutions to Eq. (10.122) should also solve Eq. (10.120) (but not necessarily the other way around). This would be true if

$$H_R(H_R\psi) = H\psi \tag{10.123}$$

Dirac quickly realized that the existence of such a linear momentum Hamiltonian was algebraically impossible if the quantum wave function describing the particle consisted of only a single complex function of space and time. Dirac's brilliant and paradigm shifting insight was that the task became possible if the equation contained not one, but a set of four independent wave functions. These wave functions could be represented in vector form using column notation. Moreover, the coefficients $\{\alpha_i\}$ and β that appear in the Dirac equation (10.122) necessarily became a set of four, 4×4, non-commuting matrices that acted on the multicomponent wave functions via matrix multiplication.[22] The four component column vector of wave functions is known as a spinor. Each component of the spinor is an independent complex function of space and time. It turns out that the Dirac equation and corresponding spinors can only have a physical interpretation if:

---

[21] The Klein Gordon equation cannot describe electrons with spin.

[22] Recall that $\boldsymbol{\alpha}$ is a spatial vector with three components. Thus each of these three components and the parameter β together constitute four matrixes.

1. the Dirac equation applies to spin $1/2$ fermions only, such as an electron.
2. the electron (and every other fundamental fermion in nature, such as muon, taon and quarks) has a partner anti-particle, with identical mass and spin but opposite charge.

Motivated by his relativistic quantum equations, Dirac predicted in 1928 the existence of the anti-particle of the electron, the so-called *positron*. The positron was discovered merely four years later in 1932 by C.D. Anderson.[23]

The discovery of the Dirac equation paved the way for quantum electrodynamics (QED) and later the standard model of particle physics (these within a framework known as quantum field theory), which correctly predicts the short distance quantum behaviour of three of the four fundamental forces of nature, namely electromagnetic, strong and weak nuclear forces. It is quite remarkable that all this started with an attempt to formulate quantum mechanics in a way that is consistent with the symmetries of special relativity.

Reference

1. D.J. Griffiths, D.F. Schroeter, *Quantum Mechanics*, 3rd edn. (Cambridge University Press, 2018)

---

[23] Anderson won the Nobel prize 1934 for his discovery of the proton. Dirac was awarded the Nobel prize a year earlier, along with E. Schrödinger, for their contributions to quantum mechanics.

Chapter 11
The Hydrogen Atom

11.1 Learning Outcomes

Conceptual

- Bohr's derivation of the hydrogen atom spectrum.
- Role of symmetry in solving the Schrödinger equation and determining quantum numbers.
- Derivation of the semi-classical spectrum from the Bohr correspondence principle.
- Hidden symmetry of the hydrogen atom and degeneracy.
- Fermions and the exclusion principle.
- Physical interpretation of hydrogen atom quantum numbers.
- Explanation of the qualitative properties of the elements in the periodic table in terms of fundamental quantum mechanics.

Acquired Skills

- Application of separation of variables to solve simple partial differential equations.
- Calculation of emission/absorption spectrum for Hydrogen.
- Labelling atomic states using notation $1s^2 2s^2 \ldots$.

11.2 Newtonian Dynamics

As shown by Rutherford (see Sect. 12.2.2 of Chap. 12 on nuclear physics), the hydrogen atom is mostly empty space. It consists of a tiny (compared to the size of the atom) proton and electron that are held together by electrostatic forces. Because the proton is much heavier than the electron, to a close approximation it stays still in the rest frame of the atom while the electron orbits around it. This is much the same as our solar system, in which the planets orbit a more massive, central Sun. In order to fully understand the quantum properties of the hydrogen atom one needs to confront the full three-dimensional dynamics. This would make the problem significantly more complex than the one dimensional ones that we considered in the previous chapter,

© The Author(s), under exclusive license to Springer Nature Switzerland AG 2022   253
G. Kunstatter and S. Das, *A First Course on Symmetry, Special Relativity and Quantum Mechanics*, Undergraduate Lecture Notes in Physics,
https://doi.org/10.1007/978-3-030-92346-4_11

but as we will see in Sect. 11.6.2, symmetry will come to the rescue and considerably simplify the quantum equations, effectively turning them into three one-dimensional Schrödinger-type equations.

The electrostatic force due to the proton on the electron with position vector **r** relative to the proton is:

$$\mathbf{F}_E = -\frac{e^2}{4\pi\epsilon_0 r^2}\frac{\mathbf{r}}{r} \tag{11.1}$$

where $r = \sqrt{x^2 + y^2 + z^2}$ is the radial distance of the electron from the nucleus and $\frac{\mathbf{r}}{r} =: \hat{r}$ is the radial unit vector. $\mathbf{F_E}$ is the force that produces the centripetal acceleration $\mathbf{a}_c$ required to keep the electron in its orbit, which we assume to be circular:

$$\mathbf{a}_c = -\frac{v^2}{r}\hat{r} \tag{11.2}$$

Using Newton's second law:

$$\mathbf{F}_E = m\mathbf{a}_c$$
$$\rightarrow \qquad \frac{e^2}{4\pi\epsilon_0 r^2} = m\frac{v^2}{r} \tag{11.3}$$

which gives the Newtonian relationship between the radius and speed of an object held in a circular orbit by electrostatic forces.

The kinetic energy of the electron is:

$$K = \frac{1}{2}mv^2 = \frac{p_x^2 + p_y^2 + p_z^2}{2m} \tag{11.4}$$

Its potential energy depends only on its radial distance from the proton:

$$V(r) = -\frac{1}{4\pi\epsilon_0}\frac{e^2}{r} \tag{11.5}$$

The dependence of $V(r)$ only on the radial distance gives rise to equations of motion for the electron that are spherically symmetric: The force on the electron always points towards the proton at the center of the atom. This observation will prove useful when solving for the wave functions of the stationary states of the Hydrogen atom (Sect. 11.6.2).

Equation (11.3) gives a relationship between the kinetic energy and the potential energy:

$$\frac{1}{2}mv^2 = \frac{1}{2}\frac{ke^2}{r} \tag{11.6}$$

where we have defined $k = \frac{1}{4\pi\epsilon_0}$. The total energy E, illustrated in Fig. 11.1, can now be written as follows:

Fig. 11.1 Plot of the relationship between the total energy (kinetic plus potential) and orbital radius of an electron bound to a proton. If the total energy is greater than zero the electron is not bound to the proton, but instead scatters off it

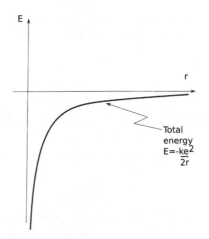

Total energy $E = -\frac{ke^2}{2r}$

$$E = \frac{1}{2}mv^2 - \frac{ke^2}{r}$$
$$= -\frac{1}{2}\frac{ke^2}{r} \tag{11.7}$$

11.3 Supplementary: Symmetries of the Hydrogen Atom

11.3.1 Spherical Symmetry

Consider the classical energy function of the Hydrogen atom, Eq. (11.7). Since the potential energy $V(r) = -ke^2/r$ depends only on the magnitude of the position vector $r = \sqrt{\mathbf{r} \cdot \mathbf{r}} = \sqrt{x^2 + y^2 + z^2}$, it is insensitive to any rotation about the origin, i.e. the proton or the centre of the force. In 3-dimensions one can do an independent rotation about any of the three axes: x, y or z. A rotation about some arbitrary axis can always be decomposed into a sequence of successive rotations around each of the x, y and z axes. As expected from the considerations of Sect. 3.2.4, the set of all such rotations forms a symmetry group with three independent parameters. The three-dimensional rotation group, known technically as $SO(3)$, plays a very important role in many areas of physics.

We learned from Noether's theorem that any continuous symmetry gives rise to a conserved quantity. So what is the conserved quantity corresponding to the continuous rotational symmetry?[1] The answer, as we shall see below, is the angular

---

[1] Note that the hydrogen atom does not possess translation symmetry, i.e. the energy function and the Hamiltonian are *not* invariant under the transformation $\mathbf{r} \to \mathbf{r} + \mathbf{d}$, for any constant vector $\mathbf{d}$. The linear momentum of the electron changes as it orbits the proton.

momentum vector **L**. We will now prove this using the techniques of Sect. 3.6. Consider a rotation around the z−axis by a small angle $\delta\theta$. This changes the coordinates of a point by

$$\delta x = -y \, \delta\theta \tag{11.8}$$

$$\delta y = x \, \delta\theta \tag{11.9}$$

$$\delta z = 0 \,, \tag{11.10}$$

which can be combined into the vector equation

$$\delta\mathbf{r} = \delta\theta \, \mathbf{k} \times \mathbf{r} \,, \tag{11.11}$$

where **k** is the unit vector along the axis of rotation, which we call the z-axis without loss of generality. Generalizing the above for a rotation by $\delta\theta$ along an arbitrary axis with unit vector **n** along it, the corresponding change is

$$\delta\mathbf{r} = \delta\theta \, \mathbf{n} \times \mathbf{r} \,. \tag{11.12}$$

Next we consider the Lagrangian $L(x^i, \dot{x}^i)$ that generates the equations of motion for the electron in the hydrogen atom. It is rotationally invariant because the potential only depends on the radial distance r to the proton.

Then its variation is given by

$$
\begin{aligned}
\delta L &= \frac{\partial L}{\partial x^i} \delta x^i + \frac{\partial L}{\partial \dot{x}^i} \delta \dot{x}^i \\
&= \delta\theta \left[\frac{\partial L}{\partial x^i} \, (\mathbf{n} \times \mathbf{r})^i + \frac{\partial L}{\partial \dot{x}^i} \, (\mathbf{n} \times \dot{\mathbf{r}})^i \right] \\
&= \delta\theta \left[\dot{p}_i \, (\mathbf{n} \times \mathbf{r})^i + p_i \, (\mathbf{n} \times \dot{\mathbf{r}})^i \right] \\
&= \delta\theta \frac{d}{dt} (p \cdot (\mathbf{n} \cdot \mathbf{r})) = \delta\theta \frac{d}{dt} (R \cdot \mathbf{r} \times p) \\
&= \delta\theta \frac{d(\mathbf{n} \cdot \mathbf{L})}{dt} \\
&= 0,
\end{aligned}
\tag{11.13}
$$

where in Eq. (11.13) we have assumed the validity of the Euler-Lagrange equations (3.30) and substituted $p_i = \frac{\partial L}{\partial \dot{x}^i}$[2] and $\frac{\partial L}{\partial x^i} = \dot{p}_i$. The last line, Eq. (11.13), follows from rotational invariance. The Einstein summation convention has been assumed. In the above, $\mathbf{L} = \mathbf{r} \times \mathbf{p}$ is the angular momentum vector (not to be confused with the scalar Lagrangian L) and $\mathbf{n} \cdot \mathbf{L}$ signifies the conserved component of the angular momentum along the rotation axis **n**. Since **n** can be oriented in any direction in space, all components of **L** and hence the vector **L** itself are conserved:

$$\frac{d\mathbf{L}}{dt} = 0. \tag{11.14}$$

---

[2] For a system with standard kinetic energy $K = \frac{1}{2}m(\dot{\mathbf{x}} \cdot \dot{\mathbf{x}})^2$ one gets as usual $\mathbf{p} = m\dot{\mathbf{x}}$.

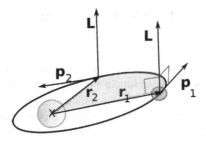

Fig. 11.2 The elliptical orbit of a planet around the sun, or an electron around a proton. The sun/proton is not at the center of the ellipse, but one of two focal points marked by an X. The position **x** and momentum **p** are not perpendicular to each other, and both change magnitude and direction as the planet moves. They stay in a fixed plane because the angular momentum **L** is perpendicular to both and stays constant. Another consequence of angular momentum conservation is that the position vector sweeps out equal areas (shaded region, in blue) in equal times along the orbit (Colour figure online)

   The conservation of angular momentum will hold, by symmetry, for any central potential, $V(r)$ not just for $V \propto 1/r$. A *central potential* is defined as one that depends only on the radial distance to the source of the corresponding force.

   Conservation of angular momentum has the same direct consequence for the classical orbit of an electron around a proton as for the orbit of a planet around the sun, since both the Coulomb potential and the Newtonian gravitational potential are proportional to $1/r$. The following discussion will make use of terminology for planetary motion but is equally valid for the orbit of an electron around a proton.

   Unless they are parallel, any two vectors in three-dimensional space define a plane. The cross product of the two vectors points in a direction perpendicular to this plane. The position vector **r** and momentum vector **p** of a planet at any point in its orbit also define a plane. This is illustrated in Fig. 11.2.

   Since the angular momentum is the cross product of **r** and **p**, its direction is perpendicular to this plane. Consequently, if the angular momentum is conserved as the planet orbits, then its direction stays constant, as does the plane defined by **r** and **p**. This implies that planets moving in *Keplerian orbits* as predicted by Newtonian gravity lie in a plane that does not change along the orbit. Keplerian orbits are said to be *planar*, and the plane in which a planet orbits is, not surprisingly, called the *orbital plane*. Note that **r** and **p** are not in general perpendicular for elliptical orbits. Finally, we reiterate that the equations describing motion in a central potential are not translationally invariant so linear momentum **p** is not conserved.

11.3.2 Accidental Symmetry of the Hydrogen Atom

The classical motion of an electron around a proton is also caused by a $1/r$ potential and hence Keplerian. The orbit takes the form of an ellipse, and angular momentum conservation requires the ellipse to stay in a fixed plane. Implied in the above discussion is that the shape of the ellipse does not change as the particle orbits. What we have not yet addressed is whether or not the ellipse itself stays in a fixed location in the plane. Remarkably, the hydrogen atom equations contain a *hidden* or *accidental* symmetry, one that is not apparent from the set up of the physical problem.[3]

It can be shown that the following set of three coordinate transformations, labelled by the subscript k, leaves the Kepler/Hydrogen atom Lagrangian invariant up to a total derivative:

$$\delta x_{(k)}^i = \epsilon \left[\dot{x}_i x_k - \frac{1}{2} x_i \dot{x}_k - \frac{1}{2} \mathbf{x} \cdot \dot{\mathbf{x}} \, \delta_{ik} \right], \tag{11.15}$$

for a fixed k.

From Noether's theorem it then follows that there are three associated constants of motion. These turn out to be the components of the vector, $\mathbf{A}$:

$$\mathbf{A} \equiv \mathbf{p} \times \mathbf{L} - m K \frac{\mathbf{r}}{r} \tag{11.16}$$

$$\frac{d\mathbf{A}}{dt} = 0, \tag{11.17}$$

where m is the mass of the particle, and K gives the strength of the potential, i.e. $V(r) = -K/r$. $\mathbf{A}$ is known as the Laplace-Runge-Lenz (LRL) vector.

This accidental symmetry and associated conserved LRL vector play an important role in determining the energies of the allowed states of the Hydrogen atom and the chemical properties of the elements in the periodic table (See Sects. 11.6.4 and 11.7).

11.4 The Bohr Atom

As discussed in Chap. 10, once we know the energy of a particle, we are able to write down its stationary states in terms of solutions to the time independent Schrödinger equation. In the present case, we need to include the momentum in all three spatial directions so that the wave function depends on all three coordinates: $\psi = \psi(x, y, z)$. The resulting three-dimensional Schrödinger equation is a complicated partial differential equation that is difficult to solve in general. In 1913, ten years before the Schrödinger equation was even formulated, Bohr was able to deduce the energy spectrum of the hydrogen atom using a combination of physical intuition and luck.

---

[3] This accidental symmetry is a property of all orbital motion in a $1/r$ potential.

He started by assuming that the electron moved in a circular orbit (not true in quantum mechanics). Bohr then went on to postulate that the magnitude of the angular momentum L of the electron was quantized:

$$L := pr = mvr = n\hbar, \qquad n = 1, 2, 3, \dots \qquad (11.18)$$

This was pure conjecture on his part, but he later justified it in terms of the de Broglie wavelength of an electron with momentum p. Specifically, he proposed that the quantum state of the electron was in the form of a standing wave on the circumference of the classical electron orbit at radius r. This in turn required that an integer number of de Broglie wavelengths fit onto the circumference, yielding the relationship.

$$\frac{h}{mv} = \frac{2\pi r}{n} \qquad (11.19)$$

A bit of algebra shows that Eqs. (11.19) and (11.18) are equivalent.

Equations (11.3) and (11.18) give two equations in the two unknowns r and v in terms of a single integer, or quantum number, n. These are easily solved. Solving for $v = n\hbar/(mr)$ in Eq. (11.19) and substituting this into Eq. (11.3) allows one to solve for r. This yields the radii r_n for the allowed electron orbits:

$$r_n = \frac{4\pi\epsilon_0\hbar^2}{e^2 m}n^2$$
$$= r_0 n^2 \qquad (11.20)$$

where r_0 is called the *Bohr radius*:

$$r_0 = \frac{4\pi\epsilon_0\hbar^2}{e^2 m} = 5.29 \times 10^{-11} m \qquad (11.21)$$

The Bohr radius r_0 sets the scale for the size of atoms to be about 1 Ånstrom (1 Å $= 10^{-10}$ m).

Substituting Eq. (11.20) into Eqs. (11.19) and (11.7), respectively yields

$$v_n = \frac{n\hbar}{mr_n} = \frac{\hbar}{mr_0 n}$$

$$E_n = \frac{1}{2}mv_n^2 - \frac{1}{4\pi\epsilon_0}\frac{e^2}{r_n}$$

$$= -\frac{me^4}{8\epsilon_0^2 h^2 n^2} \qquad (11.22)$$

$$= -13.6\,\text{eV}\frac{1}{n^2} \qquad n = 1, 2, 3, \dots \qquad (11.23)$$

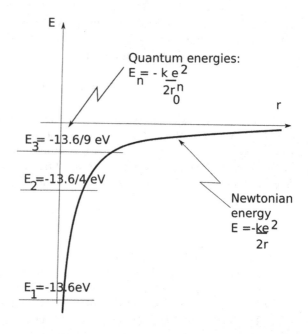

Fig. 11.3 Lowest lying quantum energy levels of the Bohr atom, superimposed on the graph of the classical energy-radius relation. Both energy and orbital radius are quantized in the Bohr model

The lowest energy state has $E_1 = -13.6\,\text{eV}$. This is the binding energy of an electron in the ground state of the hydrogen atom. It is the amount of energy that it must absorb in order to escape to infinity from its orbit (Fig. 11.3).

As $n \to \infty$, $E_n \to 0$. Energies greater than zero do not appear in the spectrum of bound states because when $E > 0$ the electron is not bound to the proton: it is energetically favourable for it to escape to infinity. In that case the electron scatters off the proton and can have any real momentum and energy.

Example 1 Semi-classical spectrum from the Bohr correspondence principle
Recall from Sect. 10.6.2 that the Bohr correspondence principle implies that for large quantum number n:

$$\frac{dE_n}{dn} = h f_{class}(E_n) \tag{11.24}$$

where f_{class} is the classical oscillation frequency of the system at energy E_n. This gives a simple first order differential equation for E_n as a function of n.

In the present case, the classical frequency is the inverse of the orbital period of the electron:

$$f_{class} = \frac{v}{2\pi r} \tag{11.25}$$

Using Eqs. (11.6) and (11.7) it is straightforward to express v and r in terms of energy. Putting these expressions into the above gives:

$$f_{class} = \frac{(-2E)^{\frac{3}{2}}}{2\pi k e^2 \sqrt{m}} \tag{11.26}$$

Putting the above into Eq. (11.20) yields the same energy spectrum Eq. (11.22) as derived by Bohr. Note that for the hydrogen atom the semi-classical spectrum turns out to be exact. This in large part explains why Bohr's somewhat naive quantization of the hydrogen atom gave the correct energy spectrum.

Exercise 1 Verify that Eq. (11.26), in conjunction with the Bohr correspondence principle, gives the correct energy spectrum for the hydrogen atom.

11.5 Emission and Absorption Spectra

The frequency and wavelength of a photon that needs to be absorbed or emitted by a hydrogen atom to take it from the state with $n = n_i$ to the state with $n = n_f$ is given by:

$$hf_\gamma = \frac{hc}{\lambda_\gamma} = E_{n_f} - E_{n_i}$$

$$= -13.6\,\text{eV} \left(\frac{1}{n_f^2} - \frac{1}{n_i^2} \right) \tag{11.27}$$

The inverse wavelength is therefore:

$$\frac{1}{\lambda_\gamma} = -\frac{13.6\,\text{eV}}{hc} \left(\frac{1}{n_f^2} - \frac{1}{n_i^2} \right)$$

$$= -\frac{13.6\,\text{eV}}{1241\,\text{eV} \cdot \text{nm}} \left(\frac{1}{n_f^2} - \frac{1}{n_i^2} \right)$$

$$= -R \left(\frac{1}{n_f^2} - \frac{1}{n_i^2} \right) \tag{11.28}$$

where

$$R = 0.01\,\text{nm}^{-1} \tag{11.29}$$

is called Rydberg's constant. When $n_f > n_i$, the hydrogen atom moves to a higher energy state by absorbing a positive energy photon. If $n_i < n_f$, the energy change is negative and a photon of the corresponding wavelength is emitted. The set of all possible wavelengths for hydrogen is called the *hydrogen spectrum*. Each allowed wavelength is called a *spectral line* because it is observed as a thin line on a spectroscope (Fig. 11.4).

Fig. 11.4 Lyman, Balmer
and Paschen Hydrogen
spectrum series

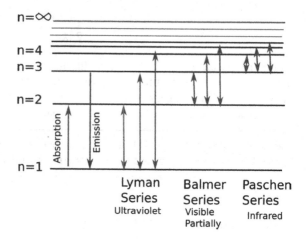

The total spectrum is classified into groups, or *series*:

- *The Lyman Series*: these are the set of lines for transitions that begin or end in the ground state, $n = 1$, which is called the *home base* of the series. If one is considering a transition from a higher state $n_i > 1$ to the home base, then the wavelength of the emitted light is:

$$\frac{1}{\lambda} = R \left(1 - \frac{1}{n_i^2} \right) \tag{11.30}$$

- *The Balmer Series* consists of the set of lines for transitions starting or ending at $n = 2$, which is the home base for this series. These transitions are in the visible region of the spectrum.
- *The Paschen Series* consists of the set of lines for transitions starting or ending at $n = 3$, which is its home base.
- *The series limit* is defined as the transition to $n = \infty$, or $E = 0$. The energy required is the *ionization energy* from the home base. This is the amount of energy required for the electron to be just barely emitted from its home base state and reach infinity with zero kinetic energy.

11.6 Three-Dimensional Hydrogen Atom

11.6.1 The Schrödinger Equation

The total Newtonian energy of the electron in the hydrogen atom, with kinetic energy written in terms of momentum, is:

$$E = KE + PE = \frac{p_x^2 + p_y^2 + p_z^2}{2m_e} - \frac{ke^2}{r} \tag{11.31}$$

Using the momentum operator $p_x = -i\hbar\partial/\partial x$ and the same for the other two coordinates, the three-dimensional time independent Schrödinger equation for the hydrogen atom is:

$$-\frac{\hbar^2}{2m_e}\left(\frac{\partial^2\psi(x,y,z)}{\partial x^2} + \frac{\partial^2\psi(x,y,z)}{\partial y^2} + \frac{\partial^2\psi(x,y,z)}{\partial z^2}\right) - \frac{ke^2}{r}\psi(x,y,z)$$
$$= E\psi(x,y,z) \tag{11.32}$$

The wave function in principle depends on all three Cartesian coordinates.

The differential operator on the left hand side of Eq. (11.32) is very important in physics. It is therefore given a name, the Laplacian, after the French mathematician Pierre-Simon de Laplace who first used it to study celestial mechanics. Since it appears very often in physics, it is given the symbol, ∇^2, for brevity. Using this notation the Schrödinger equation can be written:

$$-\frac{\hbar^2}{2m_e}\nabla^2\psi(x,y,z) - \frac{ke^2}{r}\psi(x,y,z) = E\psi(x,y,z) \tag{11.33}$$

This is in general a difficult equation to solve, but we will show in the next section that by exploiting the spherical symmetry of the equations describing the hydrogen atom, we can simplify the task immensely.

11.6.2 Solutions: Symmetry to the Rescue

Spherical Coordinates

The Laplacian operator ∇^2 in Eq. (11.32) takes different forms in different coordinates. As we have seen, an important feature of the hydrogen atom is that the potential energy in which the electron moves depends only on its radial distance r to the proton and is therefore spherically symmetric: it looks the same from all directions. This implies that the Schrödinger equation is invariant under arbitrary rotations about the origin. As discussed in Sect. 2.3.3, the presence of such a symmetry implies the existence of a particularly convenient coordinate system in which the problem simplifies immensely.

We also know that continuous symmetries such as rotational invariance imply the existence of conserved quantities. These will emerge naturally in spherical coordinates (r, θ, ϕ), which are illustrated in Fig. 11.5.

Fig. 11.5 Spherical
Coordinates: r is the radial
distance from the point P to
the origin, θ is the angle of
inclination of the position
vector of the point P relative
to the z-axis and ϕ is the
azimuthal angle, which is the
rotation relative to the x-axis
of the projection of the
position vector to the $x - y$
plane

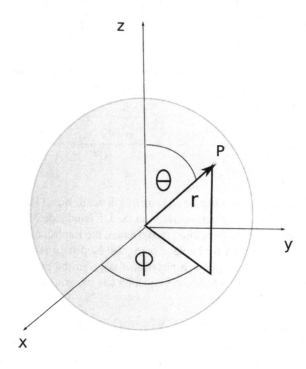

The range of r is from 0 to ∞, while θ goes from 0 to π, and ϕ from 0 to 2π. [4] In
terms of (r, θ, ϕ), the time independent Schrödinger equation is:

$$
-\frac{\hbar^2}{2m_e}\left[\frac{1}{r^2}\frac{\partial}{\partial r}\left(r^2\frac{\partial\psi(r,\theta,\phi)}{\partial r}\right) + \frac{1}{r^2\sin^2(\theta)}\frac{\partial}{\partial\theta}\left(\sin(\theta)\frac{\partial\psi(r,\theta,\phi)}{\partial\theta}\right)\right.
$$
$$
\left. + \frac{1}{r^2\sin^2(\theta)}\frac{\partial^2\psi(r,\theta,\phi)}{\partial\phi^2}\right] - \frac{ke^2}{r}\psi(r,\theta,\phi)
$$
$$
= E\psi(r,\theta,\phi) \tag{11.34}
$$

When calculating expectation values in spherical coordinates it is necessary to use
the correct integration measure and range of coordinates, so that the normalization
condition is:

$$
\langle\psi|\psi\rangle = \int_0^\infty r^2 dr \int_0^\pi d\theta\sin(\theta) \int_0^{2\pi} d\phi\psi^*(r,\theta,\phi)\psi(r,\theta,\phi)
$$
$$
= 1 \tag{11.35}
$$

---

[4] Mathematicians often use a convention in which θ and ϕ exchange roles. This is quite naturally a
potential source of confusion for physics students.

while the expectation value of an operator $\hat{O}$ is:

$$\langle\psi|\hat{O}|\psi\rangle = \int_0^\infty r^2 dr \int_0^\pi d\theta \sin(\theta) \int_0^{2\pi} d\phi \psi^*(r, \theta, \phi)\hat{O}\psi(r, \theta, \phi) \qquad (11.36)$$

Separation of Variables

Although Eq. (11.34) looks more complicated than its Cartesian counterpart Eq. (11.32), it is considerably easier to solve. Since the potential depends only on one coordinate, r, we can use a very general technique, called *separation of variables*, to find the solutions. We used this for the first time when we partially solved the time dependent Schrödinger equation in Sect. 10.7 by separating the solution into the product of two functions. The first depended only on x while the second depended only on t. In that case, the relevant symmetry was time translation invariance. The associated constant of motion that emerged was the energy, E, that appeared as the eigenvalue on the right hand side of the time independent Schrödinger equation.

In order to solve Eq. (11.34), we first multiply both sides by r^2 and then write it in the following suggestive form:

$$-\frac{\hbar^2}{2m_e}\frac{\partial}{\partial r}\left(r^2\frac{\partial\psi(r, \theta, \phi)}{\partial r}\right) - ke^2 r\psi(r, \theta, \phi) - r^2 E\psi(r, \theta, \phi)$$

$$= \frac{\hbar^2}{2m_e}\left[\frac{1}{\sin^2(\theta)}\frac{\partial^2\psi(r, \theta, \phi)}{\partial\phi^2} + \frac{1}{\sin^2(\theta)}\frac{\partial}{\partial\theta}\left(\sin(\theta)\frac{\partial\psi(r, \theta, \phi)}{\partial\theta}\right)\right]$$

$$(11.37)$$

You will notice that all the coefficients and differential operators depending on r are found on the left hand side of the equation, and all the coefficients and differential operators depending on the angles θ and ϕ are found on the right hand side. You may recall that we reduced the time dependent Schrödinger equation (10.78) by looking for stationary states $\Psi(t, x) = f(t)\psi(x)$ that were a product of a function depending on time and a function depending only on x. The separation of variables in the solution was technically possible because the explicit[5] time dependence (i.e. the time derivative) was on the left hand side of the equation, and the explicit spatial dependence (spatial derivatives and potential energy) was on the right hand side.

In the present case, we can separate the variable r from the angular variables (θ, ϕ) to write the solution in the form:

$$\psi(r, \theta, \phi) = R(r)Y(\theta, \phi) \qquad (11.38)$$

---

[5] By explicit dependence we mean that the parameter shows up in the equation, as opposed to implicit dependence where it is present within the wave function.

Exercise 2 Show that plugging the separated function Eq. (11.38) into Eq. (11.37) yields the following:

$$
-\frac{\hbar^2}{2m_e}\frac{1}{R(r)}\frac{\partial}{\partial r}\left(r^2\frac{\partial R(r)}{\partial r}\right) - ke^2r - r^2E
$$
$$
= \frac{\hbar^2}{2m_e}\left[\frac{1}{\sin^2(\theta)}\frac{1}{Y(\theta,\phi)}\frac{\partial^2 Y(\theta,\phi)}{\partial\phi^2} + \frac{1}{\sin^2(\theta)}\frac{1}{Y(\theta,\phi)}\frac{\partial}{\partial\theta}\left(\sin(\theta)\frac{\partial Y(\theta,\phi)}{\partial\theta}\right)\right]
$$

$$(11.39)$$

In Eq. (11.39) we see the miracle of separation of variables. The left hand side depends only on r, whereas the right hand side depends only on θ and ϕ. The only way the right hand side can be equal to the left hand side for all values of r and of θ and ϕ is for both sides to be constant. Thus we have:

$$
-\frac{\hbar^2}{2m_e}\frac{1}{R(r)}\frac{\partial}{\partial r}\left(r^2\frac{\partial R(r)}{\partial r}\right) - ke^2r - r^2E
$$
$$
= \frac{\hbar^2}{2m_e}\left[\frac{1}{\sin^2(\theta)}\frac{1}{Y(\theta,\phi)}\frac{\partial^2 Y(\theta,\phi)}{\partial\phi^2} + \frac{1}{\sin^2(\theta)}\frac{1}{Y(\theta,\phi)}\frac{\partial}{\partial\theta}\left(\sin(\theta)\frac{\partial Y(\theta,\phi)}{\partial\theta}\right)\right]
$$
$$
= -\frac{\hbar^2 B}{2m_e}
$$

$$(11.40)$$

where B is an arbitrary positive dimensionless constant. Negative B gives rise to solutions that are not regular on the sphere and is therefore forbidden. A more physical and intuitive reason for restricting to $B > 0$ is that $\hbar^2 B$ corresponds to the observed values of the angular momentum squared operator, which must be positive.

We have two separate equations to solve:

$$
-\frac{\hbar^2}{2m_e}\frac{1}{r^2}\frac{\partial}{\partial r}\left(r^2\frac{\partial R(r)}{\partial r}\right) - \frac{ke^2}{r}R(r) + \frac{\hbar^2 B}{2m_e r^2}R(r) = ER(r)
$$

$$(11.41)$$

$$
-\hbar^2\left[\frac{1}{\sin^2(\theta)}\frac{\partial^2 Y(\theta,\phi)}{\partial\phi^2} + \frac{1}{\sin^2(\theta)}\frac{\partial}{\partial\theta}\left(\sin(\theta)\frac{\partial Y(\theta,\phi)}{\partial\theta}\right)\right] = \hbar^2 BY(\theta,\phi)
$$

$$(11.42)$$

We can now see that a judicious use of coordinates matched to the symmetry of the situation has revealed some interesting physics. Equation (11.42) takes the form of an eigenvalue equation:

$$
\hat{L}^2 Y(\theta,\phi) = \hbar^2 BY(\theta,\phi)
$$

$$(11.43)$$

for the three-dimensional angular momentum operator:

$$\hat{L}^2 := -\hbar^2 \left[\frac{1}{\sin^2(\theta)} \frac{\partial^2}{\partial\phi^2} + \frac{1}{\sin^2(\theta)} \frac{\partial}{\partial\theta} \left(\sin(\theta) \frac{\partial}{\partial\theta} \right) \right] \qquad (11.44)$$

with eigenvalue $\hbar^2 B$. Solution of Eq. (11.43) is neither straightforward nor particularly enlightening at this stage. The important outcome is that the allowed eigenvalues must take the form:

$$B = l(l+1), \qquad \text{where} \quad l = 0, 1, 2, ..., n \qquad (11.45)$$

where n is the principal quantum number. The measurable values (i.e. *spectrum*) of the magnitude of the angular momentum squared L^2 are then:

$$L^2 = \hbar^2 l(l+1) \qquad (11.46)$$

Recall that spherical symmetry implies that the angular momentum operator commutes with the Hamiltonian operator so that the magnitude of the angular momentum is a conserved quantity. This is the reason that separation of variables allows us to find stationary states which have definite values of angular momentum squared.

There are two surprising things about the result in Eq. (11.46). First, in contrast to the Bohr atom, the allowed values of the magnitude of angular momentum $L = \hbar\sqrt{l(l+1)}$ are not equally spaced. Second, again in contrast to the Bohr atom and perhaps to physical intuition, $l = 0$ is an allowed eigenvalue, so that zero angular momentum states exist for the hydrogen atom. This is contrary to the classical picture of the atom (see Sect. 11.2) in terms of an electron orbiting the nucleus. How can an electron orbit a point source and have no angular momentum? The answer will emerge in the following analysis. Given that the potential in which the electron in the hydrogen atom moves is spherically symmetric there is nothing wrong mathematically with the existence of a spherically symmetric solution to the eigenvalue equations.

Angular Dependence

We can make progress in solving Eq. (11.42) by once again using separation of variables. We are encouraged to try this because spherical symmetry implies the presence of three associated conserved quantities, namely the three components of the angular momentum. So far we have only uncovered one of them, the magnitude of the angular momentum vector. We first multiply both sides by $\sin^2(\theta)$, and rewrite it in the form:

$$-\frac{1}{Y(\theta,\phi)} \frac{\partial^2 Y(\theta,\phi)}{\partial\phi^2} = +\frac{1}{Y(\theta,\phi)} \frac{\partial}{\partial\theta} \left(\sin(\theta) \frac{\partial Y(\theta,\phi)}{\partial\theta} \right) + l(l+1)\sin^2(\theta)$$

$$(11.47)$$

As before, all the functions and operators of ϕ appear on the left, and those depending on θ are on the right. We therefore write:

$$Y(\theta, \phi) = P(\theta)g(\phi) \tag{11.48}$$

Substituting into Eq. (11.47) yields

$$-\frac{1}{g(\phi)}\frac{d^2\,g(\phi)}{d\,\phi^2} = +\frac{1}{P(\theta)}\frac{d}{d\,\theta}\left(\sin(\theta)\frac{\partial P(\theta)}{\partial\theta}\right) + l(l+1)\sin^2(\theta) \tag{11.49}$$

Again, we see that the two sides of the equation can only be equal for all values of the coordinates θ and ϕ if they are equal to the same dimensionless constant, which we call m^2.[6]

$$-\frac{d^2\,g_m(\phi)}{d\phi^2} = m^2 g_m(\phi) \tag{11.50}$$

$$\frac{d}{d\theta}\left(\sin(\theta)\frac{d\,P_l^m(\theta)}{d\,\theta}\right) + l(l+1)\sin^2(\theta)P_l^m(\theta) = m^2 P_l^m(\theta) \tag{11.51}$$

where m^2 is an arbitrary positive constant.[7] We have labeled the eigenfunctions $P_l^m(\theta)$ and $g_m(\phi)$ to emphasize their dependence on the eigenvalues l and m. We can write Eq. (11.50)

$$\hat{L}_z^2 g_m(\phi) = \hbar^2 m^2 g_m(\phi) \tag{11.52}$$

where we have defined

$$\hat{L}_z = -i\hbar\frac{d}{d\,\phi} \tag{11.53}$$

$\hat{L}_z$ is the operator corresponding to angular momentum along the z-axis.[8] The constant m in Eq. (11.52) is called the *magnetic quantum number* because in the presence of an external magnetic field, the angular momentum of the electron, according to Maxwell's equations, aligns itself along the magnetic field. In this case $\hbar m$ gives the component of the angular momentum along the magnetic field.

Solving Eq. (11.50) is straightforward. Multiplying both sides by $g(\phi)$ one gets a harmonic oscillator equation with ϕ playing the role of "time" and m playing the role of the angular frequency.

---

[6] This is somewhat unfortunate standard notation for this constant which, for reasons given below is called the *magnetic quantum number*. It will hopefully be clear from the context whether m stands for mass, meters or magnetic quantum number.

[7] The eigenvalue m^2 must be positive because zero or negative values lead to unphysical behaviour of the wave function (see Eq. (11.54)).

[8] There is nothing special about the z-axis. We have, without loss of generality, defined it to point in the direction that the electron has a definite component of angular momentum. Spherical symmetry dictates that we can find such states.

Exercise 3 Verify that the most general solution to Eq. (11.50) is:

$$g_m(\phi) = D_m e^{im\phi} + B_m e^{-im\phi}. \tag{11.54}$$

We require g_m to be well defined for every value of ϕ. Since ϕ and $\phi + 2\pi$ are the same point on the circle, it must be true that $g_m(\phi + 2\pi) = g_m(\phi)$. For solutions of the form given in Eq. (11.54) this will be true if and only if m is an integer:

$$g_m(\phi) = B_m e^{im\phi}, \quad m = 0, \pm 1, \pm 2, \dots \tag{11.55}$$

where by allowing m to be positive or negative, we have made the second term in Eq. (11.54) redundant.

$g_m(\phi)$ is a complex-valued function with norm $|B_m|$ for all ϕ. It is therefore a pure phase, which in turn means that the probability density $\mathcal{P}$ is independent of the azimuthal angle ϕ

$$
\begin{aligned}
\mathcal{P} &:= \psi^*(r, \theta, \phi)\psi(r, \theta, \phi) \\
&= R(r)^* R(r)(P_l^m)^*(\theta) P_l^m(\theta) g_m^*(\phi) g_m(\phi) \\
&= R(r)^* R(r)(P_l^m)^*(\theta) P_l^m(\theta) B_m^* B_m
\end{aligned}
\tag{11.56}
$$

This tells us that all the allowed wave functions are axially symmetric about the z axis. The $l = 0$ solutions are spherically symmetric, which of course, also encompasses axial symmetry, but the solutions with non-zero angular momentum can only be invariant under rotations about the direction in which the angular momentum vector is pointing.

The normalization condition obtained from the probability density Eq. (11.56) splits into a factor of three integrals:

$$\langle \psi | \psi \rangle = \int_0^\infty r^2 dr\, R^*(r) R(r) \int_0^\pi d\theta\, \sin(\theta)(P_l^m)^*(\theta) P_l^m(\theta) \int_0^{2\pi} d\phi\, B_m B_m^* \tag{11.57}$$

It turns out to be convenient to normalize each factor in the solution separately. Setting the integral over ϕ to one requires:

$$B_m = \frac{1}{\sqrt{2\pi}} \quad \rightarrow \quad g_m(\phi) = \frac{1}{\sqrt{2\pi}} e^{im\phi} \tag{11.58}$$

We now turn to Eq. (11.51). The equation and its solutions depend on both the total angular momentum quantum number, l and the magnetic quantum number m. The method for solving it is rather involved, so we will just present the eigenfunctions, labelled by both quantum numbers l and m. One important feature of the solutions is that there is a mathematical and corresponding physical restriction on the magnitude of the magnetic quantum number: it must be less than the l. Solutions only exist for

$$m = 0, \pm 1, \pm 2, ..., \pm l \tag{11.59}$$

Here are the normalized functions, called the (fully-normalized) *associated Legendre polynomials* $P_l^m(\theta)$:

$$l = 0, m = 0: \quad P_0^0(\theta) = \frac{1}{\sqrt{4\pi}} \tag{11.60}$$

$$l = 1, m = 0: \quad P_1^0(\theta) = \sqrt{\frac{3}{4\pi}} \cos(\theta) \tag{11.61}$$

$$l = 1, m = \pm 1: \quad P_1^{\pm 1}(\theta) = \mp\sqrt{\frac{3}{8\pi}} \sin(\theta) \tag{11.62}$$

$$l = 2, m = 0: \quad P_2^0(\theta) = \sqrt{\frac{5}{16\pi}} \left(3\cos^2(\theta) - 1\right) \tag{11.63}$$

$$l = 2, m = \pm 1: \quad P_2^{\pm 1}(\theta) = \mp\sqrt{\frac{15}{8\pi}} \sin(\theta)\cos(\theta) \tag{11.64}$$

$$l = 2, m = \pm 2: \quad P_2^{\pm 2}(\theta) = \sqrt{\frac{15}{32\pi}} \sin^2(\theta) \tag{11.65}$$

....

The Legendre polynomials as defined above are *orthonormal*, namely:

$$\int_0^\pi d\theta \, P_l^m(\theta) P_{\tilde{l}}^{\tilde{m}} = \delta^{m\tilde{m}} \delta_{l\tilde{l}} \tag{11.66}$$

where we have used the Kronecker delta defined by[9]:

$$\delta^{m\tilde{m}} = \begin{cases} 1 & \text{when } m = \tilde{m} \\ 0 & \text{otherwise} \end{cases} \tag{11.67}$$

That is, the Legendre polynomials are normalized when $m = \tilde{m}$ and $l = \tilde{l}$, but the integrals are zero otherwise. We will discuss the physical significance of these solutions once we have presented solutions of the radial equation.

Radial Wave Function

Substituting $l(l + 1)$ for B in Eq. (11.41) yields:

$$\hat{H}_r R_{n,l}(r) = E_n R_{n,l}(r) \tag{11.68}$$

---

[9] The Kronecker delta is essentially the identity matrix written in component form.

where:

$$\hat{H}_r := -\frac{\hbar^2}{2m_e}\frac{1}{r^2}\frac{\partial}{\partial r}\left(r^2\frac{\partial}{\partial r}\right) - \frac{ke^2}{r} + \frac{\hbar^2}{2m_e}\left(\frac{l(l+1)}{r^2}\right) \qquad (11.69)$$

Equation (11.68) is the eigenvalue equation that determines the allowed values (spectrum) of the energy E_n for the hydrogen atom. We have added a subscript to the radial wave function $R_{n,l}$ to reflect its dependence on both the angular quantum number l and a new quantum number n associated with the energy eigenvalue E_n. Despite the fact that Eq. (11.68) only contains derivatives and functions of r, the angular contribution to the energy is included via the last term in Eq. (11.69), which is the increase in potential energy due to the centripetal force required to hold the electron in orbit with fixed angular momentum $\hbar^2 l(l+1)$.

Equation (11.41) is somewhat difficult, so we will skip the derivation and just present the solutions. The energy eigenvalues are labelled by a single positive integer n, often called the *principal quantum number*:

$$E_n = -\frac{m_e}{2\hbar^2}\left(\frac{e^2}{4\pi\epsilon_0}\right)^2\frac{1}{n^2} \qquad n = 1, 2, 3, ... \qquad (11.70)$$

An interesting feature of the energy spectrum given by Eq. (11.70) is that the energy eigenvalues E_n are independent of the angular momentum quantum number, l. This is not guaranteed by spherical symmetry. As we will explain below in Sect. 11.6.4, this is a consequence of the hidden symmetry of the hydrogen atom present in Sect. 11.3.

It is remarkable that the energies agree precisely with those derived by Bohr (see Sect. 11.4) despite the fundamental conceptual errors in his model. For example, in the full quantum version, the electrons do have fixed energy and angular momentum, but they do not move along orbits of fixed radius. Moreover, the allowed values of angular momentum that Bohr derived are incorrect, not least because zero angular momentum is an allowed state in the quantized hydrogen atom, but not in the Bohr atom. These zero angular momentum states are spherically symmetric, something else that is contrary to Bohr's simple picture of an electron orbiting the proton.

The radial equation only has solutions for certain values of the orbital quantum number. There are exactly n allowed values of l within a given shell (i.e. specific value of n), namely:

$$l = 0, 1, 2, ..., n - 1. \qquad (11.71)$$

The lowest, or ground state, energy is given by:

$$E_1 = -\frac{m_e}{2\hbar^2}\left(\frac{e^2}{4\pi\epsilon_0}\right)^2$$
$$= -13.6\,\text{eV}. \qquad (11.72)$$

As a consequence of Eq. (11.71), the ground state ($n = 0$) must have $l = 0$ and therefore is invariant under rotations.

Exercise 4 1. Verify that the $\frac{m_e}{2\hbar^2} \left(\frac{e^2}{4\pi\epsilon_0} \right)^2$ has units of energy, and that its numerical value is $13.6\,\text{eV}$.

We now write down some of the normalized radial wave functions $R_{n,l}$

$$n = 1, l = 0: \quad R_{1,0} = \frac{2}{r_0^{3/2}} e^{-r/r_0} \tag{11.73}$$

$$n = 2, l = 0: \quad R_{2,0} = \frac{1}{\sqrt{2} r_0^{3/2}} \left(1 - \frac{r}{2r_0} \right) e^{-r/2r_0} \tag{11.74}$$

$$l = 1: \quad R_{2,1} = \frac{1}{2\sqrt{6} r_0^{3/2}} \left(\frac{r}{r_0} \right) e^{-r/2r_0} \tag{11.75}$$

$$n = 3, l = 0: \quad R_{3,0} = \frac{2}{3\sqrt{3} r_0^{3/2}} \left(1 - \frac{2}{3} \frac{r}{r_0} + \frac{2}{27} \left(\frac{r}{r_0} \right)^2 \right) e^{-r/3r_0} \tag{11.76}$$

$$l = 1: \quad R_{3,1} = \frac{8}{27\sqrt{6} r_0^{3/2}} \left(1 - \frac{1}{6} \frac{r}{r_0} \right) \left(\frac{r}{r_0} \right) e^{-r/3r_0} \tag{11.77}$$

$$l = 2: \quad R_{3,2} = \frac{4}{81\sqrt{30} r_0^{3/2}} \left(\frac{r}{r_0} \right)^2 e^{-r/3r_0} \tag{11.78}$$

where r_0 is the Bohr radius given in Eq. (11.21).

11.6.3   Probability Densities

We have shown that the complete set of solutions to the time independent Schrödinger equation for an electron in a hydrogen atom depends on three quantum numbers: the *principal quantum number n*, the *orbital quantum number l* and the *azimuthal or magnetic quantum number m*. The full three-dimensional solutions take the form:

$$\psi_{n,l,m}(r, \theta, \phi) = R_{n,l}(r) Y_{n,l}(\theta) g_m(\phi) \tag{11.79}$$

From the quantum wave function we can define the probability per unit volume $\mathcal{P}$ of finding the electron near the point with coordinates (r, θ, ϕ). It is:

$$\mathcal{P} = \psi_{n,l,m}^*(r, \theta, \phi) \psi_{n,l,m}(r, \theta, \phi)$$

$$= \frac{1}{2\pi} R_{n,l}^2(r) Y_{n,l}^2(\theta) \tag{11.80}$$

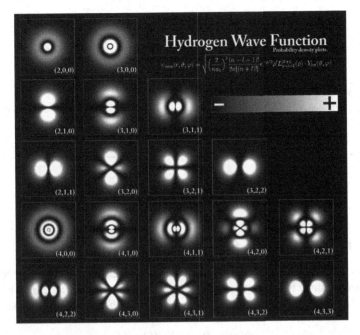

Fig. 11.6 Plots of normalized hydrogen probability densities. The first number in parenthesis is the principal quantum number, n, the second is the orbital quantum number l and the third is the magnetic quantum number m. The probability densities are identical for $\pm m$. Figure credit Copyright PoorLeno at Wikipedia

As mentioned previously, the azimuthal quantum number m disappears from the density because $g_m(\phi)$ is a complex number of constant norm. This means that the probability density is independent of ϕ and axially symmetric. The probability densities for $n = 2, 3, 4$ are shown in Fig. 11.6.

Figure 11.6 only shows the cross-section of the densities in a plane containing the z-axis. Using the axial symmetry, one can visualize the three-dimensional plots by rotating the figures about the z-axis. The probability density for states with zero angular momentum ($l = 0$) is spherically symmetric, as expected, since there is no preferred direction in this case. The probability density for the $n = 1$, (and hence $l = 0$) state looks like a spherical cloud, with magnitude maximum at $r = 0$ and decreasing as one moves away from the nucleus. Note that for $n = 2$ and higher, the $l = 0$ states have spherical "dead zones" where the electron is unlikely to be found. These are the three-dimensional analogues of the nodes in the harmonic oscillator wave function for example. The number of these dead zones increases with principal quantum number, and is in fact equal to $n - 1$, also in keeping with the nodes of the harmonic oscillator. The structure of the dead zones gets considerably more interesting as the orbital angular momentum and orbital magnetic quantum number increase.

Exercise 5 Calculate $\langle r \rangle$ for the following states:

1. $n = 1, l = 0$
2. $n = 2, l = 0$
3. $n = 2, l = 1$
4. $n = 3, l = 2$

Note: you should not have to do any angular integrals. Why?

11.6.4   Shells, Orbitals and Degeneracy

Recap of Hydrogen Atom Solutions and Energy Spectrum

To recap: the complete separable solution to the Hydrogen atom Schrödinger equation is labelled by three integers: the *principal quantum number*, n, the orbital quantum number l, and the *azimuthal*, or *magnetic quantum number m* and takes the separable form[10]:

$$\psi(\mathbf{r}) = R_{n,l}(r)Y_l^m(\theta)g_m(\phi) \tag{11.81}$$

n is the principal quantum number that determines the energy of the state according to the formula originally derived by Bohr:

$$E_n = -13.6\,\text{eV}\frac{1}{n^2} \tag{11.82}$$

States with a given value of n are called **shells**.

For each value of n, the family of solutions for $Y_l^m(\theta)$ are labelled by two quantum numbers l and m. l is the quantum number that determines the magnitude of the electron's orbital angular momentum L according to:

$$L = \sqrt{l(l+1)}\hbar \tag{11.83}$$

States within a given shell, but having differing values of l are called **sub-shells**, or **orbitals**. There are exactly n allowed values of l within a given shell (i.e. specific value of n). They are:

$$l = 0, 1, 2, ..., n-1 \tag{11.84}$$

Equation (11.83) shows that the angular momentum L is not quantized according to the same formula as assumed by Bohr and that states with $L = 0$ exist.

---

[10] Note that solutions to Eq. (11.34) of the form Eq. (11.81) exist for any spherically symmetric potential, that is any potential $V(r)$ that depends only on the radial coordinate.

The second quantum number, that labels Y_l^m, and also the azimuthal part of the wave function $g_m(\phi)$, is the magnetic quantum number m. It specifies the value of the electron's angular momentum along the z axis via:

$$\hat{L}_z = m\hbar, \quad \text{where } m = -l, -(l-1), -(l-2), .., 0, ..., (l-2), (l-1), l$$
$$(11.85)$$

There are $2l + 1$ allowed values of m within each sub-shell. For example for the $l = 2$ sub-shell, the allowed values of the magnetic quantum number are: $-2, -1, 0, 1, 2$. The number of values is $2 \times 2 + 1 = 5$, as expected from the general formula.

Hidden Symmetry and the Degeneracy of the Hydrogen Atom

As remarked earlier, the energy of each shell is independent of both the orbital quantum number, l, and the magnetic quantum number, m, something that is not expected from spherical symmetry alone. This implies that the stationary states of the hydrogen atom are highly *degenerate*, which simply means that there are many different states with the same energy. We now show how this degeneracy results from the symmetries of the system, including the hidden symmetry presented in Sect. 11.3.

We have seen in Sect. 10.9 that the quantum counterpart of a classical conservation law is the statement that the corresponding operator commutes with the Hamiltonian operator. This guarantees that the expectation value of the operator remains constant for any state as it evolves via the time dependent Schrödinger equation. It also ensures that if a state initially has a definite value for that operator (is in an eigenstate with fixed eigenvalue), then that definite value is preserved with time.

The quantum counterparts of Eq. (11.14) are the three equations (one for each component of $\hat{\mathbf{L}}$) $[\hat{\mathbf{L}}, \hat{H}] = 0$, where $\hat{\mathbf{L}}$ is an operator constructed out of the position operator $\mathbf{r}$ and momentum operator $\mathbf{p}$ as $\hat{\mathbf{L}} = \hat{\mathbf{r}} \times \hat{\mathbf{p}}$. Each component of $\hat{\mathbf{L}}$ is an operator that commutes with the Hamiltonian (i.e. $[\hat{L}_z, \hat{H}] = 0$, etc.). The components of angular momentum do not, however, commute with each other. In fact,

$$\left[\hat{L}_x, \hat{L}_y\right] = i\hbar\hat{L}_z + \text{cyclic permutations} \quad (11.86)$$

This will be examined in more detail in the context of spin in Sect. 13.4. For the time being we simply note that the largest number of mutually commuting operators is three: the Hamiltonian, the magnitude squared of angular momentum and one of the three components of spin, usually taken without loss of generality to be the z-component.

We are now in a position to better understand the spectrum of states of the Hydrogen atom. According to the discussion in Sect. 10.9, we can label all states by the quantum numbers associated with the three commuting operators, namely (n, l, m). The question is whether states with different l and m should have the same energy. The answer to this question does not require solving complicated differential equations. Instead, we start with some state $\psi_{n,l,m}(r, \theta, \phi)$ with energy E_n, angular momentum squared $\hbar^2 l(l+1)$ and component of spin along the z-axis $\hbar m$. We now act on this state with the operator $\hat{L}_y$. It commutes with the Hamiltonian so we know that it will not change the energy. It also commutes with $\hat{L}^2$, so it will not change the square of the angular momentum. $\hat{L}_y$ does not, however, commute with $\hat{L}_z$, so it follows that the result will be an eigenstate of $\hat{L}_z$ with the same energy, but different eigenvalue $\hbar m$. This state is therefore a new state that has the same energy as the original state, and we have two states that are degenerate. This implies states with the same l and n but different m are degenerate.

What about changing l? Does this produce a new state with the same energy? Since both $\hat{L}_x$ and $\hat{L}_y$ commute with $\hat{L}^2$, they will not change l, and therefore will not yield degenerate states. This is where the accidental symmetry and associated conserved LRL vector $\hat{\mathbf{A}}$ discussed in Sect. 11.3 play a role. The components of this new conserved vector operator commute with the Hamiltonian but not $\hat{L}^2$, so their action on the state $\psi_{n,l,m}(r, \theta, \phi)$ will change l but not the energy.

The astute reader will have noticed that the motion of an electron in a hydrogen atom has three degrees of freedom, namely the three components of its position. There are at most six possible independent conserved quantities: the three coordinates and their associated momenta (angular momentum is not independent). We now seem to have seven conserved quantities: the energy, three components of the angular momentum and three components of $\mathbf{A}$. Luckily, the math takes care of itself. Only one of the components of $\mathbf{A}$ is independent of the other conserved quantities, so the number of independent conserved quantities is actually five.[11]

The bottom line is that energy eigenstates with different l and different m all have the same energy. For each value of the principal quantum number n, there are n possible values of l, and for each value of l there are $2l + 1$ possible values of m. A straightforward counting of the degeneracy $d(n)$ then yields[12]

$$d(n) = \sum_{l=0}^{l} \sum_{m=-l}^{l} 1 = \sum_{l=0}^{n-1} = n^2$$

For example, if $n = 4$,

$$d(4) = 1 + 3 + 5 + 7 = 16 \qquad (11.87)$$

Inclusion of relativistic corrections to the hydrogen atom affects the above picture significantly, because they lead to terms in the corrected Hamiltonian proportional to higher powers of the momentum, such as p^4. While these terms preserve the spherical symmetry because the magnitude of momentum, p^2, is a three-dimensional scalar, they produce small changes in the energies. The hidden symmetry is broken and hence the degeneracy of the energy eigenstates as well. For example with these corrections,

---

[11] Quantum mechanically only three of the corresponding operators commute, so that three quantum numbers suffice to specify the state uniquely. See Sect. 11.6.6.

[12] The easiest way to prove this rigorously is by induction: Check that it is true for $n = 1$ and then verify explicitly that if it is true for some value of n, then it is also true for $n + 1$.

Fig. 11.7 Allowed angular momentum quantum numbers for each value of the lowest three principal quantum numbers of the Bohr atom. The total number of electrons (excluding spin) that each orbital can accommodate are also indicated in brackets

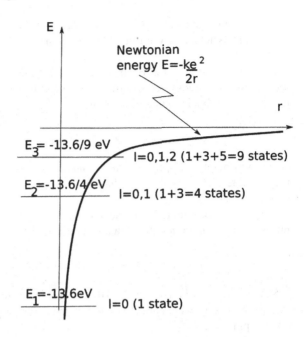

the energies depend on the orbital angular momentum. Angular momentum of course is still preserved.[13]

In the context of the Hydrogen atom the relativistic corrections to the energies of the states are observationally important because the breaking of the degeneracy affects the emission and absorption spectrum described in Sect. 11.5 in nontrivial and interesting ways.

The occupation number of the first three shells of the Hydrogen atom are illustrated in Fig. 11.7.

11.6.5 Fermions and the Spin Quantum Number

Elementary particles, such as photons, electrons and protons have a purely quantum attribute called *intrinsic spin*, or simply, spin. Spin is a form of angular momentum that is different from what we will call *orbital angular momentum*, **L**. The latter is due to motion of a particle at some finite distance from some specified axis. This is clear from the formula:

$$\mathbf{L} := \mathbf{r} \times \mathbf{p} \tag{11.88}$$

---

[13] In terms of the classical description of a central force problem, such as the hydrogen atom and planetary orbits, what results is a modification from the stationary ellipse predicted by the uncorrected equations, to a precessing ellipse. Such corrections gave rise to the famous 43 s of arc precession per century for the planet mercury, as described in Sect. 7.2.2. The planarity of the orbit remains unchanged.

As the name suggests intrinsic spin, s, is an intrinsic property of a quantum particle, that is not directly attributable to motion around an axis. It has no Newtonian counterpart.

Intrinsic spin is a vector. Each component is a quantum observable, or operator, having a discrete eigenvalue spectrum. The magnitude of spin squared has eigenvalues similar to orbital angular momentum[14]:

$$S^2 = \hbar^2 s(s+1) \tag{11.89}$$

where in principle the *spin quantum number s* can take on half-integer as well as integer values. The possibility of half-integer spin s is one of the most significant features of intrinsic spin.

Elementary particles fall into one of two classes. *Bosons*, such as the photon, have intrinsic spin quantum number s that is an integer

$$s = n \qquad n = 0, 1, 2, ... \tag{11.90}$$

Scalar particles like the famous Higgs boson have zero spin, photons have unit spin and gravitons, which are the quantum analogues of gravity waves, are thought to have spin $s = 2$.[15]

Exercise 6 Verify that $\hbar$ has the same units as angular momentum.

The other class of elementary particles consists of *fermions*. Their spin is always a half integer. That is:

$$s = \frac{1}{2}, \frac{3}{2}, \frac{5}{2}, ... \tag{11.91}$$

Electrons have intrinsic spin $s = 1/2$.

It turns out that the only values of intrinsic spin that are realizable in nature are $s = 0, \frac{1}{2}, 1, \frac{3}{2}, 2$. Higher values of spin lead to mathematical and theoretical inconsistencies.

Just like its orbital counterpart, intrinsic spin is a vector. In the case of electrons, its component along any given z-axis can have one of two possible values, $s_z = +\frac{\hbar}{2}$ or $s_z = -\frac{\hbar}{2}$. This provides a fourth quantum number called the *spin magnetic quantum number*, m_s, that is needed to fully specify the state of an electron in the hydrogen atom. The spin magnetic quantum number has two possible values $m_s = +\frac{1}{2}$ or $m_s = -\frac{1}{2}$, which correspond qualitatively to the quantum spin vector pointing along the positive, and negative z-axis. For simplicity the two values of the spin quantum number are often denoted as simply spin up ($\uparrow$) or spin down ($\downarrow$).

---

[14] Mathematical details of the spin operator will be given in Sect. 13.4.

[15] Although classical gravitational waves have recently been observed, their quantum properties, and hence the observation of gravitons, remain out of reach experimentally.

11.6.6 Summary of the 3D Hydrogen Atom

The electron in the Hydrogen atom occupies one of an infinite number of possible quantum states that are labelled or specified by the four quantum numbers (n, l, m, m_s). The energy depends only on the principal quantum number n, that labels the shell, and there are $2n^2$ different states for an electron to occupy in each shell. As n increases, the energy approaches 0 from below, while the expectation value of the orbital radius approaches ∞. The allowed electron states, including spin, in each shell up to $n = 3$ are summarized in Table 11.1.

You may have noticed at this stage that there is an apparent contradiction in that the equations for the hydrogen atom are invariant under five symmetries (time translation, three rotations and the independent component of the accidental symmetry described in Sect. 11.3.2) but the quantum states are labelled by only three quantum numbers (not counting spin), each of which corresponds to a conserved quantity associated with one of the symmetries. Why aren't there five, corresponding to the five classically conserved charges, namely energy, the three components of $\mathbf{L}$ and the single independent component of the LRL vector $\mathbf{A}$? (see Sect. 11.3.2). The answer to this question comes from the observation that the operators corresponding to the three components of angular moment, namely $\hat{L}_x$, $\hat{L}_y$ and $\hat{L}_z$ are mutually complementary. The order that they act on quantum states matters (they do not commute), so that only one of the three can have a fixed value for any given state. The operator $\hat{L}^2$, on the other hand does commute with each of the components $\hat{L}_x$, $\hat{L}_y$ and $\hat{L}_z$. One can therefore construct states with fixed $\hat{L}^2$ and one of the three components. Thus we have three independent conserved observables, namely $\hat{H}$, $\hat{L}^2$ and $\hat{L}_z$, where we have chosen, without loss of generality, $\hat{L}_z$ as the third instead of $\hat{L}_x$ or $\hat{L}_y$. The LRL vector does not commute with the three others, so it does not produce a fourth quantum number.

Table 11.1 Allowed states for hydrogen-like atoms up to $n = 3$, including spin

Shell n	1	2		3								
Subshell l	0	0	1	0	1			2				
Orbital m_l	0	0	1	0	−1	0	1	0	−1	2	1	0	−1	−2
Spin m_s	↑↓	↑↓	↑↓	↑↓	↑↓	↑↓	↑↓	↑↓	↑↓	↑↓	↑↓	↑↓	↑↓	↑↓

11.7   The Periodic Table

11.7.1   Hydrogen-Like Atoms

An element $^A_Z X$ consists of Z electrically charged protons and $N = A - Z$ neutrons.
In order to be electrically neutral, an atom must also contain Z electrons. If one
neglects the forces between the electrons, one can assume that each electron is acted
on only by the electrostatic force between it and the positively charged nucleus. In
this approximation, the Newtonian equations describing the motion of each electron
are virtually identical to those describing the electron in the hydrogen atom. The only
difference is that the electric charge $+e$ of the proton that appears in the Hydrogen
potential is replaced by the charge of the nucleus, namely Ze. The Schrödinger
equation describing the quantum state of each electron is then:

$$-\frac{\hbar^2}{2m_e}\nabla^2\psi(r,\theta,\phi) - \frac{Ze^2}{r}\psi(r,\theta,\phi) = E\psi(r,\theta,\phi) \qquad (11.92)$$

The solutions to this equation take precisely the same form as before, with the
substitution $e^2 \to Ze^2$ everywhere. Such atoms are called *hydrogen-like*.

In this case the quantum states of the hydrogen atom that we have already examined
provide the energy eigenstates that each of the Z electrons can in principle occupy.
One can imagine building hydrogen-like atoms by adding electrons one at a time
to each of the allowed states. It is necessary to fill the lowest energy states first, so
one would add electrons from the ground up in order to minimize the total energy
until all Z electrons have a "home". The chemical properties of the resulting neutral
atoms depends to a large extent on the states occupied by the outermost electrons,
those in the highest energy state with largest expectation value of radial distance r.

To understand how this happens, and why all the electrons cannot simply occupy
the ground state, we need to discuss the *Pauli exclusion principle*. Fermions are
fundamentally anti-social, at least with respect to their own kind: The Pauli exclusion
principle states that fermions cannot share a quantum state with an identical fermion.
This might seem natural in some ways, but is not true of bosons. Photons are happy
to share the same quantum state with as many other photons as they can.

For hydrogen-like atoms the Pauli exclusion principle implies that each state corresponding to specific values of the four quantum numbers (n, l, m, m_s) is occupied
by no more than one electron. This tells us how the Z electrons in a hydrogen-like
atom are distributed among the quantum states allowed by the Schrödinger equation
(11.92). The electrons occupy the lowest energy states first, filling first the $n = 0$
shell, then the $n = 1$ shell (if there are sufficient electrons) then the $n = 2$ shell and
so on. The energy of the electron in each shell depends only on the principal quantum
number n, and for each value of n, there are $2n^2$ possible states for an electron to
occupy. Figure 11.8 illustrates how electrons occupy the available states for mass
number A between one and ten.

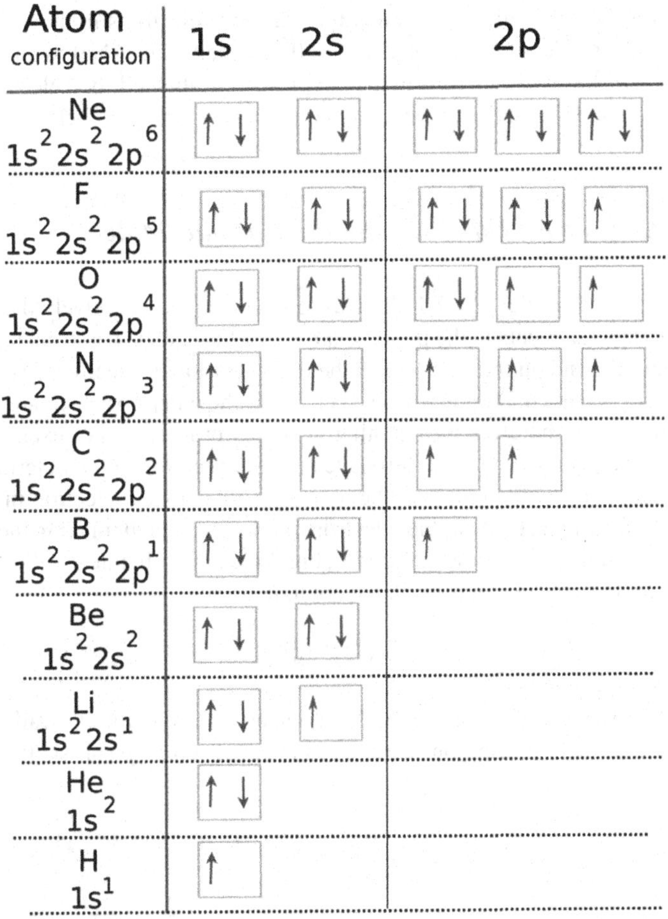

Atom configuration	1s	2s	2p		
Ne $1s^2 2s^2 2p^6$	↑ ↓	↑ ↓	↑ ↓	↑ ↓	↑ ↓
F $1s^2 2s^2 2p^5$	↑ ↓	↑ ↓	↑ ↓	↑ ↓	↑
O $1s^2 2s^2 2p^4$	↑ ↓	↑ ↓	↑ ↓	↑	↑
N $1s^2 2s^2 2p^3$	↑ ↓	↑ ↓	↑	↑	↑
C $1s^2 2s^2 2p^2$	↑ ↓	↑ ↓	↑	↑	
B $1s^2 2s^2 2p^1$	↑ ↓	↑ ↓	↑		
Be $1s^2 2s^2$	↑ ↓	↑ ↓			
Li $1s^2 2s^1$	↑ ↓	↑			
He $1s^2$	↑ ↓				
H $1s^1$	↑				

Fig. 11.8 This figure illustrates how the shells, sub-shells and orbitals are occupied by electrons for the first eight elements in the periodic table. The chemical properties are determined by how the outermost electrons are distributed in the unfilled orbitals

The configuration of electrons in an atom are described in compact form by specifying the number of electrons in each sub-shell as follows. The principal quantum number, or shell number is specified by an integer, $n = 1, 2, ...$ The sub-shells themselves are given letter names: s, p, d, f, g, h, ... for $l = 0, 1, 2, 3, 4, 5, ...$ respectively.[16] Thus, counting the two possible values of the spin quantum number in each state, the f sub-shell ($l = 4$) can hold $2 \times (2 \cdot 4 + 1) = 18$ electrons, for example. The number of electrons in each sub-shell is specified in the exponent naming the sub-shell. As indicated at the far right of Fig. (11.8), the electron configuration

---

[16] Unfortunately, chemists use the same letter, s, to denote the $l = 0$ sub-shell as we have used to denote the spin of the electron. Hopefully the usage will be clear from the context.

for lithium is $1s^2 2s^1$, which means that there are two electrons in the $n = 1, l = 0$
sub-shell, and one electron in the $n = 2, l = 0$ sub-shell. Neon (Ne), with configuration $1s^2 2s^2 2p^6$ has two electrons in the $n = 1, l = 0$ sub-shell, two electrons in the
$n = 2, l = 1$ sub-shell and six electrons in the $n = 2, l = 2$ sub-shell.

11.7.2 Chemical Properties and the Periodic Table

The considerations in Sect. 11.7.1 allow us to understand qualitatively the chemical
properties of the elements in the periodic table (see Fig. 11.9). These chemical properties largely depend on the electrons in the outermost shell, specifically on how the
sub-shells are occupied. Atoms whose electrons in the outermost shell are in similar
configurations will therefore have similar chemical properties. For example all the
elements in the column on the far left have only one electron in the outermost shell,
with zero angular momentum, just like hydrogen. This lonely electron in the outer
most orbit can easily accept another electron to fill its sub-shell, making the atoms in
the far left column, the so-called *alkalides*, highly reactive chemically. For example,
sodium (Na), has eleven electrons in the configuration:

$$1s^2 2s^8 2p^1 \tag{11.93}$$

Similarly eighteen of potassium's (K) nineteen electrons fill all available states
up to the n = 3 shell, leaving one electron in the outermost $4p$ sub-shell.

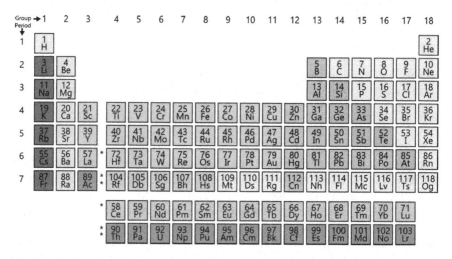

Fig. 11.9 The chemical properties of the elements in a given column are similar and can be explained
qualitatively in terms of the quantum mechanics of hydrogen-like atoms. *Credit* Author Offnfopt,
WikiCommons

The next element in the alkalide series, namely rubidium (Rb) displays a slight anomaly. The thirty-seven electrons are in the configuration:

$$1s^2 2s^2 2p^6 3s^2 3p^6 3d^{10} 4s^2 4p^6 5s^1 \tag{11.94}$$

You will notice that the 5s sub-shell has an electron in it even though the (apparently) lower energy 4d sub-shell is empty. This is where the approximation of non-interacting electrons in the model breaks down somewhat, because the difference in energy between adjacent shells gets smaller for larger principal quantum numbers. It turns out that it is energetically favourable in this case for the last (37th) electron to move to the 5th shell with zero angular momentum rather than staying in the 4th shell with higher angular momentum.

The *noble gases* on the far right on the other hand have outermost sub-shells that are full, so they do not easily accept additional electrons. They do not react chemically and are therefore classified as *inert*.

Exercise 7 How many electrons does aluminum ($Z = 13$) have in its outermost sub-shell? Is there room for more? If so, how many.

As a final comment we highlight that in this Chapter we have succeeded in applying fundamental physics and symmetry principles to explain the classification of elements in the periodic table. Quite a remarkable achievement! It is perhaps this success of quantum mechanics that prompted Rutherford to supposedly claim "All science is either physics or stamp collecting". Somewhat ironically, Rutherford received the 1908 Nobel prize in chemistry and not physics.[17]

---

[17] As quoted in Rutherford at Manchester. J. B. Birks, Ed. Heywood, London, 1962; Benjamon, New York, 1963.

Chapter 12
Nuclear Physics

12.1 Learning Outcomes

Conceptual

- Rutherford versus Thompson model of the nucleus.
- Qualitative understanding of the nucleon-nucleon potential.
- Qualitative properties of the nuclear stability curve, including implications for radioactive decay.
- Types of radioactive decay, including conservation of charge and lepton number.
- Qualitative understanding of fusion and fission in the context of binding energy per nucleon.
- A brief history of the formation of the elements.

Acquired Skills

- Calculations of decay rates.
- Solving problems involving carbon dating.
- Calculations of binding energy per nucleon.

12.2 Properties of the Nucleus

12.2.1 The Constituents

Atoms consist of negatively charged electrons that are bound to a nucleus with equal and opposite charge.[1] The nucleus is made up of positively charged protons and

---

[1] An atom runs into a police station and exclaims: 'You have to help me. I have lost an electron'. 'Are you sure?' asks the officer on duty. 'I'm positive!' replies the atom.

© The Author(s), under exclusive license to Springer Nature Switzerland AG 2022   285
G. Kunstatter and S. Das, *A First Course on Symmetry, Special Relativity and Quantum Mechanics*, Undergraduate Lecture Notes in Physics,
https://doi.org/10.1007/978-3-030-92346-4_12

Table 12.1 Particle masses

Particle	kg	Atomic mass units(u)	MeV
Electron	$9.109\,39 \times 10^{-31}$	$5.485\,79 \times 10^{-4}$	0.511
Proton	$1.672\,62 \times 10^{-27}$	1.007276	938.28
Neutron	$1.674\,93 \times 10^{-27}$	1.008665	939.57
Hydrogen atom	$1.673\,53 \times 10^{-27}$	1.007825	938.783
Helium atom	$6.644\,66 \times 10^{-27}$	5.001506	3,727.38
Carbon atom	$19.926\,50 \times 10^{-27}$	12.000000	11,177.9

uncharged neutrons.[2] The number, Z of protons and electrons is largely responsible for the chemical properties of the atom as exhibited in the periodic table, at least for hydrogen-like atoms.

The nucleus is held together by the *strong force*, whose name reflects the fact that in order to bind positively charged protons together it must be stronger than the electrostatic force that repels them from each other. Although the strong force is indeed strong, it is short ranged, which means that protons will not bind together into nuclei unless they get close enough together. As we will see, the presence of neutrons that only feel the strong force and not the electrostatic force is vital for the formation of stable nuclei. Both the protons and neutrons in the nucleus have roles to play in creating the diversity of elements on which we rely in our daily lives.

First a review of some terminology. The number, Z, of protons in a nucleus is called the *atomic number*. The sum of the atomic number and the number of neutrons, N, namely $A = Z + N$ is called the *mass number* for the rather obvious reason that it determines the total mass of the nucleus. The notation $^{A}_{Z}X$ denotes a nucleus X with mass number A and atomic number Z. We will see in Sect. 12.4.1 that the total mass is not simply the sum of the masses of the constituents. Due to the equivalence of mass and energy (see Sect. 6.7.3) one must take into account the *binding energy* when calculating the total relativistic mass. This binding energy is due primarily to the strong force that holds the nucleons together and is equal to the difference between the mass of the bound nucleons and the sum of the their individual, unbound masses. The masses of the individual nucleons and three common nuclei are given in Table 12.1.

An atomic mass unit $1\,u = 1.661 \times 10^{-27}\,\text{kg} = 931.5\,\text{MeV}/c^2$ is defined as one twelfth of the mass of the Carbon atom. It is less than the mass of an individual nucleon because the mass/energy of the Carbon atom is reduced by the binding energy.

---

[2] A neutron orders a smoothy at a juice bar and asks the price. The server says: 'For you, no charge'.

12.2.2 Structure of Nucleus

In 1909, two years after leaving a position at McGill University in Montreal to go to Manchester, England, Ernest Rutherford performed an experiment in which he fired *alpha particles*, which are positively charged Helium nuclei consisting of two protons and two neutrons (see Sect. 12.3.3), through very thin layers of gold foil. The layers were so thin in fact that the chances of the alpha particles interacting with more than one atom as they passed through the gold was very small. To his surprise, Rutherford found that many of the α particles were deflected at very large angles by the atoms in the foil.

It was known at the time that the atom consisted of positive and negative charged particles. The reason that Rutherford was surprised by his results was that the commonly accepted model of the atom consisted of a uniform distribution of positive charge, as first postulated by J.J. Thompson in 1897 (see Fig. 12.1). Such a uniform distribution of positive charge would change the trajectory of the alpha particles only slightly and could not have produced the large angle deflections observed by Rutherford.[3] Large deflections would only happen if the alpha particles were hitting very dense, almost point-like concentrations of positive charge, much like in the collision of billiard balls.

Rutherford concluded that instead of being uniformly distributed throughout the thin gold foil, the positive charge was concentrated in a very small region at the center of the atoms. His data suggested that the positive charge was confined to a roughly spherical volume, V, of size:

$$V = V_0 A \tag{12.1}$$

where A is, as before, the number of nucleons. In this model, called the *Rutherford model* of the atom, the volume of the nucleus was expected to be the volume occupied by each nucleon times the number of nucleons. The radius of a nucleus with atomic number A would then be

$$r = r_N A^{1/3} \tag{12.2}$$

where r_N is the radius of one nucleon. Rutherford measured the radius of a nucleon to be roughly $r_N \sim 1\,fermi = 10^{-15}$ m. This is 5 orders of magnitude smaller than the radius r_0 of the atom, (i.e. the Bohr radius), which is about 1 Angstrom or 10^{-10} m. Thus, Rutherford showed that atoms are mostly empty space, much like our solar system, with negatively charged electrons orbiting the central nucleus at large distances, relative to r_N. This is illustrated in Fig. 12.2.

Equation (12.2) implies that the mass density of the nucleus is

---

[3] The electrons in the gold atoms were irrelevant because of their small mass compared to the alpha particle. A collision would knock the electron out of the atom without significantly altering the trajectory of the alpha particle.

Fig. 12.1 A schematic of the Thompson model of the atom in which the positive charge is continuously distributed throughout the atom, with negatively charged electrons scattered throughout like raisins in a loaf of raisin bread, or as more common in Thompson's day, plums in a plum pudding. The model was often called the *plum pudding model* of the nucleus

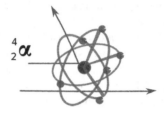

Fig. 12.2 A schematic of the Rutherford atom, not to scale. The alpha particles that hit the nucleus scatter through large angles, as observed in Rutherford's experiments

$$\rho = \frac{M}{V} \sim \frac{A \times 1\,u}{\frac{4\pi}{3}r_N^3 A} = \frac{3}{4\pi r_N^3}u/m^3 = 10^{17} kg/m^3 \qquad (12.3)$$

The mass number A has cancelled out of the final expression in Eq. (12.3). This means that the density of the nucleus is the same as the density of a single nucleon which is consistent with a picture of the nucleus consisting of A tightly packed but uncompressed spherical nucleons with little space between them. This is illustrated in Fig. 12.3.

Exercise 1 1. Osmium, $Z = 76$ is the densest stable element known. It has several stable isotopes, the most common being $^{192}_{76}Os$. The density of Os is about $23{,}000\,kg/m^3$. What fraction of a given volume V of Os is taken up by the nuclei? Is this consistent with what you would expect from the Rutherford model of the atom?

2. Compare your answer above to the fraction of the volume of the solar system taken up by the Sun. Consider the radius of the solar system to be the mean radius of Neptune's orbit, which is about 4.5 billion kilometers. The sun's radius is about 700,000 km.

12.2.3   The Nuclear Force

We now discuss in more detail the *strong force* that binds the nucleons close together without crushing them altogether. It turns out to be strongly repulsive at distances of the order of the nucleon size (about 1 fm) which gives them a hard, impenetrable

Fig. 12.3 Billiard ball
model of the nucleus
consisting of tightly packed,
incompressible protons
(blue) and neutrons (red).
(Color figure online)

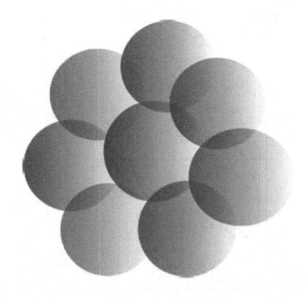

surface much like a billiard ball. The strong force is attractive out to about 3 fm
beyond which it is effectively zero. Hence it has a very short range, compared to the
electrostatic forces between two charged particles, such as protons. The situation is
summarized in Fig. 12.4 showing the potential energy as a function of distance for
a neutron-proton, and for a proton-proton. In the latter, we can see the effect of the
electrostatic repulsion at distances above 3 fm. Below that the strong force takes over.
Since the strong force only extends to a distance roughly equal to the size of two or
three nucleons, it acts more or less just between adjacent nucleons. The electrostatic
repulsion, being long range, acts on protons throughout the nucleus.

12.3 Radioactivity

12.3.1 Isotopes

Isotopes: are nuclei that have the same atomic number, Z, as each other, but different
numbers of neutrons and hence different mass number A. There may be several
isotopes for a given atomic number. Most of them are unstable and will fall apart into
more stable nuclei via radioactive decay. For a given Z, the presence of neutrons is
required in order to provide enough attractive strong forces in order to balance out
the repulsive electrostatic force between protons. If there are too few or too many
the nucleus will be unstable.

Not surprisingly, the greater the number of protons, the greater the electrostatic
repulsion between them and the greater the number of neutrons that are required to

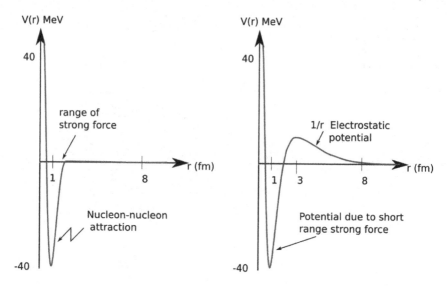

Fig. 12.4 The strong force between nucleons is attractive, but short range. The neutron-neutron and neutron-proton potential energy due to the strong force is shown in the figure on the left. In the case of proton-proton interactions, there is also a Coulomb repulsion, whose effect on the potential energy is shown on the right

provide stability. This is illustrated in Fig. 12.5: If the atomic number is too high, i.e. there are too many protons, then no amount of neutrons will be able to maintain stability and elements with $Z > 83$ are unstable against nuclear decay.

Notice that if the atomic number is small, then the stability curve approximately follows $Z = N$, so that roughly an equal number of neutrons and protons are required for stability. As the atomic number increases, more and more neutrons are required causing the stability curve to move upwards away from $Z = N$. If one moves below the stability curve by taking away a neutron or two, the strong attraction will not be sufficient and the nucleus will become unstable.

It is less obvious why adding neutrons and moving above the stability curve causes instability as well. This is due to a purely quantum force that has no classical equivalent. As we saw in Sect. 11.6.5, electrons are fermions that obey the *Pauli exclusion principle* which prevents more than one electron from occupying precisely the same quantum state. Neutrons and protons are also fermions, so they also obey the Pauli exclusion principle. In the context of atomic nuclei, the net effective is to make the neutrons want to stay as far away from each as possible. This is what gives rise to a quantum repulsive force between any two neutrons that are too close together. This force is called *Fermi repulsion*. Protons also experience Fermi repulsion but it is negligible compared to the electrostatic force. In brief, if you add too many neutrons to nuclei on the stability curve, Fermi repulsion makes the neutrons feel uncomfortable being so close to other neutrons. One of the excess neutrons will

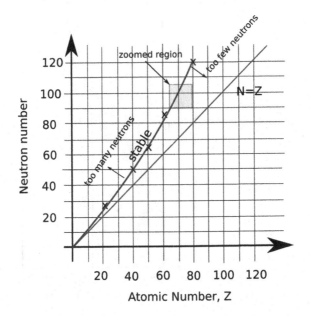

Fig. 12.5 The blue line indicates the region where stable nuclear isotopes can be found. In this region, the excess of neutrons provides just the right amount of short range strong forces to balance the long-range electrostatic repulsion between the protons. Below the line, there is an excess of protons, whereas above the line there is an excess of neutrons. As shown in Fig. 12.6, the stability line is not straight, but zig-zags, and there are a few isolated stable isotopes slightly off this line. (Color figure online)

spontaneously decay into a proton, an electron and a neutrino. This is called beta decay and will be discussed in Sect. 12.3.3.

Most stable nuclei have unstable isotopes that occur naturally but in small quantities. Unstable isotopes can also be created in the lab by inducing certain reactions. The graph in Fig. 12.6 shows the many isotopes, stable and unstable, of elements with Z between 65 and 80.

Some important features of Fig. 12.6

- The blue dots indicate stable isotopes.
- The line of stable isotopes is zig-zagged, not straight. It turns out that the most stable nuclei have even numbers of both protons and neutrons, while the least stable have odd numbers of both. You will notice that the longer stretches of the zig-zag occur for even values of protons or neutrons. The reason for this is that, just like the electrons in atomic shells (see Sect. 11.6.5), it is energetically favourable for the spin 1/2 protons and neutrons to pair up with like partners of opposite spin, in a type of nuclear "dance".
- The red region (top left) contains unstable isotopes with too many neutrons. They decay via negative beta decay.

Fig. 12.6 Unstable isotopes for $65 < Z < 80$. (Color figure online)

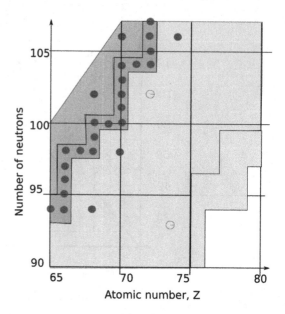

- The green region (middle) contains unstable isotopes with too many protons. They decay via positive beta decay.
- The yellow region (bottom right) and dots indicate unstable nuclei with too many protons that decay via alpha decay.

12.3.2 Neutrinos

Neutrinos are spin 1/2 particles like electrons, but with a mass that is less than 1 eV/c^2. In fact, for a long time after their discovery neutrinos were thought to be exactly massless. We know now that this is not the case, even though their mass has not yet been directly measured. There are upper and lower bounds on its value, however. The most recent laboratory bound was obtained in 2019 by a huge and very complicated laboratory experiment called Karlsruhe Tritium Neutrino (KATRIN).[4] They showed that the rest mass of the neutrino must be less than 1 eV/c^2.[5] This is 1/500,000th of the mass of the electron, which is somewhat awkward since it does not fit in very nicely with the other masses in the standard model.

Neutrinos have an interesting history, one that is steeped in mystery. In 1911 Lise Meitner and Otto Hahn performed an experiment showing that the electrons emitted in certain radioactive decays seemed to violate energy and momentum conservation.

---

[4] Aker et al. [1].

[5] The previous bound was 2eV.

In particular, their energy was not equal to the difference between the energy of the original nucleus and that of the emitted nucleus. For a while physicists thought that energy conservation and momentum conservation were violated in this process, a very disturbing prospect given its implications for time and space translation invariance of the relevant equations.

Wolfgang Pauli in 1930 was the first to make the radical proposal that the missing energy was emitted in the form of a new, unseen, uncharged low mass particle. Since the mechanism for nuclear decay was not understood at the time, it was thought that this new particle must somehow co-exist with the neutrons and protons in the nucleus prior to the decay.

It was not until 1933, a year after James Chadwick discovered the neutron, that Enrico Fermi developed a consistent, energy conserving theory of beta decay that included the emission of a neutrino, a name that he coined. The actual experimental discovery of the neutrino was made in the Nobel prize winning experiment by Clyde L. Cowan and Frederick Reines in 1956. Neutrinos do not interact very strongly with other forms of matter, making them very hard to detect. This accounted for the twenty year time lapse between prediction and discovery.

It wasn't too long after their discovery that neutrinos became the source of yet another deep puzzle called the *solar neutrino problem*. In 1968 Raymond Davis Jr. and John N. Bahcall performed an experiment designed to test the theory that nuclear fusion was the primary energy source in our Sun. If correct, this nuclear fusion should produce very many neutrinos ("solar neutrinos") some of which could be captured by their detector, which consisted of about 100,000 gallons of cleaning fluid, the primary ingredient of which was chlorine. Davis and Bahcall did detect solar neutrinos, as predicted, but the number detected was far fewer than predicted, in fact almost exactly one third the number predicted by the models. The final resolution to this problem was provided by measurements made at the Canadian Sudbury Neutrino Observatory (SNO) (http://www.sno.phy.queensu.ca/). The director of SNO-Lab, Art McDonald, shared the 2015 Nobel Prize in physics for this discovery (http://www.cbc.ca/news/technology/nobel-prize-physics-2015-1.3258178) with Takaaki Kajita, the director of the Japanese lab that also played a vital role in resolving the puzzle. They proved that neutrinos came in three flavours, much like a scoop of Neapolitan ice cream, which consists of vanilla, chocolate and strawberry.[6] The Sun was emitting the right number of neutrinos, but the neutrino detectors were only sensitive to the chocolate ones, so they saw only one third of the neutrinos passing through them.

After that brief introduction to neutrinos, we now go on to discuss the different types of radioactive decay that unstable nuclei undergo.

---

[6] The actual names of the flavours given to the neutrinos by particle physicists were: *the electron neutrino*, which was emitted in beta decay, *the muon neutrino* and *the tau neutrino*. The last two are emitted in more exotic decays involving different spin one half, charged particles, called the muon and the taon.

12.3.3   Types of Radioactive Decay

Beta decay

Beta decay refers in general to a radioactive process involving an electron or its anti-particle, the positron. The name derives from the fact that the electron was once called the "beta particle". Nuclei in the red regions of Fig. 12.6 decay via a form of beta decay that changes an excess neutron into a proton by emitting an electron and an anti-neutrino:

$$n \rightarrow p + e^- + \bar{\nu}_e \qquad (12.4)$$

This is called beta minus decay, because it emits negatively charged electrons. The mass number of the nucleus stays the same after the decay, but the atomic number increases by one.

An example of beta-minus decay is that of a radioactive isotope of phosphorus decaying to sulphur:

$$^{32}_{15}P \rightarrow ^{32}_{16}S + e^- + \bar{\nu}_e \qquad (12.5)$$

$\bar{\nu}_e$ is the electron anti-neutrino, i.e. the anti-particle of the electron neutrino. It has no electric charge, so that the total charge before and after the decay is the same, as required by charge conservation. The fact that it is an anti-neutrino and not a neutrino has to do with another conservation law, namely the conservation of *lepton number* which is an important property of weak interactions. Electrons have lepton number 1, as do neutrinos, while protons and neutrons have zero lepton number. Anti-neutrinos have lepton number -1.

Isotopes in the green region of Fig. 12.6 have too few neutrons and decay via a form of beta decay that turns a proton into a neutron by emitting a positively charged positron e^+ (anti-electron) and a neutrino:

$$p \rightarrow n + e^+ + \nu_e \qquad (12.6)$$

This keeps the mass number the same and decreases the atomic number. An example of beta plus decay is copper to nickel:

$$^{64}_{29}Cu \rightarrow ^{64}_{28}Ni + e^+ + \nu_e \qquad (12.7)$$

This process can only happen inside a nucleus. Free protons cannot decay into a neutron in this way.

Exercise 2 Are charge and lepton number conserved by the decay in Eq. (12.7)? Justify your answer.

Beta minus and *beta plus* decay are examples of *weak interactions*, which play a big role in the thermonuclear reactions inside stars. As we have seen, conservation laws are vital to determining what decays are allowed. The need to conserve lepton number provides the reason for the presence of the essentially massless neutrino (or anti-neutrino) in beta decay. The presence of extra conservation laws suggests the existence of new symmetries governing the weak interactions. A full understanding of these symmetries requires a detailed study of the theory of the electroweak interactions that describes both electromagnetic and weak interactions within a unified framework.

Alpha decay

In Fig. 12.6 the two light dots and the shaded region in the bottom right quadrant (all yellow in colour version) represent unstable nuclei that have too many protons compared to the number of neutrons. The most efficient way to dispense with these is via *alpha decay*, in which the nucleus emits a Helium nucleus (alpha particle) containing two protons and two neutrons. In alpha decay, the atomic number, Z, of the nucleus decreases by two while the mass number, A, decreases by 4:

$$_Z^A X \rightarrow _{Z-2}^{A-4} Y + _2^4 He \qquad (12.8)$$

X is called the *parent nucleus*, while Y is called the *daughter nucleus*. One example is the Uranium to Thorium decay.

$$_{88}^{226} U \rightarrow _{86}^{222} Th + _2^4 He \qquad (12.9)$$

The kinetic energy of alpha particle in such decays is between about 4 and 5 MeV.

Exercise 3 Is the decay in Eq. (12.9) relativistic? Justify your answer.

As discussed in Sect. 10.5.3 alpha decay as represented by Eq. (12.8) is an example of fission, in which one larger nucleus decays into two less massive nuclei with a release of energy (see also Sect. 12.4). A relatively simple quantum mechanics tunnelling calculation (*Gamow's theory of alpha decay*) correctly produces the lifetime of such nuclei as a function of their energy.

12.3.4 Decay Rates

Radioactive decay is a form of quantum mechanical tunnelling. One can obtain valuable quantitative information about decay rates as follows. Quantum mechanics predicts that the probability dP of an atom decaying during some time interval dt is proportional to dt:

$$dP = \lambda dt \qquad (12.10)$$

where λ is called the *decay constant*, or *disintegration constant*. If one has a large number, N of radioactive, unstable nuclei, each with probability dP of decaying in time dt, then the number of nuclei dN that are "lost" to radioactive decay in time dt is the number of nuclei present at time t times the probability of decay:

$$dN = -N(t)dP = -\lambda N(t)dt \tag{12.11}$$

Assuming that λ is constant in time, the above can be integrated to give the number of nuclei present (i.e. not yet decayed) as a function of time:

$$N(t) = N_0 e^{-\lambda(t-t_0)} \tag{12.12}$$

where N_0 is the number of nuclei present at initial time t_0. The *decay rate*, or *activity* $R(t)$ of a given sample of $N(t)$ radioactive nuclei is the number that decay per unit time:

$$R(t) := \left| \frac{dN}{dt} \right| = \lambda N(t) = \lambda N_0 e^{-\lambda(t-t_0)} \tag{12.13}$$

The activity is usually given in one of two units:

- 1 Becquerel (Bq) = 1 decay per second
- 1 Curie (Ci) = 3.7×10^{10} decays per second = 3.7×10^{10} Bq.

The *half life* $T_{1/2}$ of a radioactive element is defined as the time it takes for the number $N(t)$ to decrease by a factor of 2. Thus:

$$N(t + T_{1/2}) = \frac{N(t)}{2}$$

$$\rightarrow N_0 e^{-\lambda(t+T_{1/2}-t_0)} = \frac{1}{2} N_0 e^{-\lambda(t-t_0)} \tag{12.14}$$

Because of the exponential nature of this relationship, everything cancels on the two sides of the above equation except:

$$e^{-\lambda T_{1/2}} = \frac{1}{2} \tag{12.15}$$

which determines the half life in terms of the decay constant:

$$T_{1/2} = \frac{\ln(2)}{\lambda} \tag{12.16}$$

Using the above, one can express $N(t)$ in terms of the half-life as follows:

$$N(t) = N_0 \left(\frac{1}{2} \right)^{\frac{t}{T_{1/2}}} \tag{12.17}$$

Exercise 4 Derive Eq. (12.17) from (12.16).

Example 1 A typical banana contains 600 mg of potassium. About 0.012% of this occurs as the radioactive isotope ^{40}K, with half life of 1.25×10^9 years. What is the activity of such a banana, in Bq?

Answer: First, we need to find out the number of radioactive ^{40}K nuclei in the banana. The mass of a ^{40}K atom is:

$$M(^{40}K) = 39.96u$$
$$= 39.96u \times 1.66 \times 10^{-24} \text{ gms/u}$$
$$= 6.6 \times 10^{-23} \text{ gm} \tag{12.18}$$

So the number of radioactive nuclei is

$$N(^{40}K) = \frac{1.2 \times 10^{-4} \times .6 \text{ gms}}{6.6 \times 10^{-23} \text{ gm/nucleus}}$$
$$= 1.1 \times 10^{18} \text{ nuclei} \tag{12.19}$$

We need to find the decay constant λ in seconds^{-1}:

$$\lambda = \frac{\ln(2)}{T_{1/2}}$$
$$= \frac{\ln(2)}{1.25 \times 10^9 \text{ yrs} \times 3.15 \times 10^7 \text{ sec/yr}}$$
$$= 1.76 \times 10^{-17} \text{seconds}^{-1} \tag{12.20}$$

The activity is then:

$$R = 1.1 \times 10^{18} \text{ nuclei} \times 1.76 \times 10^{-17} \text{ seconds}^{-1}$$
$$= 19 \text{ Bq} \tag{12.21}$$

Note that ^{40}K primarily decays by beta minus decay, emitting an electron with maximum energy of 1.33 MeV. Thus, assuming you eat the whole banana the amount of energy that you can absorb per second via the beta particles is about (19×1.33) MeV = 25 MeV = 4×10^{-12} Joules. The benefits of eating the banana therefore outweigh the potential disadvantages.

12.3.5 Carbon Dating

All life on Earth is carbon-based, which means that carbon, $^{12}_{6}$C, is the fundamental element. A radioactive isotope of $^{12}_{6}$C, namely $^{14}_{6}$C, exists in the air in the following proportion:

$$\frac{N(^{14}_{6}C)}{N(^{12}_{6}C)} \sim 1.3 \times 10^{-12} \tag{12.22}$$

This ratio is kept roughly constant by the cosmic ray events in the upper atmosphere that create $^{14}_{6}C$ at roughly the same rate as are destroyed by decays. This proportion of $^{12}_{6}C$, to $^{14}_{6}C$ is thought not to have changed significantly on Earth for many years.

Any living organism, plant or animal, continuously exchanges molecules with the air thereby replenishing its content of $^{14}_{6}C$. Once the organism dies, this exchange ceases, and the amount of $^{14}_{6}C$ relative to the stable $^{12}_{6}C$ decreases as the former decays:

$$N(^{14}_{6}C)(t) = N(^{14}_{6}C)(t_0)e^{-\lambda(t-t_0)} \tag{12.23}$$

where t_0 is the time of death. Thus after the death of the organism, the ratio of $^{14}_{6}C$ to $^{12}_{6}C$ in the object decreases exponentially with time according to:

$$\frac{N(^{14}_{6}C)(t)}{N(^{12}_{6}C)(t_0)} \sim 1.3 \times 10^{-12}e^{-\lambda(t-t_0)} \tag{12.24}$$

This in turn gives an expression for the activity of $^{14}_{6}C$ as a function of time:

$$R = \lambda N(^{14}_{6}C)(t) = \lambda N(^{12}_{6}C)(t_0) \times 1.3 \times 10^{-12}e^{-\lambda(t-t_0)} \tag{12.25}$$

This gives us a remarkably simple and relatively accurate way of dating very old fossils. By measuring the activity of the fossil, and knowing the amount of stable carbon it contains, one can determine the time interval, $t - t_0$, that elapsed since the death of the corresponding animal or plant.

Exercise 5 In 2019, paleoanthropologists showed that a skull found in the 1970's in southern Greece was in fact the oldest human fossil found outside of Asia. The skull was proven by radioactive dating to be 210,000 years old. If the skull was composed primarily of carbon and weighed 2 kg, how many $^{14}_{6}C$ atoms did it contain?

12.4   Fission and Fusion

12.4.1   Binding Energy

Special relativity tells us that mass and energy are equivalent. The energy of a particle in its rest frame is given by $E = m_0 c^2$. Conversely, the total energy of a system of bound particles, such as the neutrons, protons and electrons in an atom determines the rest mass of that system, which is the mass as measured in the center of mass rest frame. We know that for such a system of particles to be bound together by conservative forces, the total potential energy of the particles must be less when they are together than when they are far apart. This in turn implies that the mass/energy of the bound nucleons must be less than the total mass of the individual particles. The difference is called the *binding energy*. More specifically, the binding energy of the nucleons in the nucleus of a neutral atom is given by the difference between the mass of Z hydrogen atoms (electron-proton pairs) plus N free neutrons, and the actual mass of the atom. The formula for binding energy is:

$$E_B = \left[Z \cdot M(H) + N \cdot M_N - M(^A_Z X) \right] \qquad (12.26)$$

where the hydrogen/neutron masses can be given either in MeV/c^2 or atomic units u to yield the binding energy in corresponding units.

Note that Eq. (12.26) applies to the neutral atom as a whole and takes into account the presence of the electrons. The electrostatic binding energy of the Z electrons to the Z protons is different when they are in an atom with atomic number Z than when they are bound in Z individual hydrogen atoms. This does not, however, contribute significantly to the total binding energy because the other binding energies are so large.

As defined, the binding energy E_B must be positive in order for the atom to be stable.

12.4.2 Binding Energy per Nucleon

The ratio E_B/A is the *binding energy per nucleon* of a given nucleus. The graph in Fig. 12.7 shows how this quantity changes with mass number A.

A higher binding energy per nucleon implies that the nucleons are more tightly bound in the nucleus, so that you would have to supply more energy to break it apart. Thus, nuclei with higher E_B/A are more stable. Nickel ($A = 62$) has the highest binding energy per nucleon. This has important consequences. Since stability is a good thing, nuclei have a tendency to change into configurations that are more stable. This can happen in two ways:

- **Fission**: Fig. 12.7 shows that for $A > 62$ it is energetically favourable for nuclei to split into smaller nuclei that have larger binding energy per nucleon. This process is called fission and can occur naturally (without external input of energy) for nuclei with large mass numbers, such as Uranium or Plutonium. This process releases energy, and is what underlies nuclear weapons and nuclear reactors.
- **Fusion**: For nuclei with $A < 62$ it is energetically favourable to join together to form nuclei with large mass number, so as to produce an element with larger binding energy per nucleon. This process, called fusion, has the potential to release huge amounts of energy. It is very difficult to realize except at the very high temperatures and pressures that occur inside the Sun. The Sun is powered by the release of energy due to fusion reactions. The heavy elements (eg iron and higher) were produced by the extreme conditions (high temperatures and pressures) that occur during supernova explosions.

Fig. 12.7 Binding energy per nucleon. Nuclei with higher binding energy per nucleon are energetically favoured. In this context nickel with mass number $A = 62$ is optimal, so nuclei with higher mass number undergo fission to produce daughter nuclei with lower A, while nuclei with lower mass number undergo fusion in order to have higher mass number A

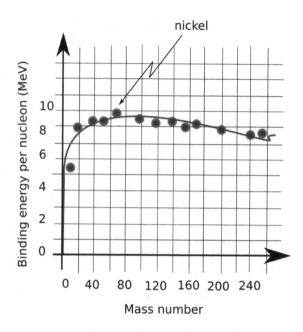

12.4.3 Formation of Elements

We have strong observational evidence from the CMBR and other observations that the Universe started out 14 billion years ago as a very small and very hot fireball before expanding into what exists today (see Sect. 7.8). The initial conditions present at that first instant in time determine everything that we observe, including the current relative density of the various types of elements. This process, which we will now briefly describe, is fairly well understood and is summarized schematically in Fig. 12.8. We start by presenting a chronology of the key events in the history of the Universe that led to the creation of the elements that we see on Earth today.[7]

History of the Universe

1. Time as we know it began with the hot Big Bang about 14 billion years ago. The Universe has been expanding and cooling (due to a gravitational redshift effect) ever since.
2. During the first 10^{-43} s the temperature of the Universe was above 10^{32} K and we know virtually nothing about the physics at such high temperatures. This is presumably the era of quantum gravity.
3. After 10^{-32} s the Universe is thought to have undergone an "inflationary period" of very rapid expansion, during which its size increased by a factor of 10^{30}. After the expansion, the Universe became a soup of elementary particles (quarks, gluons, leptons) whose physics we think we understand to be described by the "standard

---

[7] See Sect. 7.8 for more discussion of Cosmology.

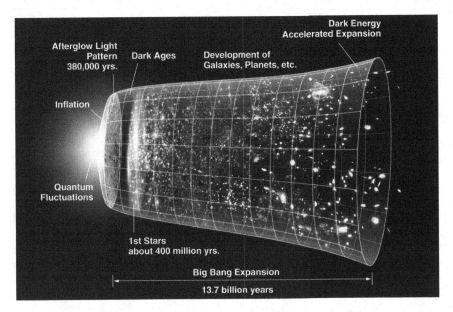

Fig. 12.8 Credit: NASA/WMAP Science Team—Original version: NASA; modified by Ryan Kaldari

model" of the strong, electromagnetic and weak interactions. The Universe at this stage was still too hot for protons and neutrons to form.

4. At around 10^{-4} s, the Universe had cooled down enough for more standard types of matter such as protons and neutrons to form. Particles of matter and anti-matter collided and annihilated each other, leaving us with the slight excess of matter over anti-matter that persists to this day. The reason for this excess is not as yet known.

5. At 1 min, the Universe was cool enough to start forming light nuclei, mostly hydrogen with a bit of helium. The Universe was too hot for the electrons to stick to the nuclei to form neutral atoms, so what existed was a form of very hot, electrically charged plasma that continually absorbed and emitted photons. This went on for about 400,000 years. During this period the photons were in thermal equilibrium with the plasma so the Universe was essentially a black body filled with radiation at fixed temperature.

6. When it was 400,000 years old, the Universe cooled down sufficiently to allow electrons and nuclei to come together to form neutral atoms that no longer absorbed or emitted radiation. The matter *decoupled* from the radiation, which for the most part continued to propagate throughout the Universe without further interactions. The expansion of the Universe caused the radiation to undergo a *gravitational redshift* and hence cool down. This residual radiation from the *decoupling era* shortly (on a cosmological scale) after the Big Bang is what we see today as the Cosmic Microwave Background Radiation. The CMBR literally

provides a snapshot of what the Universe looked like almost 14 billion years ago when the matter and radiation decoupled. See the Fig. 8.6 in Sect. 8.3.5.

7. At 150 million years the Universe cooled further and the gas (primarily hydrogen) started forming clumps that eventually ignited to form stars that converted hydrogen into helium as they burned. As the stars aged their cores got more and more dense so that fusion into heavier elements became possible. Elements with mass number up to $A \sim 62$ were now able to form. Recall that nickel, at $A = 62$ has the highest binding energy per nucleon.

8. Once stars got too old and had sufficiently depleted their thermonuclear fuel, they formed *supernovae*, huge explosions that released enough energy to produce elements even heavier than nickel. In addition to forming the heavy elements, supernova explosions also helped spread them throughout the Universe. The matter from which we are all made is thought to have mostly formed inside stars and distributed via supernovae. This process continues to this day.

9. Today humans make observations from their (relatively) infinitesimal home on Earth that enable them to deduce the 14 billion year history of our Universe with remarkable accuracy.

Reference

1. M. Aker et al., Improved upper limit on the neutrino mass from a Direct Kinematic Method by KATRIN. Phys. Rev. Lett. 123, 221802 (2019). (KATRIN Collaboration)

Chapter 13
Mysteries of the Quantum World

13.1 Learning Outcomes

Conceptual

- Physical interpretation of spin.
- Complex vector representation of spin 1/2 states.
- Bra-Ket notation.
- Definition and significance of entanglement.
- Qualitative understanding of the Einstein-Podolsky-Rosen (EPR) conundrum.
- Implications of Bell's inequalities for the EPR conundrum.
- Qubits and quantum gates.
- Why quantum computers are useful.

Acquired Skills

- Calculating inner products and probabilities for spin 1/2 states.
- Calculating expectation values of spin observables.

13.2 What Is Real?—A Quantum Conundrum

One of Einstein's objections to quantum mechanics was the idea that the physical observables such as the position and momentum of a particle are inherently uncertain. This was the view subscribed to by most other physicists of the day, including its most avid proponent, Niels Bohr. Einstein believed that quantum mechanic's inability to predict both the position and momentum with certainty was a sign that quantum mechanics was not a complete description of reality. There must, he reasoned, exist underlying truths (*hidden variables*) that would ultimately be revealed by a more

© The Author(s), under exclusive license to Springer Nature Switzerland AG 2022   303
G. Kunstatter and S. Das, *A First Course on Symmetry, Special Relativity and Quantum Mechanics*, Undergraduate Lecture Notes in Physics,
https://doi.org/10.1007/978-3-030-92346-4_13

complete theory. In a seminal paper,[1] Einstein, along with co-authors, Boris Podolsky and Nathan Rosen (EPR) attempted to rigorously prove this view by way of a thought experiment. They used one basic definition, and one physical assumption:

- **Definition:** An *element of reality* attached to a system is a quantity that can be predicted with absolute certainty (in principle at least) without interacting in any way with that system. EPR wished to associate such elements of reality with actual physical attributes possessed by the system, irrespective of whether or not we measure them.

- **Assumption:** The elements of reality of a system can only be affected by *local interactions*. By local they meant interactions that obey special relativity. Another way to state this is that events outside the past light cone of a system cannot affect the elements of reality of that system.

EPR considered an experiment involving two particles, which we will call particle 1 and particle 2, that were created in a quantum state with zero total momentum in the lab frame. Quantum mechanics requires the position and momentum of each individual particle to obey the uncertainty principle. If the position of particle 1, for example, is known with certainty, then its momentum is completely unknown. The same is true of particle 2.

Momentum conservation requires the total momentum of the two particles to be conserved. If the total momentum is zero when the particles are emitted, it must be zero at all subsequent times unless one or both particles are subjected to an external force. In terms of equations:

$$p_1 + p_2 = 0 \leftrightarrow p_1 = -p_2 \qquad (13.1)$$

where p_1 is the momentum of particle 1 and p_2 is the momentum of particle 2. This means that any time t after the particles are emitted, one can hypothetically determine the momentum p_2 of particle 2 by measuring the momentum p_1 of particle 1, and vice versa. This requires no local interaction with particle 2.

EPR go on to state that immediately after such a measurement of the momentum of particle 1, one is able to predict the momentum of particle 2 with certainty without in any way interacting with it. The momentum of particle 2 therefore satisfies the EPR definition of an element of reality. EPR then observed that one can in principle wait until the particles are separated by light years before doing the measurement on particle 1. The assumption of locality implies that such a measurement on particle 1 could not possibly affect the elements of reality of particle 2. Thus if the momentum of particle 2 is an element of reality after the measurement on particle 1, then it must also have been an element of reality before the measurement. The inevitable conclusion according to EPR was that in such a state the momentum of particle 2 is an element of reality whether or not the momentum of particle 1 is measured. The same holds for elements of reality of particle 1 before and after measurements done on particle 2.

---

[1] Einstein et al. [1].

EPR then explicitly constructed a quantum state of a pair of particles for which not only is the sum of their momentum zero, but the difference in their position is also known with certainty to be constant, say Δ:

$$x_1 - x_2 = \Delta \leftrightarrow x_1 = x_2 + \Delta \qquad (13.2)$$

such a state is possible because, although x_1 is complementary to p_1, and x_2 is complementary to p_2, the particular combinations $x_1 - x_2$ and $p_1 + p_2$ are not complementary. One can prove that $x_1 - x_2$ commutes with $p_1 + p_2$:

$$[x_1 - x_2, \; p_1 + p_2] = [x_1, \; p_1] - [x_2, \; p_2] = i\hbar - i\hbar = 0 \qquad (13.3)$$

Thus the combinations of observables $x_1 - x_2$ and $p_1 + p_2$ can both have definite values without violating the uncertainty principle. [2] The wave function for the quantum state they considered can be written as follows:

$$\psi(x_1, x_2) = \frac{1}{2\pi} \int_{-\infty}^{\infty} dp \, e^{i(x_1 - x_2 + \Delta)p/\hbar}$$
$$= \frac{1}{2\pi} \int_{-\infty}^{\infty} dp \, e^{i(x_1 + \Delta)p/\hbar} e^{-ix_2 p\hbar} \qquad (13.4)$$

This is not a state with definite x_1 or x_2. Both particles are in a linear combination (actually an infinite sum given by the integral in Eq. (13.4)) of plane waves, each of the form $e^{i(x_1 - x_2 + \Delta)p}$. As discussed in Appendix 15.2.3, the right hand side of Eq. (13.4) is just the Fourier representation of the *Dirac delta function* $\delta(x_1 - x_2 + \Delta)$. It is infinite when $x_1 - x_2 + \Delta = 0$ and zero otherwise. If one integrates $\delta(x_1 - x_2 + \Delta)$ over either x_1 or x_2 the result is unity as long as the integration covers $x_1 - x_2 + \Delta = 0$. The delta function form of the wave function guarantees that the difference between the positions of the two particles is fixed to be Δ, even though the state of each particle is a linear combination of plane waves with uncertain position.

To confirm that the EPR state Eq. (13.4) has definite total momentum, i.e. $p_1 + p_2 = 0$, we note that the integral is an infinite sum over terms, each of which is a product of pure waves with definite wave number. The first factor in the integrand of the second line in Eq. (13.4) is a pure wave that depends only on the location of particle one, with wave number $k = p/\hbar$, while the second factor is a pure wave that depends only on the location of particle 2, with wave number $k = -p/\hbar$. Each of these factors is a quantum state of fixed moment p and $-p$ respectively. A measurement of the momentum of particle 1, say, will yield some value of $p_1 = \tilde{p}$ (all are equally probably), but it will also *collapse the wave function* so that a subsequent measurement of the momentum of particle 2 will necessarily yield $p_2 = -\tilde{p}$.

---

[2] This would be a very strange combination of attributes for a pair of classical particles because zero total momentum would suggest that they were flying apart, not staying at constant distance. The EPR particles are in a highly quantum state that has no classical analogue.

Suppose one chooses to measure the position of particle 1 instead of its momentum. After the measurement, the position of particle 2 will always be measured as $x_2 = x_1 - \Delta$. According to the EPR definition, the position of particle 2 is again an element of reality after the measurement on particle 1. Evoking the assumption of locality one concludes that if the position of particle 2 was an element of reality after the measurement of the position of particle 1, it must also have been an element of reality before the measurement. Thus, the position of particle 2 is also an element of reality irrespective of what measurement is made on particle 1.

The above arguments led EPR to the conclusion that both the position of particle 2 and its momentum are simultaneously elements of reality for such a quantum state of two particles, irrespective of whether or not they are measured.[3] Since the quantum uncertainty principle does not allow particle 2 to simultaneously have both a definite position and momentum, EPR observed that quantum mechanics does not in this case provide a description of all elements of reality. They concluded that quantum mechanics must therefore be incomplete. Their hope was that there exists a more complete theory that describes both x_2 and p_2 in terms of some unknown (hence "hidden") variables.

The EPR argument is logically consistent. Their conclusion follows rigorously from the definition and assumption. Nonetheless, as described in Sects. 13.3 and 13.5, experiments have been done whose results are consistent with the predictions of quantum mechanics but **prove that the EPR conclusion is in general incorrect**. For these experiments no hidden variables theory that obeys the EPR locality assumption can possibly reproduce the confirmed predictions of quantum mechanics. Although quantum mechanics is not complete in the sense that EPR would like it to be, no such complete and local theory exists.[4] Not only is this startling, but the non-locality turns out to have very practical consequences for secure communications and quantum computation, as we will see in Sect. 13.7.

13.3   Bell's Theorem and the Nature of Quantum Reality

In the mid-1960s John Bell showed conclusively that there are physical situations involving two particles in which the predictions of quantum mechanics are not consistent with the EPR notion of a local reality. If one tries to adopt the EPR definition of an element of reality in such situations, quantum mechanics requires that the values of the elements of reality of particle 2 depend non-locally on the choice of measurement made on particle 1, no matter how great the distance between them.

We will now present a straightforward thought experiment illustrating what goes wrong when one tries to assign values to variables that are not allowed by quantum mechanics. The following thought experiment is a variation of an experiment first

---

[3] Of course the entire argument can be repeated for the elements of reality of particle 1 as revealed by measurement of the position and momentum of particle 2.

[4] A useful link on this subject can be found at https://plato.stanford.edu/entries/qt-epr/.

described by David Bohm in his textbook *Quantum Theory*, Prentice Hall, 1951, pp. 614–619 and then elucidated in a beautiful and accessible paper by D. Mermin, "Bringing home the atomic world: quantum mysteries for everyone", American Journal of Physics, 1981. The version presented here is closest in spirit to that of Kunstatter and L. Trainor in the paper "For Whom the Bell Tolls", American Journal of Physics 52, 598, 1984.

Suppose you are in a laboratory in Winnipeg, which is located at the geographic center of Canada and less than 500 km from the geographic center of North America. You have 12,000 numbered pairs of sealed envelopes each containing three pieces of paper, one green, one blue and one red. Each piece of paper can have written on it either a plus sign (+) or a minus sign (−) (see Fig. 13.1). The experiment involves sending one of the envelopes in each pair west to Alice in Vancouver, British Columbia and the other to Bob in Halifax, Nova Scotia, at opposite ends of the continent. Only one pair of envelopes is mailed per day so that there is no chance of confusion as to which pair a given envelope belongs.[5]

As soon as she receives an envelope, Alice opens it, chooses one of the three pieces of paper **at random**, looks at the sign, records the envelope number, the colour of paper she chose, and whether it contained a plus or minus sign. As soon as she looks at the paper, the envelope and the two other pieces of paper self-destruct, much as in the old Mission Impossible TV shows and movies, so what may or may not have been written on them remains forever unknown. Alice then waits to receive the next envelope and repeats the above until she has looked at all 12,000.[6] Bob does exactly the same, also randomly choosing a piece of paper in each envelope to look at.

The quantum experiment on which this thought experiment is based involves electrons and requires a deeper understanding of electron spin than we have at the moment. This experiment will be discussed in Sect. 13.5, where the predicted statistics for the various outcomes are derived. Here we will just use the corresponding quantum statistics for our thought experiment. They can be summarized as follows:

- If Alice and Bob look at the same coloured piece of paper, they should always see opposite signs.
- If Alice and Bob look at different colours, then they should see opposite signs exactly one quarter of the time. The one quarter is a prediction of the quantum version. A classical experiment would predict one half for the odds of seeing different signs on different coloured pieces of paper. Note that we are not interested in which specific signs Alice and Bob see, only whether or not they are the same for their given choice of colours.

The reason for sending the envelopes far apart before they are opened is to ensure that there can be no collusion between Alice and Bob that might affect the statistics. For example, Alice can't open her envelope first and then somehow make sure that

---

[5] We assume here, somewhat unrealistically that it takes less than one day for each envelope to be delivered.

[6] This is of course another idealization, since the experiment as described would take about thirty years, by which time the research funding would have dried up.

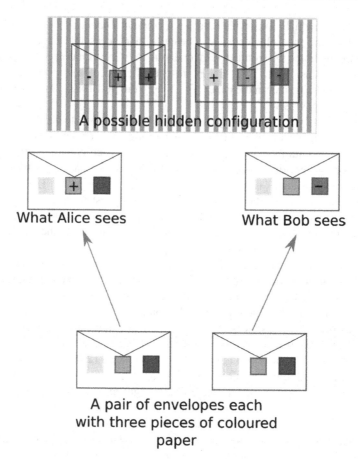

A possible hidden configuration

What Alice sees   What Bob sees

A pair of envelopes each
with three pieces of coloured
paper

Fig. 13.1 Example showing one possible pair of envelopes in the experiment. Alice and Bob each observe only one piece of paper before the envelopes are destroyed, so that the remainder of the configuration is not known. At the top is one of two possible configurations that are consistent with their observations. See Table 13.2 for all possible configurations for the two envelopes

Bob sees the opposite sign. There is of course a loophole, given that it is possible to travel faster than the speed of a mailed letter. This can in principle be circumvented by sending the envelopes light years apart in opposite directions before opening, lengthening the time of the experiment considerably. For simplicity, we just assume that both Alice and Bob are honest.

Once all the envelopes have been examined and contents recorded, Alice and Bob return to Winnipeg in order to tabulate the results, which are summarized in Table 13.1.

Let us now try to interpret the results. As predicted by theory, whenever Alice and Bob looked at the same colour, the signs were 100% correlated: if Alice saw a minus sign, Bob saw a plus and vice-versa. We can interpret this in terms of the

Table 13.1 Data collected by Alice and Bob

Colours	Total number	No. same	No. opposite
Green/Green	2000	0	2000
Blue/Blue	2000	0	2000
Red/Red	2000	0	2000
Green/Blue	2000	1500	500
Green/Red	2000	1500	500
Blue/Red	2000	1500	500

EPR argument in Sect. (13.2) as follows: given that the predicted 100% correlations have been confirmed by experiment, in future experiments when Alice looks at the red piece of paper and sees a minus sign, she can predict with certainty that if Bob looks at red in the corresponding envelope, he will see a plus sign. Given that her measurement cannot affect the elements of reality inside Bob's envelope, that plus sign must be real in that it exists independently of whether or not Bob or Alice look at the red paper in their respective envelopes. The same can be said for the signs written on the other two pieces of paper in each envelope in every pair. Using EPR's language: Given the experimental results, the three signs written on all three pieces of paper in each pair of envelopes are elements of reality whether or not they are observed, or even predictable in principle.

The above is consistent with our intuition: We naturally assume that each envelope leaves Winnipeg containing three coloured pieces of paper that have written on them a plus or minus sign. Bob's values should not be changed by what Alice observes and vice versa. Of course, we only know the values for one piece of paper in each envelope because the others were burned immediately after the observation. They are unknown to Alice and Bob and in fact unknowable after the envelope is burned, but this does not deter us from concluding that they did exist.

We can learn more about the contents of the envelopes by examining the statistics in Table 13.1 in the cases that Alice and Bob look at different coloured pieces of paper. We first consider only one of the envelopes in each pair. There are exactly eight possible ways that two signs can be distributed among the three pieces of paper. Given the correlations observed in the first three rows of Table 13.1, once the configuration of pluses and minuses is chosen for one of the envelopes, the other is uniquely determined to be the opposite. Thus, there are only eight possible configurations for both envelopes in each pair. They are listed in Table 13.2.

For later convenience the configurations have been numbered. For example $N(3)$ refers to the number of pairs of envelopes in which configuration 3 occurred, etc. Because the envelopes are burned after one colour is observed, it is impossible to know the values of the $N(i)$. We nonetheless do have some information that we can use. First, we assume that values for all $N(i)$ exist independently of what we choose

Table 13.2 The eight possible configurations for the pieces of paper in each pair of envelopes

Config.	Alice			Bob			Times observed
	Green	Blue	Red	Green	Blue	Red	
1	+	+	+	−	−	−	N(1)
2	+	+	−	−	−	+	N(2)
3	+	−	+	−	+	−	N(3)
4	+	−	−	−	+	+	N(4)
5	−	+	+	+	−	−	N(5)
6	−	+	−	+	−	+	N(6)
7	−	−	+	+	+	−	N(7)
8	−	−	−	+	+	+	N(8)

to measure because we have determined from the correlations above the contents of each envelope are elements of reality prior to any measurement. Second, the numbers on the right hand side of the table must, by definition, add up to the total number of pairs of envelopes in the experiment:

$$N(1) + N(2) + N(3) + N(4) + N(5) + N(6) + N(7) + N(8) = N_{tot} \quad (13.5)$$

Now let's look at the statistics tabulated by Alice and Bob when they looked at different colours. For green/blue, for example, they found as predicted that different signs were observed one quarter of the time. We see from Table 13.2 that the only configurations for which Alice and Bob see opposite signs for blue and green are $N(1)$, $N(2)$, $N(7)$ and $N(8)$, so whatever the number of times the individual configurations appeared, in order to be consistent with the experimental data in Table 13.1 it must be true that

$$N(1) + N(2) + N(7) + N(8) = \frac{N_{tot}}{4} \quad (13.6)$$

Recall that specific configurations are assumed to exist inside each pair of envelopes irrespective of which measurements are made by Alice and Bob.

Repeating the above for green/red, we see that configurations $N(1)$, $N(3)$, $N(6)$ and $N(8)$ give opposite signs for Alice and Bob for these colours so that:

$$N(1) + N(3) + N(6) + N(8) = \frac{N_{tot}}{4} \quad (13.7)$$

Finally, for blue/green we must have:

$$N(1) + N(4) + N(5) + N(8) = \frac{N_{tot}}{4} \quad (13.8)$$

Here comes the really interesting part. Let's add up both sides of Eqs. (13.6)–(13.8). This yields:

$$N(1) + N(2) + N(3) + N(4) + N(5) + N(6) + N(7) + N(8) + 2N(1) + 2N(8) = \frac{3N_{tot}}{4} \quad (13.9)$$

Using Eqs. (13.5) in (13.9) we see that

$$N_{tot} + 2N(1) + 2N(8) = \frac{3N_{tot}}{4}$$

$$\rightarrow \quad 2(N(1) + N(8)) = -\frac{N_{tot}}{4} \quad (13.10)$$

Hopefully you are as astounded and skeptical about the result in Eq. (13.10) as we were when we first saw it. $N(1)$ and $N(8)$ are the number of times configurations 1 and 8 were realized in the experiment, hence they are by definition non-negative integers. Two non-negative integers cannot add up to a negative number, and yet, this is the conclusion that has been forced upon us by two things: the EPR notions of reality and the measured statistics in Table 13.2.

Equation (13.10) illustrates a very deep and important result proven by John Bell in the 1960s. For classical envelopes that have contained within them the information about the quantities to be observed, there is an inequality that must be satisfied by the numbers of times different signs are observed by Alice and Bob for different colours. The problem is the 1/4 on the right hand sides of Eqs. (13.6–13.8). In a classical experiment involving real envelopes the entries are indeed real and determined before any measurement is performed, so that simple arithmetic puts a constraint on the possible values of the right hand sides of Eqs. (13.6–13.8). To derive this constraint, let's replace 1/4 by an arbitrary number, α, between 0 and 1. Equation (13.10) then yields:

$$2(N(1) + N(8)) = (3\alpha - 1)N_{tot} \quad (13.11)$$

The condition that all the $N(i)$ be positive requires:

$$\alpha > \frac{1}{3} \quad (13.12)$$

If this is not the case, then it simply does not make sense to assign values to the number of times each of the eight configurations are realized in a given run of the experiment. The only possible loophole is to give up locality, and allow the values to be affected instantaneously at a distance, or worse, retroactively, before the measurements are made.

Equation (13.12) is a specific example of a general set of inequalities proved by John Bell[7] to be conditions that must be satisfied by experimental data obtained from certain types of classical experiments similar to our envelope experiment. The

---

[7] Bell [2].

key ingredients for such experiments are that they involve two separated sub-systems (envelopes) containing multiple observable quantities whose values are *strongly correlated*. Strongly correlated refers to the fact that the sign on each piece of paper in one envelope is not independent of the sign on the same piece of paper in that envelope's partner in the experiment.

The data produced by Alice and Bob violate *Bell's inequalities*, as these conditions are called, so they cannot be obtained in a real classical experiment. In arguably one of the most profound discoveries of theoretical physics, Bell found that there were some real experiments for which quantum mechanics predicts results that violate Bell's inequalities. The predicted violations were confirmed experimentally in the early 1980s, proving that for at least these experiments, the microscopic world does not respect EPR's assumptions about the nature of reality.

So what is the feature of quantum mechanics that allows for such craziness? In order to answer this, it will be useful to learn a bit more about spin.

13.4 More on Spin

13.4.1 Overview

In this Section we provide some mathematical details about *intrinsic spin*, which we first introduced in Sect. 11.6.5. Intrinsic spin, or spin for short, is a quantum attribute that is possessed not just by electrons, but by all elementary particles, including electrons, protons, neutrons and photons. Spin is a vector quantity that, although a form of angular momentum, is not associated with orbital motion. It is an innate quantum property of the particle that has no classical analogue. We will denote the vector spin operator by $\hat{\mathbf{S}}$. As usual the hat denotes that it is a quantum operator/observable, while bold face implies that it is a vector quantity with magnitude and direction. In Cartesian coordinates:

$$\hat{\mathbf{S}} = \hat{S}_x \mathbf{i} + \hat{S}_y \mathbf{j} + \hat{S}_z \mathbf{k}. \tag{13.13}$$

The properties of the spin component operators $\hat{S}_x$, $\hat{S}_y$ and $\hat{S}_z$ will be discussed below.

Spin does contribute to the total angular momentum of a system, so that in the absence of external forces it is the sum of total spin and total orbital angular momentum that is conserved. The square of the spin operator $\hat{S}^2$ is defined as:

$$\hat{S}^2 := \hat{\mathbf{S}} \cdot \hat{\mathbf{S}}$$
$$= \hat{S}_x^2 + \hat{S}_y^2 + \hat{S}_z^2 \tag{13.14}$$

Its spectrum is similar to that of the angular momentum squared operator (see Eq. (11.46) in Sect. 11.6.1):

$$|S^2| = \hbar^2 s(s+1). \tag{13.15}$$

where the *spin quantum number* s[8] can in principle take on any positive half integer value $s = 0, 1/2, 1, 3/2, 2, \ldots$.

One important difference between orbital angular momentum and spin is that only the latter can take on half integer values. Particles with half integer spin $(1/2, 3/2, \ldots)$ are called *fermions* while particles with integer spin $(0, 1, 2, \ldots)$ are called *bosons*. Fermions have fundamentally different properties than those of bosons, the most important of which is that fermions obey the Pauli exclusion principle. As shown in Sect. 11.7.1, the Pauli exclusion principle plays a vital role in our understanding of the periodic table.

The fundamental constituents of matter, namely electrons, protons and neutrons have spin quantum number $s = 1/2$. Photons, the quantum particles associated with electromagnetism, have spin $s = 1$, while the quantum particle associated with the gravitational force is thought[9] to have spin $s = 2$. Elementary particles with integer spin are the quantum particles associated with classical forces. Elementary particles with zero spin are rare in nature. One notable exception is the famous Higgs boson discovered in 2012 at the Large Hadron Collider, dubbed by popular media to be the "god particle" because it provides an explanation for the origin of mass. The Higgs particle has spin $s = 0$. It is not completely clear whether the Higgs boson is indeed elementary or a composite particle made up of more fundamental particles, say spin 1/2 fermions.

We end this brief tour of Nature's zoo of elementary particles with the observation that, despite the (countably) infinite possible values of the spin quantum number, s, the allowed residents of the zoo (without supersymmetry) are restricted by mathematical theorems to be remarkably few in number. These theorems, dating back to the 1960s, prove that it is impossible to construct consistent theories of fundamental relativistic particles whose spins are greater than two.[10] Moreover, no more than one particle with spin 2 can exist. There is still ongoing research into whether or not spin 3/2 particles can consistently interact with gravity, either at the classical or quantum level. It is thought that there is only one way to do this, namely by putting fermions and bosons on the exact same footing. Such theories possess a very interesting but as yet unverified symmetry of nature, called supersymmetry, which requires the laws of physics to be unchanged by (a carefully chosen) set of symmetry operations that interchange fermions with bosons. Although there is no experimental evidence for the existence of supersymmetry, the conjecture by its proponents is that it is a symmetry that is *broken* in the sense detailed in Sect. 2.4.

The theory of spin provides a beautiful example of how mathematical consistency places strong constraints on what can be realized in nature. This brings to mind yet another famous quote by Einstein: "What really interests me is

---

[8] The spin quantum number will often be referred to as *spin* for short.

[9] We say "thought to have" because a complete, experimentally verified theory of quantum gravity does not as yet exist, and gravitons have not been observed in nature. Their classical counterparts, gravitational waves, have been observed numerous times since the advent of advanced LIGO in 2016.

[10] Relativistic particles with spin greater than 2 are allowed as long as they are *composites*, that is made up of two or more lower spin fundamental particles.

whether God could have created the world any differently; in other words, whether the requirement of logical simplicity admits a margin of freedom". Source: https://blogs.scientificamerican.com/observations/einsteins-famous-god-letter-is-up-for-auction/.

13.4.2 Mathematical Details

We saw in Sect. 11.6.2 that angular momentum and its components are differential operators. Spin is quite different: its components are complex valued matrices that act via matrix multiplication on complex-valued vectors living not in three-dimensional space, but an "internal" vector space that has no direct classical analogue. We will now see how this works explicitly in the case of spin 1/2.

For a particle with spin quantum number, s, the possible values of the spin components along any arbitrarily chosen axis (let's call it the z-axis without loss of generality) are quantized as follows:

$$s_z = -s\hbar, (-s+1)\hbar, \ldots, (s-1)\hbar, s\hbar$$
$$= m_s \hbar \quad \text{where} \quad m_s = -s, (-s+1), \ldots, (s-1), s \qquad (13.16)$$

In the above, s is the spin quantum number defined in Eq. (13.15) and m_s is the *magnetic spin number* analogous to the orbital magnetic quantum number m in Eq. (11.85). As described in Sect. 11.6.5 electrons have two possible values for the component of spin along any axis, namely $-\hbar/2$ and $+\hbar/2$. Particles with spin $s = 1$ normally would allow three possible values for m_s namely $-1, 0, +1$. However, because photons have no rest mass, Maxwell's theory has a special symmetry called *gauge invariance* that renders $m_s = 0$ unphysical.

Note that the maximum value of the magnitude of the component of spin along any axis is $s\hbar$. This is less than the magnitude $s(s+1)\hbar$ of the spin vector as given in Eq. (13.15). If one thinks in classical terms, this suggests that the spin vector is never fully aligned with the z-axis. This is a very strange feature of both orbital angular moment and spin. A heuristic picture sometimes used to illustrate this is given in Fig. 13.2. The spin, when thought of as a classical vector, is not aligned with the axis, but precesses around it. When the z component of spin is known, the x and y component are not fixed, but a large number of measurements at random times produce all values with equal probability and an average value of zero.

The picture in Fig. 13.2, while appealing heuristically, must, however, be taken with a large grain of salt, since it suggests that all three components of spin have specific values at any instant in time. In fact, a key and very intriguing property of spin is that the components of the spin of a particle obey a form of uncertainty principle analogous to that of position and momentum. As will be shown below, the three components of spin are complementary in the sense that no two operators commute. This means that if one knows with certainty the component of the spin of an electron along the z-axis (i.e. the particle is in an eigenstate of $\hat{S}_z$), then the components along the x and y axes are completely unknown. Since the values of the components are

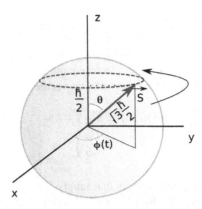

Fig. 13.2 Heuristic picture of the spin of an electron with spin $+\hbar/2$ along the z-axis. It is not aligned with the axis, but precesses around it. The x and y component are not fixed, but if measured at random times all values occur with equal probability, giving an average value of zero. The true quantum description must be different: only the values $\hbar/2$ or $-\hbar/2$ can be obtained, with equal probability so that the average is still zero

quantized this means that after one measures the z-component of spin to be, say $+\hbar/2$, there is an equal probability of a subsequent measurement along the x-axis, say, to yield $+\hbar/2$ or $-\hbar/2$. After a measurement along the x-axis is made the spin along the z-axis immediately becomes uncertain, with 50% probability of obtaining $+\hbar/2$ or $-\hbar/2$. The value of 50% follows from the equations but is also a direct consequence of the rotational invariance of the quantum equations of motion. Once one knows the z-component of spin symmetry suggests that there can be no preferred direction in the $x - y$ plane.

The mathematically and physically correct representation of the spin state of a spin $1/2$ particle is given in terms of a two dimensional complex-valued vector. One can choose basis vectors arbitrarily, but they do not, in general, point in any direction in physical space. Instead, these basis vectors map out the *quantum vector space* in which the spin vector lives. They do, nonetheless have physical interpretations in terms of ordinary three-vectors. Let us choose the following two-vectors as basis vectors[11]:

$$|+\rangle := \begin{pmatrix} 1 \\ 0 \end{pmatrix} \tag{13.17}$$

$$|-\rangle := \begin{pmatrix} 0 \\ 1 \end{pmatrix} \tag{13.18}$$

---

[11] Although this is a two-dimensional vector space, the components of the vectors are allowed to be complex so that the number of parameters is sufficient for describing spin in three dimensions.

The physical interpretation of the states $|+\rangle$ and $|-\rangle$ is that $|+\rangle$ represents the state of an electron with spin $+\hbar/2$ along the z-axis (spin up), while $|-\rangle$ represents the state of an electron with spin $-\hbar/2$ along the z-axis (spin down).

The most general spin state of an electron can be written as an arbitrary linear combination of the spin up and spin down basis vectors:

$$|\psi(a, b)\rangle = a|+\rangle + b|-\rangle \tag{13.19}$$

$$= \begin{pmatrix} a \\ b \end{pmatrix} \tag{13.20}$$

a and b are arbitrary complex numbers that satisfy $|a|^2 + |b|^2 = 1$.

Equations (13.17) and (13.18) use "bra-ket" symbols originally due to Dirac. It uses the symbol $|\rangle$ to generally denote the quantum state of a system. The $|+\rangle$ is called a ket. The notation stems from the fact that the magnitude squared of any vector $|\psi\rangle$ is written as the bracket ("bra-ket") $\langle \psi|\psi\rangle$, while the inner product of a vector $|\psi_2\rangle$ with $|\psi_1\rangle$ is denoted $\langle \psi_2|\psi_1\rangle$. It is a very elegant notation in fact. The symbol $|\psi\rangle$ represents a column vector in a particular coordinate system as in Eq. (13.20). The bra symbol $\langle \psi|$ represents the complex transpose of the same vector, namely the row vector:

$$\langle \psi| := (a^*, b^*) \tag{13.21}$$

Using this notation the inner product of any two-vectors $|\psi_1\rangle$ and $|\psi_2\rangle$:

$$|\psi_1\rangle = a_1|+\rangle + b_1|-\rangle \tag{13.22}$$
$$|\psi_2\rangle = a_2|+\rangle + b_2|-\rangle \tag{13.23}$$

is the matrix multiplication of the complex transpose of $|\psi_2\rangle$ with the column vector $|\psi_1\rangle$, namely

$$\langle \psi_2|\psi_1\rangle = (a_2^*, b_2^*) \begin{pmatrix} a_1 \\ b_1 \end{pmatrix}$$
$$= a_2^* a_1 + b_2^* b_1 \tag{13.24}$$

The magnitude squared of a vector $\langle \psi|\psi\rangle$ is then:

$$\langle \psi|\psi\rangle = a^* a + b^* b = |a|^2 + |b|^2 \tag{13.25}$$

Note that Eq. (13.24) implies that:

$$\langle \psi_2|\psi_1\rangle = \langle \psi_1|\psi_2\rangle^* \tag{13.26}$$

The above rules define in general the inner products and magnitudes of vectors in a two-dimensional complex vector space. The normalization condition on the quantum states implies that spin one half fermion states form a two-dimensional space of

complex vectors with unit norm. The space of states is then a complex analogue of
the set of all unit vectors in three space. In the latter case these unit vectors span
a two dimensional sphere. The symmetries of a two dimensional sphere consist of
rotations about any of three coordinate axes. This is the same as the symmetries of
the space spanned by the quantum spin states of the electron.

Using the above definitions of inner product and magnitude, the two vectors $|+\rangle$
and $|-\rangle$ form an *orthonormal basis* for the space of spin 1/2 states. That is:

$$||+\rangle|^2 := \langle +|+\rangle = 1$$
$$||-\rangle|^2 := \langle -|-\rangle = 1$$
$$\langle +|-\rangle = 0 \tag{13.27}$$

We now discuss in more detail the physical interpretation of the general spin state
Eq. (13.20). $|a|^2 = a^*a$ gives the probability of the electron's spin being up, while
$|b|^2 = b^*b$ gives the probability of it being down. When the z-component of spin
is measured, the only possible outcomes are either up or down. The state must be
normalized ($|a|^2 + |b|^2 = 1$), which imposes one condition on the four real numbers
that determine the two complex numbers a and b. This leaves three real parameters.

The spin state can be specified in terms of a useful set of parameters χ, θ, ϕ by
writing the complex components a and b of the two-vector as follows:

$$a = e^{i\chi}\cos(\theta/2) \tag{13.28}$$
$$b = e^{i\chi}e^{i\phi}\sin(\theta/2) \tag{13.29}$$

with χ, ϕ, θ all real.

Exercise 1 Verify that the expression for a and b in Eq. (13.29) identically satisfy
$|a|^2 + |b|^2 = 1$.

In terms of χ, ϕ and θ, the most general spin state of an electron takes the form:

$$|\psi(\theta, \phi, \chi)\rangle = e^{i\chi}\left(\cos(\theta/2)|+\rangle_z + e^{i\phi}\sin(\theta/2)|-\rangle_z\right)$$
$$= e^{i\chi}\begin{pmatrix} \cos\theta/2 \\ e^{i\phi}\sin\theta/2 \end{pmatrix}. \tag{13.30}$$

In the above, we have added a subscript z to the basis vectors to emphasize that they
point up and down along the z-axis. The overall factor, $e^{i\chi}$ affects neither the direction
nor magnitude of the complex vector $|\psi\rangle$. It contributes only an overall phase that
can be dropped without loss of generality, as was the case for the Schrödinger wave-
functions in Sect. 9.2.2. *Relative phases are, however, important when adding two
spin state vectors*, as is evident from the second term in Eq. (13.30). We will when
appropriate henceforth drop **overall** phases when discussing spin states.

You will have noticed that the spin state of a single electron in Eq. (13.30) is
specified by two real parameters, namely the angles θ and ϕ, and the overall, physically irrelevant phase χ. Although we are using complex valued vectors to represent

them, spin states such as $|\psi(\theta, \phi)\rangle$ live in a vector space that has, in effect two real dimensions. As illustrated in Fig. 13.3, the states are in one to one correspondence with the points on a two dimensional sphere in three dimensional space. $|\psi(\theta, \phi)\rangle$ represents the spin state of an electron that has spin up along the axis of three-space that makes an angle θ with respect to the z-axis, and ϕ with respect to the x-axis, as shown in Fig. 13.3. Thus the angles in the spin states correspond precisely to the angles that locate a point on a sphere of fixed radius as in Fig. 13.2.

We note that if a particle is known to be in a state $|\psi(\theta, \phi)\rangle$ the probability of measuring its spin to be up along a different axis defined by $|\psi(\theta', \phi')\rangle$ is given by:

$$P := \left|\langle \psi(\theta', \phi')|\psi(\theta, \phi)\rangle\right|^2$$
$$= \cos(\theta'/2)\cos(\theta/2) + e^{i(\phi-\phi')}\sin(\theta'/2)\sin(\theta/2). \qquad (13.31)$$

Example 1 1. Show that the general spin state $\psi(\theta, \phi, \chi)$ in Eq. (13.30) is normalized for all values of θ, ϕ, χ.
Solution: The probability P in Eq. (13.31) must equal 1 if $\theta' = \theta$ and $\phi' = \phi$ since in this case it represents the norm squared of the state $|\psi(\theta, \phi)\rangle$. This is indeed the case, since:

$$\cos(\theta/2)\cos(\theta/2) + e^{i(\phi-\phi)}\sin(\theta/2)\sin(\theta/2) = \cos^2(\theta/2) + \sin^2(\theta/2) = 1.$$
$$(13.32)$$

2. Consider the spin state for an electron that corresponds to spin up in the x-direction.

   (a) Express this state as a linear combination of spin vectors pointing up and down along the z-axis, i.e. the basis vectors $|\pm\rangle_z$.

Fig. 13.3 Diagram of generic spin one half state specified by two angles on the unit sphere

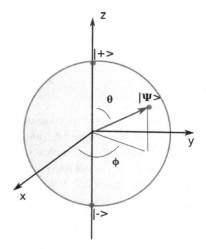

Solution:
The point on the unit two sphere lying on the positive x-axis has spherical coordinates $\theta = \pi/2$ and $\phi = 0$. Using Eq. (13.30) with these angles, we get:

$$|+\rangle_x = \cos(\pi/4)|+\rangle_z + e^{i0}\sin(\pi/4)|-\rangle_z$$
$$= \frac{1}{\sqrt{2}}|+\rangle_z + \frac{1}{\sqrt{2}}|-\rangle_z. \tag{13.33}$$

Similarly, the point on the unit two sphere lying on the negative x-axis has spherical coordinates $\theta = \pi/2$ and $\phi = \pi$ so:

$$|-\rangle_x = \cos(\pi/4)|+\rangle_z + e^{i\pi}\sin(\pi/4)|-\rangle_z$$
$$= \frac{1}{\sqrt{2}}|+\rangle_z - \frac{1}{\sqrt{2}}|-\rangle_z. \tag{13.34}$$

(b) What is the probability that an electron with spin up along the x-axis will be measured with spin up along the z-axis? What is the probability that it will be measured with spin down along the z-axis?
Solution: Using Eq. (13.31) and the orthonormality of the basis vectors $|\pm\rangle_z$, we get:

$$P = {}_z\langle+|+\rangle_x$$
$$= \left| \frac{1}{\sqrt{2}}{}_z\langle+|+\rangle_z + \frac{1}{\sqrt{2}}{}_z\langle+|-\rangle_z \right|^2$$
$$= \frac{1}{2}. \tag{13.35}$$

By symmetry this probability must also be $\frac{1}{2}$ for spin down along the x-axis.

Exercise 2 1. Repeat Example 1 for electrons in states with:

(a) Spin down along the x-axis.
(b) Spin up along the y-axis, and spin down along the y-axis.
(c) Spin up along an axis in the $y - z$ plane rotated $120°$ clockwise with respect to the z-axis and spin down along the same axis.

2. Suppose an electron is known to have spin up along the z-axis. What is the probability of measuring spin up along an axis rotated clockwise away from the z-axis by an angle of $120°$? What is the probability of measuring spin down along the same axis?

Finally, a note on time evolution of spin states and physical observables. As in the case of wave functions, spin states also obey a time dependent Schrödinger equations.

Instead of it being a differential equation, however, it is a matrix equation. For a spin vector

$$|\psi\rangle = \begin{pmatrix} a \\ b \end{pmatrix} \tag{13.36}$$

the time dependent Schrödinger equation takes the form:

$$i\hbar\frac{\partial}{\partial t}|\psi\rangle = \hat{H}|\psi\rangle$$

$$i\hbar\begin{pmatrix} \frac{\partial a}{\partial t} \\ \frac{\partial b}{\partial t} \end{pmatrix} = \hat{H}\begin{pmatrix} a \\ b \end{pmatrix} \tag{13.37}$$

where

$$\hat{H} = \begin{bmatrix} h_{11} & h_{12} \\ h_{21} & h_{22} \end{bmatrix} \tag{13.38}$$

is a two dimensional complex matrix. In order for the spin state to be preserved under time evolution, $\hat{H}$ must be a *Hermitian* matrix. That is $\hat{H}$ must be the same as its Hermitian conjugate $\hat{H}^\dagger$:

$$\hat{H}^\dagger := \begin{bmatrix} h_{11}^* & h_{21}^* \\ h_{12}^* & h_{22}^* \end{bmatrix} = \begin{bmatrix} h_{11} & h_{12} \\ h_{21} & h_{22} \end{bmatrix} =: \hat{H} \tag{13.39}$$

Comparing the components of Eqs. (13.38) and (13.39), we see that this requires:

$$h_{11} = h_{11}^* \text{ (real)} \tag{13.40}$$
$$h_{12} = h_{21}^* \tag{13.41}$$
$$h_{22} = h_{22}^* \text{ (real)} \tag{13.42}$$

Note that complex conjugation of Eq. (13.41) implies that $h_{21} = h_{12}^*$, so this need not be listed separately.

We can say a bit more about the solution to the spin version of the time dependent Schrödinger equation Eq. (13.37). It must take the form:

$$|\psi(t)\rangle = \hat{U}(t, t_0)|\psi(t_0)\rangle \tag{13.43}$$

where $|\psi(t_0)\rangle$ is the state vector at some initial time t_0 and $\hat{U}(t, t_0)$ is a unitary matrix that converts the initial state $|\psi(t_0)\rangle$ to the final state $|\psi(t)\rangle$. A unitary matrix is one whose Hermitian conjugate is equal to its inverse:

$$\hat{U}^\dagger = \hat{U}^{-1} \tag{13.44}$$

Such a unitary transformation plays the role of a rotation: it takes a spin vector from one point on the unit two sphere to another point on the unit two sphere.

Exercise 3 Prove that the norm of a spin vector is preserved under an arbitrary unitary transformation.

As you have probably discerned, not least by the notation, $\hat{H}$ in Eq. (13.38) is the energy operator, or Hamiltonian, associated with the spin state. The Hamiltonian that appears in the time dependent Schrodinger equation in Eq. (10.78) is a differential operator. In the present case it is a linear matrix operator whose (real) eigenvalues correspond to the allowed values of energy that one can measure for the given spin state. For example, the energy, or Hamiltonian operator for an electron in a classical magnetic field **B** is:

$$\hat{H} = -\frac{e}{m_e c} \hat{\mathbf{S}} \cdot \mathbf{B}$$
$$= -\frac{e}{m_e c} \left(\hat{S}_x B_x + \hat{S}_y B_y + \hat{S}_z B_z \right). \tag{13.45}$$

The minus sign signifies that the spin of the electron is happiest (has lowest energy) when it is lined up with the magnetic field.

In order for the eigenvalues to be real it is necessary for the Hamiltonian to be a Hermitian matrix. We can now write the linear operators that correspond to the components of spin in the basis that we have been using:

$$\hat{S}_x = \frac{\hbar}{2} \begin{bmatrix} 0 & 1 \\ 1 & 0 \end{bmatrix} \tag{13.46}$$

$$\hat{S}_y = \frac{\hbar}{2} \begin{bmatrix} 0 & -i \\ i & 0 \end{bmatrix} \tag{13.47}$$

$$\hat{S}_z = \frac{\hbar}{2} \begin{bmatrix} 1 & 0 \\ 0 & -1 \end{bmatrix} \tag{13.48}$$

These are Hermitian matrices as required. The linear combination with real coefficients of Hermitian matrices is also Hermitian, so that $\hat{H}$ in Eq. (13.45) provides a suitable Hamiltonian operator with real eigenvalues.

There are many interesting things one can do with and learn about spin matrices. For the present we limit ourselves to the observation that the three components of spin are complementary in the sense that the order in which they operate on a spin state matters: they do not commute (see Sect. 10.8).

Exercise 4 1. Verify that $\hat{S}_x$, $\hat{S}_y$ and $\hat{S}_z$ are hermitian.
2. Show that the spin operators defined in Eqs. (13.46)–(13.48) satisfy the following:

$$[\hat{S}_x, \hat{S}_y]|\psi\rangle := \left[\hat{S}_x \hat{S}_y - \hat{S}_y \hat{S}_x \right] |\psi\rangle$$
$$= i\hbar \hat{S}_z |\psi\rangle \tag{13.49}$$

when acting on any spin vector $|\psi\rangle$.

We end on a mathematical note. The spin states Eq. (13.30) provide an alternative, but equivalent, description of the points on a two dimensional sphere. The angles θ and ϕ, whose values determine specific spin states can also be used to locate points on the two sphere. Even the angle χ in the phase that we have been dropping has a geometrical interpretation. Recall that there are three independent rotations that leave a sphere invariant. θ gives a rotation off the z-axis that can without loss of generality be considered a rotation *about* the x-axis. ϕ then gives a subsequent rotation away from the x axis about the z axis. These two operations move the point at the north pole of the unit sphere, for example, (i.e. on the z-axis) to a new point on the unit two sphere which we will call P. The spin transformations with same values of θ and ϕ do the analogous thing to spin states. They take a state that is an eigenstate of $\hat{S}_z$ with eigenvalue $+\hbar/2$ and rotate it to an eigenstate of the component of spin pointing along the axis that goes through the point P. We know there is a rotation about a third axis that we can do and still stay on the two sphere. In terms of angles, this is a rotation χ about the new direction that the z-axis is pointing. It rotates all the other points on the sphere, without changing the location of the point under consideration. This is in keeping with the interpretation of χ for the spin states: it is a change of phase that does not change the state.

Rotations of the two dimensional sphere in three dimensional space can be implemented by three real matrices that preserve the length of three-vectors. We now see that the same set of symmetry operations can also be represented using transformations of two dimensional complex vectors with unit norm. In this case the transformations take the form of two by two complex matrices that preserve the norm of two dimensional complex vectors. A general transformation $U(\theta, \phi, \chi)$ depends on the three angles (θ, ϕ, χ) and takes the form:

$$U(\theta, \phi, \chi) = e^{i\chi} \begin{bmatrix} \cos(\theta/2) & e^{i\phi}\sin(\theta/2) \\ -e^{i\phi}\sin(\theta/2) & \cos(\theta/2) \end{bmatrix} \tag{13.50}$$

Such a matrix is said to be *unitary*, in that:

$$U^{\dagger}U = \mathbb{1} \tag{13.51}$$

where $U^{\dagger}$ is the complex transpose of U. You can verify for yourself that such *unitary transformations* preserve the norm of complex two-vectors.

We have learned therefore that unitary transformations on complex two-vectors (i.e. spin states) provide a different representation of precisely the same group of symmetry operators as do rotations of three-vectors. The fact that symmetry groups have different representations plays an important role in both physics and mathematics. This is yet another example of physics and mathematics as different sides of the same coin.

13.4.3   Summary

Given the large amount of information we have just covered, it is worth summarizing the properties of electron spin:

- Spin is a purely quantum observable with no classical counterpart and is quite different in nature from orbital angular momentum.
- It is the sum of the total spin and total orbital angular momentum that is conserved in general interactions.
- The spin state of an electron is represented by a two dimensional complex vector of unit norm. Since the overall phase is not relevant physically, spin states are in one to one correspondence with points on the unit sphere, labelled by two angles θ and ϕ, which correspond to their namesakes in spherical coordinates.
- We can define two basis vectors that correspond to spin states $|+\rangle$ and $|-\rangle$ that have positive and negative spin, respectively, along a given, arbitrarily chosen axis in three space. Without loss of generality, one normally chooses the z-axis.
- A spin state must preserve its norm under time evolution. Thus any time evolution can be represented by the action of a unitary 2×2 complex matrix on the spin vector.
- Physical observables, such as the components of spin, are represented by two dimensional complex *Hermitian* matrices.

13.5 Experimental Confirmation of Quantum Weirdness

In Sect. 13.3, a thought experiment was presented in which the statistical predictions made by theory violated Bell's inequalities and were therefore inconsistent with the common sense view that the envelopes contained the values of all possible observables irrespective of whether or not they were measured. Of course, predictions for a physical macroscopic (i.e. classical) experiment will always obey Bell's inequality Eq. (13.12). Amazingly, the microscopic world described by quantum mechanics behaves very differently. As mentioned briefly in Sect. 13.3, experiments involving microscopic particles that violate Bell's inequalities do exist, showing that Nature does not obey the EPR notion of a locality reality.

The following experiment is completely analogous to the envelope experiment. Instead of envelopes, it involves a pair of electrons in a quantum state with total spin equal to zero that fly apart in opposite directions towards two distant detectors. Let's call the particles electron 1 and electron 2 and the detectors Alice and Bob. The role of coloured pieces of paper is played by the components of spin of the electrons along each of three different axes.[12] The first axis, which we will call green, is the z-axis. The second axis, which we call blue is in the $x - z$ plane ($y = 0$) rotated 120° clockwise with respect to the z-axis. The third, or red, axis is also in the $x - z$ plane at 120° counterclockwise from the z-axis. In terms of the spherical coordinates in Fig. 4.8, $\theta = 120°$, $\phi = 0$ for the blue axis, while $\theta = -120°$, $\phi = 0$ for the red axis. The experiment is illustrated schematically in Fig. 13.4.

---

[12] Although the two detectors are far apart, it is possible for them to agree on a common Cartesian coordinate system before moving apart to start the experiment.

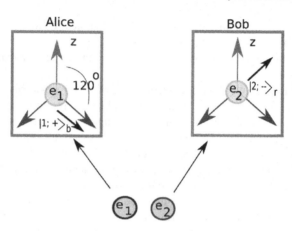

Fig. 13.4 In the microscopic version of the envelope experiment, two electrons are created in a combined quantum state of total spin zero and enter two distant detectors. The detectors can measure the spin of their respective electrons along one and only one of three axes, which we call green, blue and red. The decision as to which axis to measure is made after the electrons are created, but before they arrive at the detectors. In the example illustrated Alice measures spin up along blue, while Bob measures spin down along red

The discussion in Sect. 13.4 allows us to calculate the predictions of quantum mechanics for the correlations listed in Table 13.1. From when the particles are first emitted until a measurement of spin is made, all axes are equivalent, so the probabilities of measuring spin up or spin down along any axis is 1/2. Suppose Alice measures the spin of electron 1 along the z (green) axis. There will be a 50–50 chance that she will measure spin up. If Alice does measure spin up, then immediately after her measurement electron 1 is in state $|1, +\rangle_g$. Since the total spin is zero a measurement of the spin of particle 2 along the z axis must yield spin down with 100% certainty. This in turn requires particle 2 to be in the state $|1, -\rangle_g$ immediately after Alice's measurement on particle 1.

In Exercise 2 of this Chapter we worked out that a particle in a state with spin up along the z-axis will be measured to have spin up along an axis rotated by an angle 120° in either direction with probability $\cos^2(120°/2) = 1/4$. In particular, once Alice has measured her particle to have spin up along the z-axis, the probability that a subsequent measurement of her particle will yield spin up along an axis rotated by 120° with respect to the z-axis will be 1/4. Recall that Bob's particle is anti-correlated with Alice's so the probability for his particle to be spin down along a given axis is equal to the probability of Alice's particle to be spin up along the same axis. As a consequence, if Alice measures spin up along the z-axis, we know the probability that Bob will measure spin down along an axis rotated by 120° relative to Alice's z-axis must be $\cos^2(120°/2) = 1/4$.

We emphasize that the green, blue and red axes are arranged symmetrically in the $x - z$ plane, so that if Alice first measures along the blue axis, the probability of Bob measuring spin down along either the green or red axis is also one quarter, and so on. Thus, the electron spin experiment provides precisely the statistics that we are looking for: the probability of anti-correlation is 100% along the same axis (colour), while the probability for anti-correlation along different axes (colours) is 25%. These statistics violate Bell's inequality Eq. (13.12) implying that it is impossible to assign values to the spins of both particles along each of the three axes in a way that is consistent with the EPR notion of an objective, local reality.

What is really going on here? Before either measurement, the possibility of measuring spin up or down along any axis is 1/2 for both Alice and Bob. Alice's measurement changes these statistics for Bob as soon as she makes the measurement. The probability of measuring spin down along the z-axis becomes one. This is explained by noting that the total spin of the two particles when they were first emitted was zero. Much more striking is the fact that the probability of measuring spin down along any axis at 120° compared to Alice's axis becomes one quarter. Bell's inequalities prove that this cannot be explained by properties possessed by the electrons when they were first emitted. Does this mean that Alice's measurement changes the values of the spin of particle 2 along different axes instantaneously at any distance? This is one possible interpretation, but it is definitely unpalatable to physicists who, like Einstein, believe that information cannot propagate faster than the speed of light. Changing the spin along a given axis from one specific value to another instantaneously at long distances would require the knowledge of Alice's spin measurement to propagate at infinite speed.

Another interpretation that is perhaps not less puzzling but certainly more consistent with special relativity is the following: The states of the electrons contain all the information that exists about the properties of their spins, so that they do not have definite values until measured. Alice's measurement does not change one specific value of electron 2 to another, since specific values do not exist. Instead, her measurement changes, non-locally, the relative probabilities of various outcomes for Bob's measurement.[13]

The second interpretation, like the first, is non-local, but only in a probabilistic sense: no information need be transmitted from Alice to Bob. From Bob's point of view, as long as he is unaware of Alice's result for any given pair of electrons, measurement of spin along any axis can and will yield spin up half the time and spin down half the time. He has no way of knowing Alice's choice of axis until they come together to compare results. Once they do, the mysterious correlations of 1/4 that violated Bell's inequalities are revealed. Thus, the correlations do not violate the letter of the laws of special relativity because no real information was transmitted when Alice made her measurement. In a sense, what Alice changes with her measurement are Bob's potentialities, not realities.

Example 2 Bob's measurements can't provide information about what axes Alice is measuring. For example suppose Alice measures the green axis and Bob measures

---

[13] Although we are only considering non-relativistic quantum mechanics, the situation becomes even weirder in special relativity, in which different observers disagree on who did their measurements first, Alice or Bob.

the blue axis for 100 pairs of electrons. Show that Bob measures spin up and spin down along the blue axis an equal number of times, so that he has no way of knowing from his results alone what measurements Alice is doing, or vice versa, no matter how many electron pairs are measured.

Solution: Given that quantum mechanics predicts that Alice should get spin up and spin down along a given axis an equal number of times, Alice will measure spin up along the green axis 50 times and spin down 50 times. When Alice measures spin up, Bob must get spin down for the corresponding electron $1/4 \times 50 \sim 12$ times and spin up $3/4 \times 50 \sim 38$ times. When Alice measures down, Bob will get down $3/4$ of the time and up $1/4$ of the time, so he will get down 38 times and up 12 times. After 100 measurements he will get down about $12 + 38 = 50$ times and up $38 + 12 = 50$ times. Thus Bob has no way of knowing from his results what Alice is doing.

The first experimental verification that quantum mechanics violates Bell's inequalities was obtained in a beautiful, ground-breaking experiment by Alain Aspect and collaborators in 1982, using pairs of photons.[14] Many subsequent experiments were done to close loop-holes, such as possible collusion between Alice and Bob, by placing the detectors very far apart, thereby ensuring that the choice of axes was made well after any possibility of interactions existed. For example, a very elaborate experiment involving electron pairs with measuring devices 1.3 km apart was done in 2015. As of 2020 the record is a remarkable 1200 km. See Scientific American, June 15, 2017 for a pedagogical description.[15]

These experiments have verified beyond any reasonable doubt that Nature obeys the predictions of quantum mechanics and that the EPR view of reality must be abandoned, however appealing it might seem. Quantum mechanics' "spooky action at a distance" is a property of the real world. To understand better how this works and how it can lead to mind-bending revolutions in encryption and computing, we will briefly discuss the notion of entanglement in quantum mechanics.

13.6 Entanglement: The Key to Unlocking Quantum Weirdness

The EPR conundrum and the violation of Bell's inequalities have at their roots the phenomenon of *entanglement*. Two or more particles are said to be entangled if the quantum states describing them cannot be separated by writing the total wave function as a simple product of two or more distinct states, each depending only on variables associated with one of the particles. As an example, consider the wave function for the EPR state in Eq. (13.4):

---

[14] *Experimental Realization of Einstein-Podolsky-Rosen-Bohm Gedanken experiment: A New Violation of Bell's Inequalities*, Alain Aspect, Philippe Grangier, and Gerard Roger Phys. Rev. Lett. 49, 91, 1982.

[15] The original scientific article describing the experiment is: Yin et al. [3].

$$\psi(x_1, x_2) = \frac{1}{2\pi} \int_{-\infty}^{\infty} dp\, e^{i(x_1+\Delta)p} e^{-ix_2 p}$$

$$= \frac{1}{2\pi} \int_{-\infty}^{\infty} dp\, \psi_1(x_1; p)\psi_2(x_2; -p) \qquad (13.52)$$

where

$$\psi_1(x_1; p) = e^{i(x_1+\Delta)p} \qquad (13.53)$$
$$\psi_2(x_2; -p) = e^{-ix_2 p} \qquad (13.54)$$

This is in effect an infinite sum (i.e. integral) of terms each of which is a product of a wave function for particle 1 and a wave function for particle 2. The infinite sum over this product of plane waves yields a Dirac delta function $\delta(x_1 - x_2 + \Delta)$, as described in Sect. 15.2.3. There is no way to write the wave function in the form:

$$\psi(x_1, x_2) = \psi(x_1)\psi(x_2) \qquad (13.55)$$

so the state is said to be entangled.

In each term of the sum, the particle 1 part of the wave function $\psi_1(x_1; p) = e^{i(x_1+\Delta)p}$ is a plane wave with wave number p. The particle 2 part $\psi_2(x_2; -p) = e^{-ix_2 p}$ is a plane wave with wave number $-p$. Because it is a sum over the product of such terms, each with different wave number, the overall wave function does not have a definite value of momentum for either particle. However, if one measures the momentum of particle 1, and obtains $p = p_0$, say, then the wave function collapses to the corresponding specific term in the infinite sum. Hence a subsequent measurement of the momentum of particle 2 must yield $p = -p_0$. Consequently, the EPR conundrum described in Sect. 13.2 is directly related to the fact that the coordinates, and momenta of particles 1 and 2 are inseparably linked.

Entanglement is a bit easier to see in the context of spin 1/2 particles. Consider two electrons. We know the state of each on its own can be represented by a complex vector of unit norm as in Eq. (13.30). It is certainly possible to construct quantum states describing two electrons in which the states are completely independent of each other. Consider the two one-particle states:

$$|\psi(1)\rangle = a_1|+; 1\rangle_z + b_1|-; 1\rangle_z \qquad (13.56)$$
$$|\psi(2)\rangle = a_2|+; 2\rangle_z + b_2|-; 2\rangle_z \qquad (13.57)$$

In the above, we have added a subscript z to the kets to specify the axis along which we are defining spin up, and inserted the particle number into the relevant kets to make it clear to which particle we are referring. The following is said to be a *direct product* of the two individual one particle states:

$$|\Psi(1,2)\rangle = |\psi(1)\rangle \otimes |\psi(2)\rangle$$
$$= (a_1|+; 1\rangle_z + b_1|-; 1\rangle_z) \otimes (a_2|+; 2\rangle_z + b_2|-; 2\rangle_z)$$
$$= a_1 a_2|+; 1\rangle_z \otimes |+; 2\rangle_z + a_1 b_2|+; 1\rangle_z \otimes |-; 2\rangle_z$$
$$+ b_1 a_2|-; 1\rangle_z \otimes |+; 2\rangle_z + b_1 b_2|-; 1\rangle_z \otimes |-; 2\rangle_z; \quad (13.58)$$

For product states such as Eq. (13.58), measurements on electron 1 leave the state of electron 2 unchanged, and vice versa. For example, measuring the spin of particle 1 gives + with probability $|a_1|^2$ and collapses the state vector so that subsequent measurements of the same quantity give the same result, i.e. only the first term in the state of particle 1 remains. The state of particle 2, and hence the outcomes of future measurements of particle 2, are completely unaffected.

The symbol $\otimes$ reminds us that each electron state is a vector in its own 2 dimension complex vector space so that this is not an ordinary cross product of two-vectors living in the same space. The result lives in a vector space with $4 = 2 \times 2$ complex dimensions, i.e. the product of the dimensions of each of the two-vectors. As can be seen in the last two lines of Eq. (13.58), there are four independent basis vectors, namely

$$|+; 1\rangle_z \otimes |+; 2\rangle_z; \qquad |+; 1\rangle_z \otimes |-; 2\rangle_z;$$
$$|-; 1\rangle_z \otimes |+; 2\rangle_z; \qquad |-; 1\rangle_z \otimes |-; 2\rangle_z. \quad (13.59)$$

A general two electron spin state is a normalized linear combination of the above four independent basis vectors:

$$|\Psi(1,2)\rangle = A_{++}|+; 1\rangle_z \otimes |+; 2\rangle_z + A_{+-}|+; 1\rangle_z \otimes |-; 2\rangle_z$$
$$+ A_{-+}|-; 1\rangle_z \otimes |+; 2\rangle_z + A_{--}|-; 1\rangle_z \otimes |-; 2\rangle_z; \quad (13.60)$$

where normalization requires the sum of the magnitudes of the coefficients to be one:

$$|A_{++}|^2 + |A_{+-}|^2 + |A_{-+}|^2 + |A_{--}|^2 = 1 \quad (13.61)$$

The quantum state in Eq. (13.60) contains four arbitrary complex parameters that are constrained by one normalization condition and have one redundant overall phase. Thus three arbitrary complex parameters (two real parameters per complex number for six parameters in all) are required to specify a general state. The product state given in Eq. (13.58) has fewer parameters. Each of the individual states is specified by one complex, or two real parameters, so there are four parameters in total, five if you only require one normalization condition. There are consequently many more non-product states than product states.

Exercise 5 In the case of two electrons there is one extra complex parameter or two extra real parameters for non-product states as compared to the product states. How many extra parameters are there for 3 electron states? How many extra parameters are there for n electron states?

Just as in the EPR state in Eq. (13.52), the two particles are said to be entangled if the coefficients in the sum in Eq. (13.60) are such that it can not be split into a product as in Eq. (13.58). An example of such a state is:

$$\Psi(1, 2) = \frac{1}{\sqrt{2}} (|+; 1\rangle_z |-; 2\rangle_z - |-; 1\rangle_z |+; 2\rangle_z). \tag{13.62}$$

Equation (13.62) is a state in which the total combined spin of the two particles is zero. If one measures the state of particle 1 to be $|+; 1\rangle_z$, then the state vector collapses to just the first term and the second term is eliminated from the state. A subsequent measurement must reveal that particle 2 is in state $|-; 2\rangle_z$. As long as the particles undergo no forces or interactions, this entanglement persists, no matter how far apart they are. In this sense, the measurement of the spin of particle 1 along the z-axis allows one to predict with certainty the value of the spin of particle 2 along the same axis, and vice versa. As such, one might conclude as did EPR that the spins of both particles along the z-axes must be elements of reality that are merely revealed by measurement, even if one cannot predict the value of either until at least one of the particles has been measured.[16]

Exercise 6 Consider the following two-electron state:

$$\Psi(1, 2) = A \left(3|00\rangle + \frac{1}{2}|01\rangle + 2e^{-i\frac{\pi}{6}}|11\rangle \right). \tag{13.63}$$

1. Find the normalization constant A
2. Is it a product state? Justify your answer.

At this stage you should ask: what is special about the z-axis that we have used to define our basis vectors? As required by the symmetry of the laws of physics under rotations, the answer is of course: nothing. We can express the spin zero state in terms of spin up and down along the x-axis for example:

$$\Psi(1, 2) = \frac{1}{\sqrt{2}} (|+; 1\rangle_z |-; 2\rangle_z - |-; 1\rangle_z |+; 2\rangle_z)$$

$$= -\frac{1}{\sqrt{2}} (|+; 1\rangle_x |-; 2\rangle_x - |-; 1\rangle_x |+; 2\rangle_x). \tag{13.64}$$

It is the same form as before up to an overall minus sign, which is a physically irrelevant phase. This may seem odd, but it is required by the rotational invariance of the quantum state, which after all has zero spin and hence no preferred direction.

---

[16] So far this is no surprise since it is merely a consequence of angular momentum conservation. The total spin is zero, so that the total spin along any axis must also be zero.

Example 3 Prove Eq. (13.64).
Solution

$$|+\rangle_x = \cos(\pi/4)|+\rangle_z + e^{i0}\sin(\pi/4)|-\rangle_z$$
$$= \frac{1}{\sqrt{2}}|+\rangle_z + \frac{1}{\sqrt{2}}|-\rangle_z \tag{13.65}$$

$$|-\rangle_x = \cos(\pi/4)|+\rangle_z + e^{i\pi}\sin(\pi/4)|-\rangle_z$$
$$= \frac{1}{\sqrt{2}}|+\rangle_z - \frac{1}{\sqrt{2}}|-\rangle_z. \tag{13.66}$$

$$|\Psi(1,2)\rangle = \frac{1}{\sqrt{2}}\left(|+;1\rangle_x|-;2\rangle_x - |-;1\rangle_x|+;2\rangle_x\right)$$

$$= \frac{1}{\sqrt{2}}\left(\left[\frac{1}{\sqrt{2}}|+;1\rangle_z + \frac{1}{\sqrt{2}}|-;1\rangle_z\right]\left[\frac{1}{\sqrt{2}}|+;2\rangle_z - \frac{1}{\sqrt{2}}|-;2\rangle_z|\right]\right.$$
$$\left. - \left[\frac{1}{\sqrt{2}}|+;1\rangle_z - \frac{1}{\sqrt{2}}|-;1\rangle_z\right]\left[\frac{1}{\sqrt{2}}|+;2\rangle_z + \frac{1}{\sqrt{2}}|-;2\rangle_z\right]\right)$$

$$= \frac{1}{\sqrt{2}}\left(|+;1\rangle_z|-;2\rangle_z - |-;1\rangle_z|+;2\rangle_z\right). \tag{13.67}$$

which corresponds to Eq. (13.64) as required.

Exercise 7 1. Complete the algebra to show that Eq. (13.64) follows as the last line of Eq. (13.67).
2. Show that $\Psi(1,2)$ takes the same form when expressed in terms of basis vectors $|+\rangle_n$ and $|-\rangle_n$ that point up and down along an arbitrary axis given by:

$$|+\rangle_n = e^{i\chi}\left(\cos(\theta/2)|+\rangle_z + e^{i\phi}\sin(\theta/2)|-\rangle_z\right). \tag{13.68}$$
$$|-\rangle_n = e^{i\chi}\left(\sin(\theta/2)|+\rangle_z - e^{i\phi}\cos(\theta/2)|-\rangle_z\right). \tag{13.69}$$

3. Show that the expressions for $|\pm\rangle_n$ for a vector in the $x - z$ plane that is at an angle $\theta = \pm 120°$ clockwise relative to the z-axis are (dropping the overall phase):

$$|+\rangle_{\pm 120°} = \left(\frac{1}{2}|+\rangle_z \pm \frac{\sqrt{3}}{2}|-\rangle_z\right). \tag{13.70}$$

$$|-\rangle_{\pm 120°} = \left(\mp\frac{\sqrt{3}}{2}|+\rangle_z + \frac{1}{2}|-\rangle_z\right). \tag{13.71}$$

We now return to the pair of electrons in a spin zero state that is given by Eq. (13.62) and see how quantum mechanics predicts the observed correlations of the spins along the green, blue and red axes. First, we see from the results of Exercise 7 that an observation by Alice of the spin up of particle 1 along an axis 120° will select $|+\rangle_{\pm 120°}$. Because of the anti-correlation, Bob's particle will, after Alice's measurement, be in the state $|-\rangle_{\pm 120°}$. The probability of Bob subsequently measuring spin down along the z-axis is therefore $1/4$, as advertised.

This calculation highlights two key factors that give rise to the spooky action at a distance. First, the entanglement of the two particles implies that Alice's measurement immediately collapses Bob's state vector as well as her own. Secondly, the properties of the spin vectors produces the result $1/4$, which is less than $1/3$ and violates Bell's inequality. The strange prediction of $1/4$ is only possible because the spin state of Bob's electron just before his measurement is the linear superposition of the spin up and spin down along the z-axis.

In brief, EPR proved using a specific entangled two particle state that quantum mechanics cannot be both local and complete, in the hopes of inspiring the search for a more complete theory. Unfortunately for EPR, the experimentally observed violations of Bell's inequalities prove that no such local and complete description of microscopic reality can possibly exist. It is fair to say that this is one of the most profound results in physics to date.

13.7 Quantum Computation: Entanglement as a Resource

Entanglement is to some extent a source of mystery, but as we will now see it is also a very valuable property of quantum mechanics that can be exploited to provide more secure encryption and exponentially faster computations.[17]

13.7.1 Quantum Recap

In the last few chapters, we saw that the classical and quantum worlds behave very differently. Classical mechanics operates at the macroscopic length scales we experience in our daily lives. It is very successful at predicting the positions and momenta of particles and their interaction with electric and magnetic fields, if charged. Quantum mechanics on the other hand, works at small (sub-atomic) distances. Each possible quantum state of a particle is determined by a wave function from which position and momentum probabilities are computed. The spin of a particle is determined by its spin state vector, which for spin one half, is two dimensional. Quantum mechanics predicts only probabilities for physical observables such as position, momentum and spin. Moreover, there exist complementary variables, including x, p and the components of spin along different axes, for which the uncertainty principle implies that precise knowledge of one variable renders the values of the complementary variable(s) completely uncertain.

---

[17] We are very grateful to Jonathan Ziprick for helpful comments and suggestions on this Section, and for sharing with us the notes for his course on Quantum Computation for Computer Scientists.

The complete quantum description of a particle with spin is given in the simplest case by the direct product of its wave function with its spin state vector[18]:

$$|\Psi(t)\rangle = \psi(x, t) \otimes |\chi(t)\rangle \tag{13.72}$$

As in Sect. 13.6 the direct product implies that the space in which the total state $|\Psi\rangle$ lives is enlarged to contain both sets of variables. In the rest of this chapter we use the term *state vector* to refer to the complete state $|\Psi\rangle$ that describes all relevant degrees of freedom in the system. However, in most cases we will not consider both the position and spin of a particle simultaneously.

We briefly summarize the general properties of the state vector:

1. State vectors in general depend on time. As in the time dependent Schrödinger equation (TDSE), Eq. (10.78), the time derivative of the state is determined by the action of a suitable Hamiltonian operator. Time evolution must preserve normalization in which case it is called *unitary evolution*. In other words there will always be a unitary operator that transforms the initial state $|\Psi\rangle_I$ into the final state $|\Psi\rangle_F$.
2. State vectors can be added to produce a new state vector. That is, if $|\Psi_1\rangle$ and $|\Psi_2\rangle$ are state vectors corresponding to two distinct normalized states of the particle, then there exists another possible state for the particle, described by the sum

$$|\Psi\rangle = \alpha|\Psi_1\rangle + \beta|\Psi_2\rangle, \tag{13.73}$$

The wave function in Eq. (13.73) is called a *linear superposition* of states $|\Psi_1\rangle$ and $|\Psi_2\rangle$. Since the time dependent Schrödinger equation is linear in the state vector (wave function), it is necessarily true that if $|\Psi_1\rangle$ and $|\Psi_2\rangle$ each solve the TDSE then so does an arbitrary linear superposition.
3. State vectors of composite, or multi-particle systems can be entangled. That is, for a system S consisting of two components (subsystems) a and b, there exist states of the complete system whose state vector Ψ_S *cannot* be factorized into a direct product of state vectors $|\Psi_a\rangle$ and $|\Psi_b\rangle$ that involve only variables associated with a and b. That is:

$$|\Psi_S\rangle \neq |\Psi_a\rangle \otimes |\Psi_b\rangle. \tag{13.74}$$

Such non-product states are said to be *entangled*.
4. The measurement of any observable associated with a quantum system yields one of a set of allowed values that are called eigenvalues (see Sect. 10.8). The set of complete eigenvalues for a given operator is called the spectrum of that operator. The probability of observing a particular (eigen-)value can be calculated from the state vector. After the measurement the state vector collapses in the sense that

---

[18] In general the spin degrees of freedom of a particle can be entangled with the wavefunction. One example occurs in the *Stern-Gerlach* experiment in which an electron interacts with a constant magnetic field that deflects it in one direction when its spin is aligned with the magnetic field and in the opposite direction when its spin is anti-aligned.

repeated measurements of the same observable must yield the same value. At this stage the state vector is said to be in an eigenstate of that observable.

The above properties all play a crucial role in distinguishing quantum computers from their classical counterparts.

13.7.2 Classical Computers, in Brief

A classical computer can be broadly separated into two parts:

1. *Hardware*: This must include a register that stores data, or information, usually in binary form. Each piece of information is called a *bit* which can take on one of two values, either 0 or 1. Each configuration or classical state of a computer with 1000 bits, say, would consist of a string of 1000 0's or 1's.
   There must also exist within the hardware a mechanism for manipulating the bits. For the most part, this can only be done one bit at a time, although modern multi-processor units and GPU's[19] are able to coordinate several inter-related computations simultaneously, so that in effect more than one bit is being manipulated at a time.
2. *Software*: This provides the algorithm, or computer code, that instructs the hardware how to manipulate the data in a specified sequence of steps designed to take some known input and produce the initially unknown desired output. The software also determines what new data should be stored and what can be discarded.

13.7.3 What Is a Quantum Computer?

The quantum counterpart of a classical computer works in much the same way in principle.[20] The crucial difference is that instead of using bits to register and manipulate data, quantum computers use *qubits*. Qubits are microscopic particles whose quantum states produce one of two possible values for some observable when measured. The quantum properties of qubits allow them to carry a great deal more information than classical bits. The easiest way to visualize qubits is to think of them as spin 1/2 electrons but other physical systems with two quantum states can also be

---

[19] GPU's or graphics processing units are a type of central processing unit (CPU) that were originally designed for gaming systems to process images as rapidly as possible. Since then they have been used more and more by scientists because they are able to speed up certain types of numerical calculations as well.

[20] There exists another type of quantum computer, called an "annealing quantum computer", which was developed and is being marketed by a Canadian company called D-Wave. Annealing quantum computers are adapted to specific tasks, namely those that can be posed as an energy minimization problem. Universal quantum computers are closer in spirit and functionality to classical universal computers.

used. Examples include photons that can have one of two polarizations perpendicular to their line of motion, and atoms that effectively have only two energy states.

The key characteristic of a qubit is that its quantum state can be described by a normalized two component complex vector, just like the spin state of an electron. Quantum computer code provides the algorithm for manipulating these qubits, not just individually, but also collectively so as to create entanglement between them. In effect, entanglement provides a uniquely quantum mechanical resource that allows for immense increases in computational speed achievable by quantum computers.

A universal quantum computer consists of three components:

1. *The hardware*: an n-qubit register along with a set of operations, called *gates*, that can be applied in sequence to the quantum state of the n-qubit register.
2. *The software*: A quantum algorithm for taking a specified initial state of the register and evolving it via a set of unitary (probability preserving) operations in a specific order to produce a desired end state.
3. *The measurement* of some observable associated with the final quantum state that extracts the desired answer and allows one to record it on a classical register (e.g. classical computer memory or even a piece of paper).

The reader will have noted that measurement is a third component in quantum computation, whereas for classical computers there is no such extra step. The relevant data is obtained in the course of the computation and recorded as needed. The separation between calculation and measurement is a key attribute of quantum computers. The algorithm, or software, evolves the qubits according to the rules of quantum mechanics. The evolution is unitary. The end state is in general a superposition of quantum states that each contain potential answers to different questions. To extract the specific answer that one wants, one must perform a measurement that picks out the relevant term in the linear superposition. As discussed further in Sect. 13.8, measurement is a tricky business in quantum mechanics since it is in general distinct from unitary evolution as described by the time dependent Schrödinger equation, for example. Probability is not strictly conserved by the measurement process.[21] We will see more clearly how all this works in a couple of examples further on.

The miracle of quantum computers lies in a couple of key properties: First, qubits can store immensely more information than ordinary binary bits. The configuration, or state, of n bits in a classical computer is specified by a string of n 0's and 1's. There are therefore 2^n possible different configurations for the state of the classical computer. In the case of 4 bits, for example, there are $2^4 = 16$ configurations, such as (0000), (0001), (1010), etc. The quantum state of a collection of n qubits, on the other hand, is in general a normalized linear superposition of all 2^n possible binary states that form the basis vectors for the n qubit state. For 4 qubits there are sixteen terms possible:

$$|\Psi\rangle = a_1|0000\rangle + a_2|0001\rangle.....a_{16}|1111\rangle \tag{13.75}$$

---

[21] There is an interpretation of quantum mechanics in which measurement is a unitary process. The cost of this simplification is that measurements cause the Universe to split into many branches, each containing one of the possible outcomes of the measurement. See Sect. 13.8.

where $a_1, a_2 \ldots a_{16}$ are arbitrary complex numbers subject to the normalization condition $|a_1|^2 + |a_2|^2 + \ldots = 1$, and as usual there is an arbitrary overall phase. Thus, in a very real sense, each state of the quantum computer is a linear combination of all possible classical computer configurations and therefore in some sense stores all 2^n binary digits at once. Moreover, one can design quantum algorithms whose interactions simultaneously affect all 2^n terms in the linear superposition.

The potential for fast computation is staggering. For certain problems, quantum computers are able to perform calculations exponentially faster than ordinary computers, thereby having the potential to solve problems that are virtually unsolvable with classical computers. One example of something quantum computers are much better at than classical computers is factoring very large integers into a product of prime numbers. For example the number 60 factors into the prime numbers $3 \times 2 \times 2 \times 5$. Quantifying the speedup is somewhat difficult, but the classical computing time and resources required to solve this seemingly simple problem grows exponentially with the size of the number. The inherent difficulty in factoring large numbers provides the basis for the encryption codes[22] protecting bank accountants and other important databases.

In 1982, Richard Feynman[23] provided an abstract model of a quantum computer. The real breakthrough was made in 1985 by David Deutsch,[24] who proved that in principle *any* process could be accurately modelled by such a quantum computer. Many quantum algorithms have been devised since Deutsch's work, awaiting the day that quantum computation becomes technologically viable.

One of the earliest and most important quantum algorithms was discovered by Peter Shor, who worked at AT&T's Bell Laboratories as a research and computer scientist at the time.[25] He wrote a quantum code that could factor very large numbers (with tens of thousands of digits) exponentially faster than any ordinary computer, which means that the computation time grew as some power of the number of digits, instead of as an exponential, as with classical computers. Once Shor's algorithm is successfully implemented on a large enough working quantum computer all electronic information (both personal and financial) will be at risk. Needless to say many governments and universities have been researching quantum computing in a big way, with amazing results.

As of 2020, functioning quantum computers not only exist, but even more impressively, are available for use by the general research community on-line. Luckily for our personal data, however, these computers do not as yet have enough qubits to tackle the factorization of the very large numbers required to break current encryption codes. However, in order to keep this data private for the foreseeable future, action needs to be taken as soon as possible. One source of hope is that the very properties of quantum computers that will allow us to one day easily break classical

---

[22] These codes use a method known as RSA (Rivest-Shamir-Adleman) encryption.

[23] Feynman [4].

[24] Deutsch [5].

[25] Shor [6].

encryption codes can also be used to create unbreakable *quantum encryption codes*. This is currently an area of intense research.

13.7.4   Examples of Quantum Algorithms

The best way to make sense of all of this is with examples:

Example 4   The Deutsch Algorithm: Constant or Balanced?
    This is the simplest example of a non-trivial, potentially useful quantum algorithm. It was first proposed by David Deutsch and therefore called the *Deutsch algorithm*.

1. *The problem*
   Consider a *binary function* of the form: $f : \{0, 1\} \to \{0, 1\}$. It has two possible outputs $\{0, 1\}$ for each of the two possible inputs $\{0, 1\}$.
   Such a function can take four possible forms:

$$
\begin{aligned}
f(0) = 0 \;\; &\text{and} \;\; f(1) = 0 \\
f(0) = 1 \;\; &\text{and} \;\; f(1) = 0 \\
f(0) = 0 \;\; &\text{and} \;\; f(1) = 1 \\
f(0) = 1 \;\; &\text{and} \;\; f(1) = 1
\end{aligned}
\tag{13.76}
$$

   Which form a particular function takes can be determined by measuring both $f(0)$ and $f(1)$. We can group the possible functions according to whether the output is the same for both inputs, or whether the output is different for both inputs:

   - If $f(0) = f(1) = 0$ or 1, then f is known as a *constant function*.
   - $f(0) \neq f(1)$ with $f(0) = 0$ or 1, then f is known as a *balanced function*.

   Two of the possible functions are constant and two are balanced.
   Suppose we do not care about the specific values of the function, but would just like to know whether a given function f is constant or balanced. The only way to do this is to do two measurements in order to determine the values $f(0)$ and $f(1)$ and then check whether they are equal or not.
   Deutsch's algorithm presents the quantum version of such a measurement and presents a simple example of how entanglement makes it more efficient.

2. *The hardware*
   The register consists of two qubits. One stores the quantum information about the function f. The second is an auxiliary qubit $|x\rangle$ that essentially stores the value of the input parameter x where $x = 0, 1$. Such auxiliary qubits are frequently needed in quantum algorithms in order to extract the answer to the problem, as we will see in Sect. 13.7.5.
   Next, we need to specify the operations (called *gates* in quantum computing literature) that the quantum computer must perform in order to implement Deutsch's

algorithm. There are in fact two gates required. The first is a unitary operation on a single qubit. It is known as the *Hadamard transformation* H and creates new states by summing $|0\rangle$, $|1\rangle$:

$$H|0\rangle = (|0\rangle + |1\rangle)/\sqrt{2} \qquad\qquad (13.77)$$

$$H|1\rangle = (|0\rangle - |1\rangle)/\sqrt{2} \qquad\qquad (13.78)$$

For electrons, the Hadamard gate takes a qubit in state $|0\rangle$ whose spin is up along the z-axis and rotates it into a state that is up along the x-axis. Similarly H takes a qubit whose spin is down along the z-axis ($|1\rangle$) and rotates it so it has spin down along the x-axis. (See Eq. (13.33)). The $\sqrt{2}$'s in the denominators on the right hand side of Eq. (13.78) ensure that the final states are normalized, or in other words, that the Hadamard gate is a unitary operator.

The second gate U_f operates on two qubits as follows:

$$U_f|x\rangle|y\rangle = |x\rangle|y + f(x)\rangle \qquad\qquad (13.79)$$

where $|x\rangle$ is the auxiliary qubit in the above. U_f adds the value of the function for the argument x in the first qubit to the previous value of the second qubit. Recall that the qubits can only be in states $|0\rangle$ or $|1\rangle$ and that the value of $f(x)$ is either zero or one. Thus, the final state of the second qubit is $|0\rangle$ whenever $y + f(x)$ is either zero or two, and $|1\rangle$ whenever $y + f(x) = 1$. The former happens when $y = 0$ and $f(x) = 0$ or $y = 1$ and $f(x) = 1$, i.e. when y and $f(x)$ are the same. The latter happens when y and $f(x)$ are different. We say that the value of the second qubit after the operation is $y + f(x)$ *mod 2*.

It is common in quantum computing literature to write the Hadamard operation and U_f operations in the form

$$|0\rangle \rightarrow H \rightarrow (|0\rangle + |1\rangle)/\sqrt{2} \qquad\qquad (13.80)$$

$$|x\rangle|y\rangle \rightarrow U_f \rightarrow |x\rangle|y + f(x)\rangle \qquad\qquad (13.81)$$

and similarly for the other operations or gates.

3. *The algorithm*

(a) Start with the initial two qubit state (the input):

$$|\Psi_1\rangle = |0\rangle|1\rangle \qquad\qquad (13.82)$$

Note that the two qubits are not yet entangled.

(b) Perform the Hadamard operations on both qubits to get:

$$|\Psi_2\rangle \equiv H|\Psi_1\rangle$$
$$= \left[(|0\rangle + |1\rangle)/\sqrt{2}\right]\left[(|0\rangle - |1\rangle)/\sqrt{2}\right]$$
$$= \frac{1}{2}[|0\rangle|0\rangle - |0\rangle|1\rangle + |1\rangle|0\rangle - |1\rangle|1\rangle] \qquad (13.83)$$

Note that the two qubits are still not entangled.

(c) Act on $|\Psi_2\rangle$ with U_f:

$$|\Psi_3\rangle \equiv U_f|\Psi_2\rangle$$
$$= U_f\,[|0\rangle|0\rangle - |0\rangle|1\rangle + |1\rangle|0\rangle - |1\rangle|1\rangle]\,/2$$
$$= \frac{1}{2}\,(|0\rangle|0 + f(0)\rangle - |0\rangle|1 + f(0)\rangle)$$
$$+ \frac{1}{2}\,(|1\rangle|0 + f(1)\rangle - |1\rangle|1 + f(1)\rangle)$$

$$(13.84)$$

The two qubits are now entangled. There are two possible outcomes for $|\Psi_3\rangle$ at this point. If f is constant so that $f(0) = f(1)$ then this operation yields:

$$|\Psi_3\rangle = [|0\rangle + |1\rangle]\,[|f(0)\rangle - |1 + f(0)\rangle]\,/2 \qquad (13.85)$$

On the other hand if f is balanced then $f(1) = f(0) + 1$. This means that if $f(0) = 0$, $f(1) = 1$, while if $f(0) = 1$, then $f(1) = 0$. Thus, one gets:

$$|\Psi_3\rangle = [|0\rangle - |1\rangle]\,[|f(0)\rangle - |1 + f(0)\rangle]\,/2 \qquad (13.86)$$

This is the crucial step in the computation. Note that the second qubit has the same value irrespective of whether the function is constant or balanced, but the auxiliary qubit is in a different state for the two possible outcomes.

(d) Act with the Hadamard operation on the first qubit:

$$|\Psi_4\rangle \equiv H|\Psi_3\rangle$$
$$= |0\rangle\,[|f(0)\rangle - |1 + f(0)\rangle]\,/\sqrt{2} \quad \text{if } f \text{ is constant}$$

$$(13.87)$$

$$= |1\rangle\,[|f(0)\rangle - |f(1)\rangle]\,/\sqrt{2} \qquad \text{if } f \text{ is balanced.}$$

$$(13.88)$$

This step puts the auxilliary qubit back into spin up or spin down with the outcome depending on whether the function is constant or balanced.

4. The final step is *measurement:* One now simply measures the first qubit. If the result is $|0\rangle$, then one concludes that the function f is constant. If the result is $|1\rangle$, then it is balanced.

This is quite remarkable. The quantum algorithm requires only *one* observation in order to arrive at the desired answer, as opposed to *two* for classical computation. While this may seem like a trivial improvement, we recall first that it is still a quantum miracle, unattainable by classical means, and second that it becomes much more significant for functions involving a large number of qubits. For example one can consider a function of n qubits. $f : \{0, 1\}^n \rightarrow \{f(0), f(1)\}^n$. In this case one needs to do $2n$ measurements classically to determine whether or not it is balanced or constant. However, there exists an extension of the Deutsch algorithm, called the Deutsch-Jozsa algorithm,[26] for which a single quantum measurement reveals whether f is constant (i.e. all outputs are the same) or balanced (half of the outputs are 0 and the other half 1)! This is a huge advantage for the very large values of n (millions or billions) that are typical for classical computers.

Example 5 The Grover Search Algorithm
The problem in this case is to search for a marked element in an unstructured database of N elements (a needle in a haystack, as it were). This task is required very often in coding, so a successful quantum search algorithm could in principle be used in conjunction with an ordinary computer to speed up classical code, or to do specific tasks on its own. Classically, one expects on average to check about half the data base before finding the desired element. This requires, $N/2$ measurements. Using qubits and quantum gates, the search can be done in $\alpha\sqrt{N}$ steps, where α is a number of order unity. This is a quadratic improvement, highly significant for very large haystacks. The algorithm that does this is known as the Grover search algorithm after its inventor.[27]

13.7.5 Two Important Technical Details

Having demonstrated some of the things that a quantum computer can do that a classical computer cannot, we now discuss a couple of subtle points of quantum computation itself. The first is the problem of the inherent reversibility of quantum gates.

13.7.5.1 Reversibility

A classical computer manipulates a series of binary inputs in a circuit in the form of 0's (no current flowing) and 1's (current flowing) and gives outputs, also as a binary string. It does this by what are known as 'gates'. These gates perform logical operations, and are the building blocks of the simplest adding machines as well as the most sophisticated computers. The gates take in one or more bits as input and give out one bit or more as output. One example of such a gate is the *NOT gate*, which

Table 13.3 NOT gate

Input	Output
0	1
1	0

reverses the input, meaning it converts a 0 to 1 and 1 to 0. The terminology can be understood as follows: the binary bit can represent the truth value of some logical or mathematical statement in the code. The value 0 signifies FALSE, while 1 signifies TRUE. The NOT gate changes FALSE to NOT FALSE = TRUE, and TRUE to NOT TRUE = FALSE. The NOT gate is most easily represented by a 'truth table', which for the NOT gate looks like Table 13.3.

Next consider a logical operation in which the input consists of two bits and the output bit is such that it is zero for all inputs *except* when both input bits are 1. This corresponds to the AND gate, because an output of 1, signifies that the statements represented by the two bits are both TRUE, whereas an output of 0 tells us that at least one of the two statements is FALSE. The truth table for the AND gate is given in Table 13.4.

Note an important difference between the two gates. The NOT gate is *reversible*. This means that the output column of the first table uniquely determines the input column. The output is in one-to-one correspondence with the input. This gate can therefore be run backwards, or equivalently, a second NOT gate with the output column as its input will reproduce the first input column as its output.

The AND gate, on the other hand, gives the same output for several inputs. Thus, if one has access to just the output column, then there is no way of unambiguously reconstructing the input column. For example, the entry 0 in the second (output) column may result from any of the following input bits: {0, 0}, {0, 1} {1, 0}. Such a gate is therefore *irreversible*. Such irreversible gates are required by all non-trivial computational algorithms.

Now let us now examine the corresponding situation for a quantum computer, where 0 and 1 are replaced by qubit states $|0\rangle$ and $|1\rangle$, respectively. All operations on a qubit $|i\rangle$ are implemented by quantum gates that are unitary, i.e. norm preserving. Such gates are reversible by definition since every unitary matrix has an inverse, namely its hermitian adjoint. The quantum version of the NOT gate can be constructed as follows. If one represents the qubit states $|0\rangle$ and $|1\rangle$ as the column vectors

$$\begin{bmatrix} 1 \\ 0 \end{bmatrix}, \begin{bmatrix} 0 \\ 1 \end{bmatrix}, \tag{13.89}$$

then the quantum NOT gate can be represented as the matrix

$$U_{NOT} = \begin{pmatrix} 0 & 1 \\ 1 & 0 \end{pmatrix}. \tag{13.90}$$

It can be easily verified that $U_{NOT}^{\dagger} = U_{NOT}^{-1}$.

Table 13.4 AND gate

Input	Output
{0, 0}	0
{0, 1}	0
{1, 0}	0
{1, 1}	1

Table 13.5 Quantum AND gate is a subset of the above

Input	Output
{0, 0, 0}	**{0, 0, 0}**
{0, 0, 1}	{0, 0, 1}
{0, 1, 0}	**{0, 1, 0}**
{0, 1, 1}	{0, 1, 1}
{1, 0, 0}	**{1, 0, 0}**
{1, 0, 1}	{1, 0, 1}
{1, 1, 0}	**{1, 1, 1}**
{1, 1, 1}	{1, 1, 0}

How does one construct a quantum version of the irreversible AND gate? The answer lies in 'enlarging' the state of the quantum computer by adding extra qubits and then reducing it at the end by making a suitable measurement. Enlarging the quantum computer state allows one to introduce higher dimensional unitary operators, or gates (i.e. bigger matrices). These bigger gates are of course reversible. Irreversibility is introduced by measuring the state of the auxilliary qubit(s) once the quantum computation is finished, and then throwing out all outputs for which the auxilliary qubits do not have the desired values. This is in fact the reason that the auxiliary qubit was required for the Deutsch algorithm (see Example 4). The final measurement of the auxilliary qubit collapsed its wave function and provided the desired answer.

To see how the quantum AND gate (known as the Toffoli gate) works, we need to consider a system of three qubits. The outputs for all possible inputs are shown in Table 13.5. The third qubit in each entry corresponds to the auxiliary qubit. The outputs are determined as follows: The first two qubits are left unchanged in all cases, while the auxiliary qubit is flipped if and only if the other two entry qubits are 1. A careful examination shows that the input and output entries are in one to one correspondence: the input uniquely determines the output and vice versa. The corresponding gate is therefore reversible.

Let us focus on the subset of entries in Table 13.5, indicated in bold, which have 0 as the input for the third, auxiliary qubit. These four inputs have an output of the form $\{a, b, x\}$, with $x = 0$ unless a and b are both 1. Thus, the data with input 0 and output 1 for the auxiliary qubit constitute the data for the classical AND gate. This then is how to implement irreversible gates in a quantum computer: add one or more auxilliary qubits. Restrict the input data to particular values of the auxilliary qubit and then measure the auxilliary qubit to ensure it has the desired value. Again we stress that the measurement is what makes the process irreversible and allows, as in the present case, the construction of an irreversible AND gate.

13.7.5.2 Quantum Communication and Error Correction

Classical information or data is manipulated by classical computers, transferred from one internal memory device to another and transmitted from one place to another, e.g. via the internet, again as a string of 0's and 1's. In order to make sure that data is not corrupted during any of these processes, several copies of each string can be made. This ensures that even if some of the bits do get corrupted in transit, the majority will continue to represent the correct configuration so that an error correction can be made by fixing the corrupted (non-conforming) bit.

For example if one starts with a three bit computer in the configuration {001}, one adds bits and makes two duplicates of each configuration so that

$$\{001\} \rightarrow \boxed{\text{copy}} \rightarrow [\{001\}\{001\}\{001\}] \qquad (13.91)$$

If under transferral one of the nine bits gets corrupted (i.e. flipped), the new configuration could be: [{001}{011}{001}]. The middle bit of the middle copy is the odd one out, and with very high probability it is the corrupted one. One can then safely flip this back to restore the correct configuration: [{001}{001}{001}]. This procedure assumes that the chance of a bit getting corrupted in one set is small, so that the probability of two in the same set being corrupted is even smaller. If the chance of corruption is a bit higher, one simply needs to add more duplicates, and always take the majority to be correct. By adding more and more bits, it is possible to make the chance of corruption of the final data set as small as one would like.

Errors can occur in quantum computation as well, where in addition to random or human errors, there is also something called *environmental decoherence*, that tends destroy entanglement and hence information. This is just a fancy way of saying that it is difficult to keep all the qubits from interacting with degrees of freedom that are not a part of the computation. Such interactions tend to transfer the entanglement to these other degrees of freedom thereby destroying the entanglement of the qubits in the hardware. One way to minimize this decoherence is to isolate the hardware qubits from their environment as much as possible and cool down the computer to near absolute zero to minimize energy transfer. Even if this is done, errors occur.

Unfortunately, the apparently simple procedure of duplicating the qubits, and making sure that the copies remain faithful, as in the classical computer, cannot be carried over directly to quantum computation. There is a very deep but easy to prove theorem in quantum mechanics called the *no-cloning theorem*, which makes it literally impossible to copy via a unitary operation the general state of a qubit onto another qubit while retaining the original one in its original state.

Proof of No-Cloning Theorem

The proof is by contradiction. Consider a qubit, call it qubit 1, in the state:

$$|\psi\rangle_1 = a|0\rangle_1 + b|1\rangle_1 \tag{13.92}$$

with $|a|^2 + |b|^2 = 1$. Now assume that there exists an unitary 'cloning operator' U_C that takes a second qubit, c initially in state $|0\rangle_c$, say, into an exact copy of the state of qubit 1 without changing the state of the first qubit. In terms of equations:

$$U_C|\psi\rangle_1 |0\rangle_c = |\psi\rangle_1|\psi\rangle_c$$
$$= (a|0\rangle_1 + b|1\rangle_1) \otimes (a|0\rangle_c + b|1\rangle_c)$$
$$= (a|0\rangle_1 \otimes a|0\rangle_c + a|0\rangle_1 \otimes b|1\rangle_c + b|1\rangle_1 \otimes a|0\rangle_c + b|1\rangle_1 \otimes b|1\rangle_c)$$
$$\tag{13.93}$$

However, when the same operator acts on the individual basis qubits in the expansion,

$$U_C |\psi\rangle_1|0\rangle_c = U_C(a |0\rangle_1|0\rangle_c + b |1\rangle_1|0\rangle_c)$$
$$= a (U_C |0\rangle_1|0\rangle_c) + b (U_c|1\rangle_1|0\rangle_c)$$
$$= a |0\rangle_1|0\rangle_c + b |1\rangle_1|1\rangle_c . \tag{13.94}$$

Thus the two routes give different answers, except when a or b are 0.

To make things worse, not only is duplicating states impossible, but trying to do occasional checks on the state vector for its integrity is also impossible because as soon as $|\psi\rangle$ is measured, it collapses into one of the eigenstates, losing all information.[28] Luckily, using a clever combination of three steps one can nonetheless correct errors that creep into the state of the quantum computer. These steps are roughly.

(a) Enlarge the space of qubits by duplicating all the bases state $|0\rangle$ and $|1\rangle$ to basis states consisting, say, of three qubits: $|000\rangle$ and $|111\rangle$.

(b) Represent the desired qubit states in terms of these enlarged basis states. The state $|\psi\rangle = a|0\rangle + b|1\rangle$ would be represented in the present case by:

$$|\psi\rangle = a|000\rangle + b|111\rangle \tag{13.95}$$

(c) Apply the operations required to encode the quantum information on these states.

(d) Monitor the individual qubits in the basis states to check if the qubits in each of them are still the same. This can be done without destroying the quantum information. If two of the qubits in an enlarged basis state are different, then correct the one that has most likely been corrupted.

---

[28] From this apparently bad feature of quantum information in the context of error correction, one can extract something very good for data encryption and secure transmission. The no-cloning theorem is the basis for quantum encryption codes. By sending data via qubits, one insures that the information contained in the qubits is destroyed as soon as a potential eavesdropper tries to measure one of the qubits.

The above procedure, called quantum error correction, complicates the codes and increases the resources required considerably, but not to the extent of eliminating the exponential benefits quantum computers provide for solving some very difficult, if not impossible, problems.

13.8   Interpretations of Quantum Mechanics: What Does It All Mean?

Quantum mechanics tells us that the state of a particle is described by a normalized, complex valued wave function. The magnitude squared of the wave function at each point reveals the probability per unit length of finding the particle at a particular point along the real line when measured. The magnitude squared of the Fourier transform of the wave function tells us the probability per unit wave number, k, of measuring the momentum of a particle to be $\hbar k$. Because of the mathematical relationship between the wave function and its Fourier transform, quantum mechanics cannot describe individual particles with both specific position and momentum. However, if a particle is in a state with well defined momentum, so that its position is uncertain and we decide to measure the position, the position of the particle directly after the measurement will be known, and the momentum will be uncertain. This is described as wave function collapse.

Does the uncertainty in the position before the measurement mean that the particle did not have a position until after the measurement? Or did the particle have both definite position and definite momentum before and after the measurement, but the quantum wave function was unable to describe it? This is in essence what Bohr, Einstein and many physicists and philosophers debated at the turn of the 20th century. Amazingly, as we saw in the Section on Bell's Theorem (Sect. 13.3), this rather philosophical debate was partially resolved via ground-breaking theoretical input combined with painstaking experiments. In particular, we found that at least for certain experiments there was no way of assigning values to all non-commuting observables for specific entangled states that satisfy the predictions of quantum mechanics without violating locality.

There nonetheless still exist many related, unanswered questions. What exactly is measurement? Since wave function collapse is non-unitary (doesn't preserve probabilities) it cannot be described within the framework of quantum mechanics. This is in effect another way of asking how particle position, for example, "comes into being" during the measurement process. Alternatively, perhaps particles do in some sense have well-defined positions and momenta, despite the implications of the violations of Bell's theorem. Clearly this latter view would require some form of non-locality. Debate still exists regarding these issues, perhaps in part because of an inherent need to understand what it all means and perhaps also in part to try to pave the way for the development of a more fundamental theory.

13.8.1   The Copenhagen Interpretation

Bohr and his disciples asserted that particles do not have positions or momenta until those attributes are measured. This implies that there is in effect no particle in the usual sense, only the particle wave function. Bohr took the view that one should not even ask questions about the position and momentum of a particle except as part of the measurement process. One need only worry about the probabilities of outcomes in order to make predictions for physical experiments that involve a large number of similarly prepared particles. Bohr's pragmatic approach, sometimes referred to as "shut up and calculate", has enabled many great discoveries but to some physicists at least, it appears less than satisfactory.

Apart from the conceptual issues that Einstein among others had with Bohr's approach, there is a serious technical problem. Bohr's world is divided into two apparently distinct parts, namely the microscopic world described as a quantum wave function, and the classical world that describes the measuring apparatus. It doesn't describe how the act of measurement results in the position of the particle going from completely uncertain to completely certain. In this sense quantum mechanics as described by Bohr can be said to be incomplete, although not in the way suggested in the EPR paper.

There have been attempts to produce wave function collapse dynamically by adding terms to the Schrödinger equation that are non-linear in the wave function, or by invoking the concept of *environmental decoherence*, as described in Sect. 13.7.5. In brief, this approach hopes to explain wave function collapse as a consequence of quantum interactions of the system with the rather "dirty" environment in which the measurement is performed. Environmental decoherence therefore stays completely within the framework of standard quantum mechanics, and is an ongoing subject of research.

13.8.2   The Many-Worlds Interpretation

One way to solve the problem associated with wave function collapse is to assume that wave functions do not collapse. This is the contention of the *many-worlds interpretation* of quantum mechanics, first put forward by Hugh Everett and advocated for effectively by Bryce DeWitt. The many-worlds interpretation does away with the dichotomy between the microscopic quantum system in an experiment and the classical measuring apparatus by suggesting that both are quantum subsystems of the experiment that are described by a wave function. A measuring apparatus behaves more or less classically because of its scale relative to Planck's constant, making quantum fluctuations irrelevant to its behaviour. In this view, a measurement is an interaction between the microscopic subsystem and the measuring apparatus that produces an entangled state of both.

As an example, suppose an electron is produced in a state that is a linear combination of spin up and spin down along some axis:

$$|\psi\rangle_e = \frac{1}{\sqrt{2}}\left(|+\rangle + |-\rangle\right) \tag{13.96}$$

The measuring apparatus, a black box with a dial, is initially in a state in which the dial is pointing straight up. If the dial moves to the right, the electron is measured to be spin up, whereas if the dial moves to the left, it is measured to be spin down. In the Bohr interpretation, the measurement collapses the wave function of the electron and the dial moves either to the left or to the right.

If the measuring apparatus is quantum, then according to the previous paragraph, it can be described by the state: $|\uparrow\rangle_M$. The complete quantum state of the electron + apparatus is therefore:

$$|\Psi\rangle_{\text{before}} = |\psi\rangle_e \otimes |\uparrow\rangle_M \tag{13.97}$$

The interaction between the electron and the measuring apparatus causes their respective wave functions to be entangled:

$$|\Psi\rangle_{\text{after}} = \frac{1}{\sqrt{2}}\left(|+\rangle_e \otimes |\rightarrow\rangle_M + |-\rangle_e \otimes |\leftarrow\rangle_M\right) \tag{13.98}$$

After the measurement, both possible outcomes are equally "real".[29] There are then two copies of the classical apparatus, one of which sees spin up, the other sees spin down. Note that the above argument works irrespective of the nature of the measuring apparatus. It could be a Stern-Gerlach detector that uses a magnetic field to deflect spin up electrons in one direction and spin down in another, or it could be a physics professor who records the results registered by the Stern-Gerlach detector. This interpretation appears to solve in principle both the conceptual and technical problems associated with measurement.

There is of course a cost. The many-worlds interpretation posits a Universe consisting of a single, immense (in the sense of describing many degrees of freedom) wave function that evolves according to one complicated Schrödinger equation. It is continuously splitting into branches that split into more branches, every time subsystems interact or a measurement is made. Nature therefore consists of an immeasurable number of parallel "Universes", each of which contains one particular outcome of every interaction or measurement. All possibilities are realized and their relative numbers are consistent with the statistical predictions of quantum mechanics. In principle the many-worlds view is indistinguishable experimentally from the Bohr interpretation. Do either provide a satisfying world view? This is a matter of taste.

---

[29] One can think of each branch existing independently in a separate Universe or world, hence the term *Many Worlds*.

13.8.3 Hidden Variables and Non-locality

Both the Copenhagen and many-worlds interpretations of quantum mechanics take the view that the wave function provides a complete description of the state of individual particles or systems of particles. Complementary observables such as position and momentum or spin along two different axes cannot simultaneously have precise values for any physical state. And yet, measurement of either produces a precise value. The question then is whether the values of observables, especially those that can't be predicted with certainty, are in a sense created by the measurement process, or whether the measurement simply reveals a previously unknown but real attribute of the particle. In the Copenhagen interpretation, the former is true: observables do not have values until they are measured by a classical system. The many-worlds interpretation on the other hand does not allow for classical systems. Measurements entangle the state of the observed system with that of the quantum observing apparatus, so all measurable values are equally real, or not real, depending on your point of view.

There are those, like EPR, who would like the values of observables such as position and momentum to be attributes of the particles themselves, independent of whether or not we choose to measure them. If individual particles have well defined values of both position and momentum, something not describable by a quantum state, how should we interpret the wave function? The *statistical interpretation* of quantum mechanics stipulates that individual electrons, or pairs of electrons in an experiment, do carry specific values of both position and momentum, say, or all spins along all three axes. The corresponding wave function, or quantum state, does not however apply to individual particles or individual systems of particles in an experiment. Instead it provides information about the statistics that emerge when performing measurements on a large number of identical systems, called *ensembles*, that are all prepared in the same quantum state. Quantum mechanics does put a powerful constraint on the observables associated with individual systems: each and every measured value of an observable must correspond to one of the eigenvalues of that observable (Sect. 10.8).

According to the statistical interpretation, one can think of the beam of electrons in the double-slit experiment (Sect. 8.4.1) as an ensemble of electrons each with a well defined position and momentum. These complementary variables cannot, of course, be known simultaneously. The relevant measurement for the electron double slit experiment involves locating each electron in the ensemble on the screen on the far side of the two slits. In this view, the wave function carries no information about the position and momentum of each individual electron. Instead it tells you how these values are distributed among members of the ensemble. According to the uncertainty principle, if the dispersion of the values of position, say, of the particles in the beam is zero, i.e. they all have the same position, then the values of momentum in the beam are maximally distributed over the allowed values, and *vice versa*. The statistical interpretation is therefore able to reconcile the interference pattern on the screen with our intuitive desire to allow individual electrons to pass through only one slit at a time. It is the wave function that passes through both slits and undergoes interference, thereby affecting the statistics of the electron locations when they hit the screen.

The statistical interpretation is no less viable than the Copenhagen or many-worlds, as long as it can be shown that it does not change any of the verified predictions of quantum mechanics. The question that interests us here is whether or not the statistical interpretation is helpful to those who would like attributes of physical systems to have a reality that is independent of measurement. Does it allow one to construct an underlying theory that in principle assigns values to all observables, both commuting and non-commuting, associated with the individual particles in an ensemble (or beam) for all possible experiments? Since quantum mechanics does not allow these values to be measured or known simultaneously, such hypothetical theories are dubbed *hidden variable theories*.

In our discussion of Bell's Inequalities in Sect. 13.3 we saw that for at least one specific experiment involving a pair of electrons in a particular quantum state (total spin zero), it appeared to be impossible to assign values to the three (non-commuting) components of the spin of each electron without giving up either locality or basic arithmetic. The predictions of quantum mechanics violated Bell's inequalities. As shown by Bell, these inequalities must be satisfied by the statistical outcomes of any experiment that allows consistent assignment of hidden variables to individual elements in the corresponding ensemble of similarly prepared systems.

Does this mean that hidden variable theories are dead? The short answer is yes, unless one is willing to allow the hidden variables to be changed via action at a distance. In order to gain a deeper understanding of the features of quantum mechanics that condemn hidden variable theories to this fate, we will describe a somewhat more complicated and less known "no-go theorem", also due to Bell.[30] This theorem forbids the assignment of hidden variables based only on fundamental properties of certain quantum observables without relying on the properties of any particular quantum state.

First some history: One of the first rigorous attempts to construct a "no-go" theorem for hidden variable theories was made in 1933 by the famous mathematician John Von Neumann.[31] The Von Neumann theorem played a large role in discouraging research into hidden variable theories for many years. However, any theorem starts with a set of assumptions. It was argued by John Bell thirty-three years later that one of Von Neumann's assumptions was, in Bell's words "silly". To get a handle on Bell's criticism, we note that not all quantum observables for a system are independent. It is often the case that one of the operators associated with an observable can be written in terms of two others. Such relations are called *operator identities*, because they are identically satisfied by the operators themselves and do not depend on what states they act upon.

Von Neumann mistakenly assumed that in any viable hidden variables theory, the values of complementary observables assigned to individual particles in an ensemble

[30] Bell [9].

[31] Von Neumann [10].

(i.e. those that quantum mechanics does not allow to be measured simultaneously) must satisfy equations identical in form to the identities satisfied by the corresponding operators.[32] Bell realized that because the values of complementary observables for individual particles cannot be known simultaneously, this restriction was not needed physically. Even if the individual values did violate such operator equations, we would never know. More significantly, Bell showed that if this unnecessary restriction on complementary observables were removed, then hidden variable theories were not only possible in principle, but already existed, albeit for a very limited set of toy models.

Bell then went on to prove his own rigorous "no-go theorem"[33] based only on assumptions that were physically sensible. Bell's theorem showed unambiguously under what conditions hidden variable theories could exist. The proof was quite complicated, but David Mermin[34] has provided a much simpler and remarkably elegant version, which we now describe.

Consider a spin 1/2 particle such as an electron. We know that the three components of spin can only take on one of two values $\pm\hbar/2$. We also know that the operator corresponding to the magnitude squared of the spin vector is:

$$\hat{S}^2 = \hat{S}_x^2 + \hat{S}_y^2 + \hat{S}_z^2 \tag{13.99}$$

For a spin 1/2 particle, $\hat{S}^2$ has the value $s(s+1)\hbar = 3\hbar/4$, where the spin quantum number $s = 1/2$. We will need the following relations between the spin operators:

$$\hat{S}_x\hat{S}_y - \hat{S}_y\hat{S}_x = i\hbar\hat{S}_z \quad +\text{cyclic} \tag{13.100}$$

$$\hat{S}_x\hat{S}_y + \hat{S}_y\hat{S}_x = 0 \quad +\text{cyclic} \tag{13.101}$$

$$\Rightarrow \hat{S}_x\hat{S}_y = -\hat{S}_y\hat{S}_x = \frac{\hbar}{2}\hat{S}_z \quad +\text{cyclic} \tag{13.102}$$

$$\hat{S}_x^2 = \hat{S}_y^2 = \hat{S}_z^2 = \left(\frac{\hbar}{2}\right)^2 \mathbb{1} \tag{13.103}$$

The above are algebraic identities obeyed by the quantum spin operators defined in Eqs. (13.46), (13.47) and (13.48) in Sect. 13.4. A formal proof can be obtained purely from symmetry considerations, but this is beyond the scope of this text. Instead, we use the fact that the spin operators are represented as 2×2 dimensional complex matrices that act on the two dimensional complex spin states (vectors) described in Sect. 13.4 by matrix multiplications.

Exercise 8 Show using matrix multiplication that $\hat{S}_x$, $\hat{S}_y$ and $\hat{S}_z$ as defined in Eqs. (13.46), (13.47) and (13.48) obey the identities in Eqs. (13.100)–(13.103).

---

[32] We will see examples of such identities a bit further on in this section.

[33] We refer to this no-go theorem as Bell's theorem. It is important to distinguish this from Bell's derivation of his famous Bell's inequalities. Interestingly, the latter was published first even though the former was completed earlier.

[34] David Mermin [11].

Because the components of the spin vector operator $\hat{\mathbf{S}}$ do not commute (Eq. (13.100)), no quantum states exist for which all three components of spin, or in fact any two, have specific values. On the other hand the total spin squared operator $\hat{S}^2$ commutes with each component:

$$0 = \hat{S}^2 \hat{S}_x - \hat{S}_x \hat{S}^2$$
$$= \hat{S}^2 \hat{S}_y - \hat{S}_y \hat{S}^2$$
$$= \hat{S}^2 \hat{S}_z - \hat{S}_z \hat{S}^2 \tag{13.104}$$

so that each electron in an ensemble can be prepared in a state that has a definite value for $\hat{S}^2$, namely $\frac{3\hbar^2}{4}$ and **one** of the components. In the language of ensembles, the measurement of both $\hat{S}^2$ and $\hat{S}_x$, say, on a large number of electrons in such a state will yield specific values with zero dispersion (rms deviation).

Exercise 9 Prove that
$$\hat{S}^2 \hat{S}_x - \hat{S}_x \hat{S}^2 = 0 \tag{13.105}$$

The analogous relations for the y and z components of spin follow by symmetry.

Hidden variable theories would like to assign values to all three spin components for each electron in a beam or ensemble. Can this be done in a way that is consistent with identities Eqs. 13.100–13.103? The answer is no (as shown in Von Neumann's proof), but we now know that this is not a problem because we can never measure all three simultaneously.

The goal of Bell's theorem was to put constraints on possible Hidden Variables theories by providing examples of situations in which consistent assignment of values for all quantum observables to all members of the relevant ensemble is not possible, even if one requires the values of only commuting observables to obey the relevant operator identities. In order to do this one needs to consider systems with two particles or more, so that one has more observables, commuting as well as non-commuting, at one's disposal. We therefore start with two spin one half particles as in our Bell inequality experiment in Sect. 13.3. These electrons have associated with them two independent sets of spin components $\{S_x^{(i)}, S_y^{(i)}, S_z^{(i)}, i = 1, 2\}$. The spin operators of particle 1 all commute with the spin operators of particle 2. The components in each set obey the identities Eqs. (13.100)–(13.103). From these six operators we can construct several sets of operators such that every operator commutes with every other operator in a given set. Consider for example, the set of three operators $\{\hat{S}_x^{(1)}, \hat{S}_x^{(2)}, \hat{X}(1, 2)\}$ where $\hat{X}(1, 2) := \hat{S}_x^{(1)} \hat{S}_x^{(2)}$. Since all three commute with each other, we can measure the values of all three of these observables simultaneously for any given two electron state, and via this measurement process create a quantum state for which all three values are precise. According to the statistical interpretation, the values of all three observables are the same for all electron pairs in the corresponding ensemble.

Interestingly, if you multiply these three operators, the relations Eqs. (13.100–13.103) imply that you get the identity operator times a numerical factor:

$$\hat{S}_x^{(1)} \hat{S}_x^{(2)} \hat{X}(1, 2) \equiv \frac{\hbar^4}{2^4} \mathbb{1} \tag{13.106}$$

Equation (13.106) states that the action in succession of these three operators leaves any state unchanged. Since the three operators commute, the order of operations is irrelevant.

Now consider an ensemble of electron pairs, all prepared in the same state such that it is possible to assign specific values to each electron pair for all three observables in Eq. (13.106) because they commute. Equation (13.106) implies that the values we assign them are not independent because, for example, we can obtain a value for $\hat{X}(1, 2)$ for a single pair of electrons by measuring only $\hat{S}_x^{(1)}$ and $\hat{S}_x^{(2)}$. A subsequent measurement of $\hat{X}(1, 2)$ must produce the inverse of the product of these values, since multiplying the product of the three measurements must give the value 1.

This then provides a set of commuting observables that is relevant to Bell's version of Von Neumann's theorem. Bell quite sensibly required the values of commuting observables assigned to each element of an ensemble of similarly prepared systems to obey the same identities as their respective operators.[35] Otherwise the predictions of the theory would not be consistent with experiment.

We now note that there exists another set of three commuting operators, namely

$$\{\hat{X}(1, 2), \ \hat{Y}(1, 2), \ \hat{Z}(1, 2)\}. \tag{13.107}$$

where $\hat{Y}(1, 2) := \hat{S}_y^{(1)} \hat{S}_y^{(2)}$ and $\hat{Z}(1, 2) := \hat{S}_z^{(1)} \hat{S}_z^{(2)}$. They satisfy the identity:

$$\hat{X}(1, 2)\hat{Y}(1, 2)\hat{Z}(1, 2) = -\frac{\hbar^6}{2^6} \mathbb{1} \tag{13.108}$$

Acting on any state with the three operators on the left of Eq. (13.108) yields minus the same state back again, irrespective of the order of operation.

Exercise 10 Using the individual single particle identities Eqs. (13.100)–(13.103) prove the following additional identities:

1.

$$\hat{S}_x^{(1)} \hat{S}_x^{(2)} \hat{X}(1, 2) = \left(\frac{\hbar}{2}\right)^4 \mathbb{1} \tag{13.109}$$

2. $\hat{X}(1, 2)$, $\hat{Y}(1, 2)$ and $\hat{Z}(1, 2)$ are mutually commuting.
3.

$$\hat{X}(1, 2)\hat{Y}(1, 2)\hat{Z}(1, 2) = -\left(\frac{\hbar}{2}\right)^6 \mathbb{1} \tag{13.110}$$

As before, all three operators $\hat{X}(1, 2)$, $\hat{Y}(1, 2)$, and $\hat{Z}(1, 2)$ can be determined simultaneously but the identity fixes the value of one of them in terms of the other two.

---

[35] Recall that Von Neumann's mistake was requiring this of noncommuting operators as well.

Now suppose there exists a pair of particles whose value of $\hat{X}(1, 2)$ we wish to measure. We can do this in one of two ways: either by measuring $\hat{S}_x^{(1)}$ and $\hat{S}_x^{(2)}$ or by measuring $\hat{Y}(1, 2)$ and $\hat{Z}(1, 2)$. These are two different experiments. A Hidden Variable Theory would require the measured value of $\hat{X}(1, 2)$ to be a property of the pair of particles that is independent of how we choose to measure it. Hence, both experiments must yield the same value. Moreover, the values of the other two pairs of observables assigned to each member of the ensemble must also respect the identities. In the present case, there is no problem consistently assigning values to all five observables. For example, suppose $\hat{X}(1, 2) = +\frac{\hbar^2}{4}\mathbb{1}$, then $\hat{S}_x^{(1)}$ and $\hat{S}_x^{(2)}$ must both be plus or minus, while $\hat{Y}(1, 2)$ and $\hat{Z}(1, 2)$ must have opposite signs.

If the above were the only set of commuting observables obeying such identities, then there would be no problem. In fact, one can construct nine of them, as laid out in Table 13.6.[36] where

$$\hat{Z}(1, 2) = \hat{X}(1, 2)\hat{Y}(1, 2) \qquad (13.111)$$

$$\hat{V}(1, 2) = \hat{S}_x^{(1)}\hat{S}_y^{(2)} \qquad (13.112)$$

$$\hat{W}(1, 2) = \hat{S}_y^{(1)}\hat{S}_x^{(2)} \qquad (13.113)$$

and the order of operators on the right hand sides of Eq. (13.111)–(13.113) does not matter since the relevant operators commute. You can verify that the three observables in each row and each column of Table 13.6 commute with each other and can therefore can be measured simultaneously. As verified in Exercise 10, the product of the three observables in all three rows gives a positive number times the identity operator, while the product of observables in the columns on the left also gives a positive number times the identity operator. The product of the three observables in the column on the right gives a negative number times the identity operator. Thus there are six identities in all satisfied by the nine operators.

In preparing for the *coup de grace*, we recall that the basic tenets of a Hidden Variables approach require that the values of all nine observables in the table be specifiable (but not necessarily observable simultaneously) for each pair of particles in an ensemble of similarly prepared systems. We now see that the values assigned to each pair must satisfy all six identities in the rows and columns of Table 13.6. The essence of Bell's theorem in this particular context is that:

It is impossible to assign values for all nine operators in Table 13.6 to a single electron pair without violating at least one of the six identities.

The proof is straightforward. The product of values of the nine observables in all three rows must give a positive number, since the product of the operators in each row is the unity operator. The product of the same nine observables obtained from the three columns must yield a negative number, since the product of two of the rows

---

Table 13.6 Nine observables associated with two electrons

$\hat{S}_x^{(1)}$	$\hat{S}_x^{(2)}$	$\hat{X}(1,2)$
$\hat{S}_y^{(2)}$	$\hat{S}_y^{(1)}$	$\hat{Y}(1,2)$
$\hat{V}(1,2)$	$\hat{W}(1,2)$	$\hat{Z}(1,2)$

give a positive number while the product of the third gives a negative number. There is no way for nine real numbers to satisfy both these requirements simultaneously.

It seems that hidden variables are dead. And yet, there is, as always, a loop-hole.[37] Each of the nine observables in Table13.6 can be measured in one of two ways: either by measuring the other two values in the corresponding row, or the two values in the corresponding column. We assumed in the previous argument that the value for this observable that emerged via the row measurement must be the same as the value that emerged via the column measurement. However, only one set of measurements can be done on an element of an ensemble, and in fact the two sets require quite different experimental setups. Hidden variable theories can avoid the *coup de grace* administered by John Bell, if they allow the value of the observable assigned to a given element of the ensemble to be different depending on which of the two experiments is performed. With this assumption it would not be necessary to satisfy all the identities simultaneously for each element of the ensemble, only the identities that were relevant to the measurement. This solution is given the name *contextuality*, because the hidden variables depend on the context of the complete experiment, and not just on the state creation.

Hopefully contextuality worries you because it seems to go against the spirit, if not the letter, of the principles behind the search for hidden variable theories. To make matters worse, Mermin showed in his 1993 article that it is possible to construct experiments in which contextuality, if accepted, leads invariably to locality. [38]

The bottom line: Hidden variable theories that assign an underlying reality to non-commuting observables associated with individual particles in an ensemble and are consistent with the predictions of quantum mechanics can only be constructed at substantial cost. One must give up non-contextuality, and in some instances locality.

This is pretty much the same conclusion we arrived at via Bell's inequalities in Sect. 13.3. The difference is that here we relied only on properties of the quantum operators and not on the creation of any particular quantum state. It is therefore more broadly applicable than Bell's inequalities, which rely on knowing the predictions of quantum mechanics for specific types of experiments. Mermin's beautiful proof also helps isolate the key features of quantum mechanics that are at the heart of

---

[37] John Bell famously said in his paper entitled "On the impossible pilot wave", Foundations of Physics 12, pp. 989–999 (1982), that "What is proved by impossibility proofs is lack of imagination". Clearly his own proof in this regard was not meant as the final word either.

[38] In order to prove that locality must be violated in some cases, it is necessary to consider three particles, not two, and restrict consideration of the states on which the operators act to a subset of all possible states. We encourage the reader to look at Mermin's 1993 article.

quantum weirdness: quantum observables are operators (differential or matrix) that behave very differently[39] than their observed values (eigenvalues). This is why the nine operators can obey all six identities, but there exists no set of eigenvalues that can do so.

References

1. A. Einstein, B. Podolski, N. Rosen, Can the quantum-mechanical description of physical reality be complete? Phys. Rev. **47**, 777–780 (1935)
2. J. Bell, On the Einstein-Podosky-Rosen Paradox. Physics **1**, 195–200 (1964)
3. J. Yin et al., Satellite-based entanglement distribution over 1200 km. Science **356**, 1140–1144 (2017)
4. R.P. Feynman, Int. J. Theor. Phys. **21**, 467 (1982)
5. D. Deutsch, Proc. Roy. Soc. London, Ser. A **400**, 97 (1985)
6. P.W. Shor, Algorithms for quantum computation: Discrete logarithms and factoring, in *Proceedings of the 35th Annual Symposium on Foundations of Computer Science*, IEEE Computer Society Press, vol. 10 (1994)
7. D. Deutsch, R. Jozsa, Rapid solutions of problems by quantum computation. Proceedings of the Royal Society of London A **439**(1907), 553–558 (1992)
8. L.K. Grover, A fast quantum mechanical algorithm for database search, in *Proceedings, 28th Annual ACM Symposium on the Theory of Computing* (1996), p. 212
9. J. Bell, On the problem of hidden variables in quantum mechanics. Rev. Modern Phys. **38**, 447–452 (1966)
10. J. Von Neumann, *Mathematische Grundlagen der Quantenmechanik* (Springer, Berlin, 1932)
11. N. David Mermin, Hidden variables and the two theorems of John Bell. Rev. Modern Phys. **65**, 803 (1993)

---

[39] In technical terms, they obey a different algebra.

Chapter 14
Conclusions

It is hoped that the reader has developed a greater appreciation for the two foundations of physics, namely special relativity and quantum mechanics. At the risk of over-extending a metaphor, if special relativity and quantum mechanics form the concrete foundation of the edifice that is twenty-first century physics, then symmetry is the rebar, or reinforcing steel, that threads its way throughout, providing durability and strength, as well as simplicity of form.

We now summarize a few key points

- The predictions of special relativity and quantum mechanics have been well tested over the years in their respective regimes. The accuracy of such tests continues to increase and not one violation has yet been confirmed.
- Both special relativity and quantum mechanics have lead to great technical advances that have changed the world, albeit not always for the better. Without special relativity, GPS satellites would not be nearly accurate enough to guide us to our destinations, nor would we have nuclear power plants or atomic bombs. The applications of quantum mechanics are even more ubiquitous. As we have seen, it lies at the heart of our understanding of the periodic table and hence most of chemistry. In addition, the digital age could just as easily be dubbed the quantum age. Quantum mechanics has lead to the CCD, which is the basis of digital cameras, as well as the transistor, the fundamental building block of smart phones, computers and virtually all the digital technology on which we currently rely. In addition, quantum computers utilizing quantum entanglement as a resource are poised to revolutionize computation and encryption.
- Equally important are the deep insights that symmetry provides about how nature operates at its most basic level. Noether's theorem, first derived within the context of Newtonian mechanics, shows that symmetry lies at the origins of conserved quantities such as energy, momentum and angular momentum. In addition, a single symmetry assumption, namely that the laws of physics must be the same for all

G. Kunstatter and S. Das, *A First Course on Symmetry, Special Relativity and Quantum Mechanics*, Undergraduate Lecture Notes in Physics, https://doi.org/10.1007/978-3-030-92346-4_14

observers, independent of their frame of reference, leads invariably to special relativity and Einstein's theory of gravity.

- Special relativity and quantum mechanics can be combined into a single theoretical framework called *relativistic quantum mechanics* and its multi-particle counterpart relativistic quantum field theory. These theories have had great success in explaining the diversity of phenomena at the microscopic level. For example, Dirac's quest to unite quantum mechanics with special relativity led him to predict the existence of positrons (anti-electrons).

- One thing not generally emphasized is that special relativity and quantum mechanics are more than theories per se: they are frameworks for constructing theories. In essence, special relativity requires whatever theory one constructs to be invariant under certain spacetime symmetry operations. Similarly, the underlying structure of quantum mechanics is quite general and can be rigorously formulated without specifying the degrees of freedom being considered or the dynamics (Hamiltonian) that describes the time evolution.

- Nature often behaves in ways that appear strange and counter-intuitive:

  - Special relativity tells us that in order for the laws of physics to be the same for all inertial observers, space and time must fuse into a single spacetime continuum. The time elapsed between events and the lengths of objects are different as measured by observers in different states of motion.
  - General relativity tells us that in order for the laws of physics to be the same in all frames of reference (accelerating as well as non-accelerating), spacetime becomes a dynamical participant that is affected by and affects the motion of matter and energy. As Shakespeare so astutely observed: "All the world is a stage...".[1] What he didn't know was that the stage itself was also a player.
  - Observers in accelerated frames and strong gravitational fields age more slowly than those in inertial frames.
  - Time as we know it appears to have had a beginning at the Big Bang singularity just over fourteen billion years ago. Time also appears to end at the singularities that lurk at the center of all black holes.
  - The microscopic world described by quantum mechanics is probabilistic. Quantum states provide statistics instead of precise values for the outcomes of measurements. Inherent in its very structure is an apparent breakdown of the Newtonian world view of an objective, local reality. Spooky action at a distance has been verified experimentally.
  - General relativity and quantum mechanics appear to be incompatible. There exists as yet no complete, falsifiable theory of *quantum gravity* that incorporates both. One of the more promising candidates, namely string theory, describes a ten dimensional Universe consisting of quantized vibrating strings. It is truly an embarrassment of riches in that it contains enough flexibility to describe an unimaginably large collection of different Universes that may or may not include one resembling our own.

---

[1] *As You Like It*, act 2, scene 7.

Clearly there is still a great deal that we do not understand about how the Universe works. What is needed in order to make further progress is guidance from a startling new experimental result that contradicts the predictions of either general relativity or quantum mechanics, or perhaps both. Alternatively, progress might be made via powerful new reasoning such as that which led to special and general relativity. There are strong reasons to hope that a new revelation may soon be forthcoming: immense strides have recently been made in quantum mechanics (vis. quantum computing), astronomy (new generations of telescopes and satellites), astrophysics and cosmology (observations of black holes and deciphering the CMBR) and gravitational waves (a new window on the Universe)

Stay tuned...

Chapter 15
Appendix: Mathematical Background

15.1 Complex Numbers

The quantum description of the state of a particle at any instant in time is given by a wave function $\psi(x, t)$ that assigns a complex number to each point x in space at each time t. It is therefore useful to review the properties of complex numbers:

1. A complex number can be written as:

$$z = a + ib \qquad (15.1)$$

where

$$i^2 = -1 \qquad (15.2)$$

and a, b are real numbers. a is the real part of z, b is its imaginary part.

2. Complex numbers can be represented as a point or vector on the two-dimensional complex plane, with the x-axis giving the real part and the y-axis giving the imaginary part. This is illustrated in Fig. 15.1.
   The wave function can be split into a real and imaginary part as well:

$$\psi(x, t) = \psi_R(x, t) + i\,\psi_I(x, t) \qquad (15.3)$$

3. **Euler Formula**: As Fig. 15.1 suggests, one can represent complex numbers in terms of polar coordinates (ρ, θ). From the diagram we can see that

$$
\begin{aligned}
a &= \rho \cos(\theta) \\
b &= \rho \sin(\theta) \\
\rightarrow z &= \rho(\cos(\theta) + i \sin(\theta))
\end{aligned}
\qquad (15.4)
$$

© The Author(s), under exclusive license to Springer Nature Switzerland AG 2022
G. Kunstatter and S. Das, *A First Course on Symmetry, Special Relativity and Quantum Mechanics*, Undergraduate Lecture Notes in Physics,
https://doi.org/10.1007/978-3-030-92346-4_15

Fig. 15.1 Complex plane

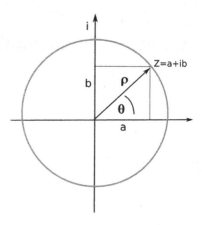

where

$$\rho = \sqrt{a^2 + b^2}$$
$$\theta = \cos^{-1}\left(\frac{a}{\sqrt{a^2 + b^2}}\right) \tag{15.5}$$

What makes this interesting is the existence of *Euler's formula*, which states that:

$$e^{i\theta} = \cos(\theta) + i\sin(\theta) \tag{15.6}$$

The proof is straightforward. Taylor expansions of the exponential, cosine and sine functions in the above show that the cosine function provides all the even power terms (θ^n, n even) in the exponential while the sine function provides the odd power terms.

Using Euler's formula in Eq. (15.4), we can write any complex number in terms of a magnitude ρ and phase θ:

$$z = \rho e^{i\theta} \tag{15.7}$$

4. **Complex conjugation**: To complex conjugate any expression containing complex numbers, simply replace i by $-i$ wherever it appears.

$$z^* = a - ib = \rho e^{-i\theta} \tag{15.8}$$

5. **Norm (magnitude) of a complex number**: The magnitude of a complex number is defined as:

$$|z| = \sqrt{z^*z} = \sqrt{a^2 + b^2} = \rho \tag{15.9}$$

15.2 Probabilities and Expectation Values

15.2.1 Discrete Distributions

- Consider a set of student test marks M_i, $i = 1, 2, 3, ...N$ obtained on a term test written by N students. One calculates the average mark on a test as follows:

$$\langle M \rangle = \frac{\sum_{i=1}^{N} M_i}{N} \tag{15.10}$$

where M_i is the mark of the ith student and N is the total number of students. Note that here you are summing over all students.

- An equivalent way to calculate the average is to ask how many times N_a a particular mark M_a occurred in the set of all the marks. This provides an alternative formula for the average:

$$\langle M \rangle = \frac{\sum_a N_a M_a}{N}$$

$$= \sum_a M_a \frac{N_a}{N}$$

$$= \sum_a M_a P_a \tag{15.11}$$

In the above $a = 1, 2, ...q$, where q is the number of different marks that are in principle possible to obtain. If no student obtained a particular mark, M_3 say, then $N_3 = 0$. If the class is large enough then this is unlikely. Note that in the above you are summing over all possible marks, not over all the students. We have defined:

$$P_a := \frac{N_a}{N} \tag{15.12}$$

P_a gives the fraction of times a particular mark appeared in the distribution. It only gives the probability in the limit that N gets very large.

- Equation (15.11) gives us a general formula for calculating the *average value* or *expectation value* or *mean value* of any quantity x, with possible values $\{x_a, a = 1..N\}$ and probability distribution P_a:

$$\langle x \rangle = \sum_{a=1}^{N} x_a P_a \tag{15.13}$$

We can also calculate the average value for the same distribution of any function $f(x)$ of x:

$$\langle f(x) \rangle = \sum_{a=1}^{N} f(x_a) P_a \tag{15.14}$$

- The *standard deviation* of a variable x for a given probability distribution tells us the *width* of the probability distribution. We often use the terms width and standard deviation interchangeably. The standard deviation gives an idea of the degree to which the values in the distribution are spread out amongst all possible values. It is defined as:

$$\Delta x_{sd} := \sqrt{\langle (x - \langle x \rangle)^2 \rangle} \tag{15.15}$$

$$= \sqrt{|\langle x^2 \rangle - |\langle x \rangle|^2|} . \tag{15.16}$$

Δx_{sd} and $(\Delta x_{sd})^2$ are also known as the standard deviation and variance respectively of the probability distribution.

Exercise 15.1 Derive Eq. (15.16) from Eq. (15.15).

15.2.2 Continuous Probability Distributions

When a quantity takes its values on the real line, i.e. it is a *continuous* variable, the sums in the above expressions must be replaced by integrals. Nothing else changes. In the context of quantum mechanics, consider a large number of particles, each prepared in a state described by the same wave function $\psi(x)$ at some time t prior to measurement. The probability of measuring the position of the particle to be in the infinitesimal range $x \to x + dx$ is given by:

$$P(x \to x + dx) = \psi^*(x)\psi(x)dx =: \mathcal{P}(x)dx \tag{15.17}$$

where

$$\mathcal{P}(x) := \psi^*(x)\psi(x) \tag{15.18}$$

is called the probability density associated with the wave function $\psi(x)$.

Multiple measurements of position, position squared, or in fact any function $f(x)$ of the position of different representatives of these identically prepared particles will yield the following average values, also called *expectation values*:

$$\langle x \rangle = \int_{-\infty}^{\infty} dx \, \psi^*(x) x \psi(x)$$

$$\langle x^2 \rangle = \int_{-\infty}^{\infty} dx \, x^2 \psi^*(x) \psi(x)$$

$$\langle f(x) \rangle = \int_{-\infty}^{\infty} dx \, f(x) \psi^*(x) \psi(x)$$

$$(15.19)$$

The standard deviation in the observed values of any function $f(x)$ is

$$f_{sd} = \sqrt{\langle (f(x) - \langle f(x) \rangle)^2 \rangle}$$
$$= \sqrt{|\langle f^2(x) \rangle - \langle f(x) \rangle^2|}$$

$$(15.20)$$

Example 15.1 Gaussian wave function

Suppose a large number of particles are prepared in a state with wave function:

$$\psi(x) = A e^{-(x-x_0)^2/(2b^2)}$$

$$(15.21)$$

where A is a complex number.

1. Plot the probability distribution $\mathcal{P}(x)$ for $b = 1/2$ and $x_0 = 2$.
   Solution:
   See Fig. 15.2.
2. What is the value of the constant A?
   Solution:
   Normalize the wave function:

$$1 = \int_{-\infty}^{\infty} dx \, \psi^*(x) \psi(x)$$
$$= A^* A \int_{-\infty}^{\infty} dx \, e^{-(x-x_0)^2/(2b^2)} e^{-(x-x_0)^2/(2b^2)}$$
$$= |A|^2 \int_{-\infty}^{\infty} dx \, e^{-(x-x_0)^2/(b^2)}$$
$$= |A|^2 \int_{-\infty}^{\infty} dy \, b \, e^{-y^2} \quad \text{where } y := (x - x_0)/b$$
$$= |A|^2 b \sqrt{\pi}$$

$$(15.22)$$

where we have used the basic integral:

$$\int_{-\infty}^{\infty} dy \, e^{-\frac{y^2}{2}} = \sqrt{2\pi}$$

$$(15.23)$$

Fig. 15.2 Gaussian
probability distribution

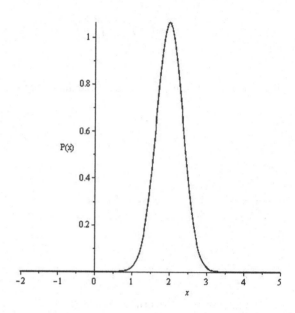

The normalization condition therefore requires:

$$|A|^2 = \frac{1}{b\sqrt{\pi}} , \tag{15.24}$$

and the normalized wave function is:

$$\psi(x) = \frac{1}{\sqrt{b\sqrt{\pi}}} e^{-(x-x_0)^2/(2b^2)} \tag{15.25}$$

The probability distribution is then:

$$\begin{aligned} \mathcal{P}(x) &:= \psi^*(x)\psi(x) \\ &= \frac{1}{b\sqrt{\pi}} e^{-(x-x_0)^2/(b^2)} \end{aligned} \tag{15.26}$$

Note that the normalization condition does not determine the phase of the complex number A, only its magnitude. Here we assume that it has vanishing phase so it is real.

3. What is the most probable value of the outcome of the measurement of the particle's position?

Solution:

The most probable value occurs at the value of x where:

$$\frac{d\mathcal{P}(x)}{dx} = 0$$

$$= \frac{d(\psi^*(x)\psi(x))}{dx}$$

$$= \frac{d\left(|A|^2 e^{-(x-x_0)^2/b^2}\right)}{dx}$$

$$= -\frac{2|A|^2}{b^2}(x - x_0)e^{-(x-x_0)^2/b^2} . \tag{15.27}$$

Therefore the most probable value is $x = x_0$. The most probable value in this case is equal to the average x_0 (see below) because the probability distribution is symmetric about x_0.

4. What will the average value be of the particle's measured position?
   Solution:

$$\langle x \rangle = \int_{-\infty}^{\infty} dx\, x\, \psi^*(x)\psi(x)$$

$$= \frac{1}{b\sqrt{\pi}} \int_{-\infty}^{\infty} dx\, x\, e^{-(x-x_0)^2/(b^2)}$$

$$= \frac{1}{b\sqrt{\pi}} \int_{-\infty}^{\infty} dx\, (x - x_0)\, e^{-(x-x_0)^2/(b^2)}$$

$$+ x_0 \frac{1}{b\sqrt{\pi}} \int_{-\infty}^{\infty} dx\, e^{-(x-x_0)^2/(b^2)}$$

$$= 0 + x_0 \tag{15.28}$$

The first term in the third line of Eq. (15.28) is zero by symmetry of the integral, the second term gives x_0 because the wave function is normalized.

5. What is $\langle x^2 \rangle$ for this state?
   Solution:

$$\langle x^2 \rangle = \int_{-\infty}^{\infty} dx\, x^2 \psi^*(x)\psi(x)$$

$$= \frac{1}{b\sqrt{\pi}} \int_{-\infty}^{\infty} ds\, (x - x_0)^2 e^{-(x-x_0)^2/(b^2)}$$

$$+ \frac{1}{b\sqrt{\pi}} \int_{-\infty}^{\infty} dx\, (2xx_0 - x_0^2)e^{-(x-x_0)^2/(b^2)}$$

$$= \frac{1}{b\sqrt{\pi}} \int_{-\infty}^{\infty} dy\, y^2\, e^{-y^2/(b^2)} + x_0^2$$

$$= \frac{b^2}{2} + x_0^2 \tag{15.29}$$

6. Calculate the standard deviation.

   Solution:

   The standard deviation gives an idea of the "width" of a probability distribution, or equivalently the uncertainty in the resulting measurements of x. For the Gaussian probability distribution above:

$$\Delta x_{sd} = \sqrt{|\langle x^2\rangle - |\langle x\rangle|^2|} \tag{15.30}$$

$$= \sqrt{\frac{b^2}{2} + x_0^2 - x_0^2} = \frac{b}{\sqrt{2}} \tag{15.31}$$

15.2.3 Dirac Delta Function

Consider a particle in a state with wave function Eq. (15.21) and hence probability distribution Eq. (15.26). If we let the parameter b get small, the width of the probability distribution given in Eq. (15.31) goes to zero. This means that in this limit, the probability distribution is zero everywhere except at $x = x_0$. As b gets small, the normalization constant $|A|^2$ in Eq. (15.24) gets very large, as does the peak value of the distribution. The probability distribution is normalized for all values of b, so it stays normalized in the limit that $b \to 0$. Putting all this together, we realize that in the limit that $b \to 0$, the probability distribution Eq. (15.26) predicts that a measurement of the position of the particle will yield x_0 with probability one.

The limit of the Gaussian probability distribution is called the *Dirac delta function* after P. A. M. Dirac:

$$\delta(x - x_0) := lim_{b \to 0}\frac{1}{b\sqrt{\pi}}e^{-(x-x_0)^2/(b^2)} \tag{15.32}$$

The Dirac delta function is zero everywhere, except at $x = x_0$, where it is infinite. The other important feature of the Dirac delta function is that the integral

$$\int dx\, \delta(x - x_0) = 1 \tag{15.33}$$

as long as the integration region covers x_0. It is zero otherwise.

Figure 15.3 illustrates the limiting procedure defined above.

The Dirac delta function is not quite a function because its value at $x = x_0$ diverges, but it turns out to be a very useful construction both mathematically and physically. We are into an area of mathematics whose development was in large part motivated by physics. $\delta(x - x_0)$ is called a "distribution". Distributions are not defined as functions, but instead they are defined in terms of their action via integration on true functions.

Fig. 15.3 Dirac delta
function as the infinitely thin
limit of a Gaussian

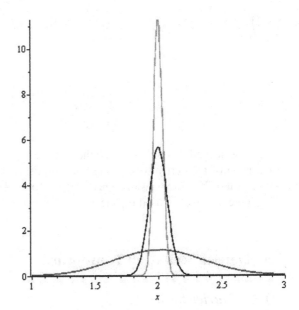

The limiting procedure in Eq. (15.32) provides one of many explicit realizations
of the Dirac delta function as a limit of an ordinary function. While useful, such
realizations are not required to rigorously define the Dirac delta function.

Here is a summary of some of the properties of the Dirac delta function:

- It is non-vanishing only at one point:

$$\delta(x - x_0) = 0, \qquad x \neq x_0 \tag{15.34}$$

- Its integral across that point is one.

$$\int_{x_0-b}^{x_0+a} dx\, \delta(x - x_0) = 1, \qquad \text{for all } a, b > 0 \tag{15.35}$$

It can also be shown that the above conditions can more or less be replaced by the
single condition:

$$\int_{x_0-b}^{x_0+a} dx\, \delta(x - x_0) f(x) = f(x_0), \qquad \text{for any function } f(x) \tag{15.36}$$

The easiest way to prove this is to do a Taylor series expansion of $f(x)$ around
x_0:

$$\int_{x_0-b}^{x_0+a} dx\, \delta(x - x_0) f(x) = \int_{x_0-b}^{x_0+a} dx\, \delta(x - x_0)\Big[f(x_0) + f'(x_0)(x - x_0)$$

$$+ \frac{1}{2} f''(x_0)(x - x_0)^2 + ...\Big]$$

$$= f(x_0) \int_{x_0-b}^{x_0+a} dx\, \delta(x - x_0)$$

$$= f(x_0) \qquad\qquad (15.37)$$

To get the second line we have used the fact that $\delta(x - x_0)$ vanishes when $x \neq x_0$ (and none of the derivatives of $f(x)$ are infinite at x_0), so that all higher order terms in the Taylor expansion vanish, leaving only the first term, namely $f(x_0)$. To get the last line we used Eq. (15.35).

15.3 Fourier Series and Transforms

15.3.1 Fourier Series

Any function of position $y(x)$ that is periodic in x with period L can be written as an infinite sum of pure waves, known as the *Fourier Series* for the function $y(x)$:

$$y(x) = \frac{1}{2} a_0 + \sum_{n=1}^{\infty} \left(a_n \cos\left(\frac{2\pi n x}{L}\right) + b_n \sin\left(\frac{2\pi n x}{L}\right) \right)$$

$$= \frac{1}{2} a_0 + \sum_{n=0}^{\infty} (a_n \cos(k_n x) + b_n \sin(k_n x)) \qquad (15.38)$$

where we have defined:

$$k_n := nk = \frac{2\pi n}{L} \qquad\qquad (15.39)$$

The coefficients can be obtained as follows:

$$a_0 = \frac{2}{L} \int_{x_0}^{x_0+L} d\tilde{x}\, y(\tilde{x}) \qquad\qquad (15.40)$$

$$a_n = \frac{2}{L} \int_{x_0}^{x_0+L} d\tilde{x}\, y(\tilde{x}) \cos\left(\frac{2\pi n}{L}\tilde{x}\right), \qquad n = 1, 2, 3, ... \qquad (15.41)$$

$$b_n = \frac{2}{L} \int_{x_0}^{x_0+L} d\tilde{x}\, y(\tilde{x}) \sin\left(\frac{2\pi n}{L}\tilde{x}\right), \qquad n = 1, 2, 3, ... \qquad (15.42)$$

Note:

- To calculate the coefficients you can integrate over any complete period.
- If the function is symmetric about $x = 0$, i.e if $f(x) = f(-x)$ then the coefficients of the sine terms in the series will vanish.
- If the function is anti-symmetric about $x = 0$, i.e. if $f(x) = -f(-x)$ then the coefficients of the cosine terms in the series will vanish.
- If you wish to approximate a non-periodic function in a finite interval, say $[-L/2, L/2]$, then just pretend it is periodic, with the finite interval corresponding to one full period, and carry on from there. In other words, you can *extend* the function from the interval so as to make it periodic.
- Often there exists a clever extension of a function that makes it either even or odd, thereby reducing the calculation of the Fourier coefficients to either the cosine or sine coefficients, respectively.
- Fourier series basically work via constructive and destructive interference. By adding periodic functions with different wavelengths you can arrange for the peaks and valleys to cancel and add at the right places so that the desired function is produced.
- If the right hand side of Eq. (15.38) is truncated to a finite number of terms (as required for example when evaluating the series using a computer), then one gets an approximation of the periodic function $y(x)$. The accuracy of the approximation gets better as more terms are included.

Example 15.2 Equation (15.43) shows the first six cosine terms in the Fourier series approximation to the Gaussian function: $f(x) = e^{-8x^2}$, $\quad -2 \le x \le 2$

$$
\begin{aligned}
f(x) = {} & .157 + .2900768785 \cos((1/2)\pi x) \\
& + .230 \cos(\pi x) + .157 \cos((3/2)\pi x) \\
& + 0.091 \cos(2\pi x) + 0.0456 \cos((5/2)\pi x) \quad\quad (15.43)
\end{aligned}
$$

Note that because the function is symmetric about $x \to -x$, the Fourier decomposition only contains cosine functions. The constant term is called the zeroth order approximation. Figure 15.4 illustrates the zeroth order (constant) term, the first order approximation (sum of first two terms in the expansion) and third approximation (sum of first four terms) to the Gaussian. One can see from the Figure that the interference between the successive terms of different wavelengths causes destructive interference where the Gaussian is close to vanishing and constructive interference where the Gaussian is at a maximum.

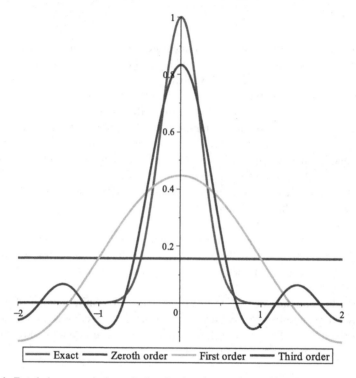

Fig. 15.4 Zeroth (constant, in brown), first (cosine, in green) and third (in blue) order Fourier approximations of the Gaussian (in red) $f(x) = e^{-8x^2}$. Terms in series and the figure were generated by Maple 16. (Color figure online)

15.3.1.1 Complex Form of Fourier Series

Exercise 15.2 Show using Euler's formula that the Fourier series above can be written in complex form:

$$f(x) = \sum_{n=-\infty}^{\infty} c_n e^{ik_n x} \qquad (15.44)$$

$$\text{where} \quad k_n = \frac{2\pi n}{L} \qquad (15.45)$$

$$c_n = \frac{a_n - i b_n}{2} \qquad (15.46)$$

$$c_{-n} = c_n^* = \frac{a_n + i b_n}{2} \qquad (15.47)$$

15.3.2 Fourier Transforms

It is possible to extend the Fourier series to approximate functions that are not periodic by taking $L \rightarrow \infty$ (*carefully*).[1] The only requirement on the function $f(x)$ is that the following integral:

$$\int_{-\infty}^{\infty} |f(x)|dx \tag{15.48}$$

be finite. If this is the case, then one has the following expression for $f(x)$:

$$f(x) = \frac{1}{\sqrt{2\pi}} \int_{-\infty}^{\infty} dk\, F(k)e^{ikx} \tag{15.49}$$

$$\text{where} \quad F(k) = \frac{1}{\sqrt{2\pi}} \int_{-\infty}^{\infty} dx\, f(x)e^{-ikx} \tag{15.50}$$

- The basic difference between Fourier series and Fourier transforms is that instead of a discrete (but infinite) set of fundamental modes with wave numbers $k_n = 2\pi n/\lambda$, there are now an uncountable infinity of modes, so you have to integrate over them instead of doing a sum. The bottom line as far as quantum mechanics goes is that you can write any normalizable wave function as an infinite sum of pure waves.
- $f(x)$ and its Fourier transform $F(k)$ can be thought of as two different, but *complementary* descriptions of the same function.

Example 15.3 The Fourier transform of

$$f(x) = \begin{cases} Ae^{-\lambda x} & \text{when } x \geq 0 \\ 0 & \text{when } x < 0 \end{cases} \tag{15.51}$$

is

$$\begin{aligned}
F(k) &= \frac{1}{\sqrt{2\pi}} \int_{-\infty}^{0} dx \cdot 0 \cdot e^{-ikx} + \frac{A}{\sqrt{2\pi}} \int_{0}^{\infty} dx\, e^{-\lambda x} e^{-ikx} \\
&= 0 + \frac{A}{\sqrt{2\pi}} \left[-\frac{e^{-(\lambda+ik)x}}{\lambda+ik} \right]_{0}^{\infty} \\
&= \frac{A}{\sqrt{2\pi}(\lambda+ik)} \tag{15.52}
\end{aligned}$$

Example 15.4 The Fourier transform of the normalized Gaussian wave function in Eq. (15.25) centred on the origin $x_0 = 0$:

---

[1] The derivation can be found, for example in Chap. 12 of the book *Mathematical Methods for Physics and Engineering: A Comprehensive Guide Paperback* by K. F. Riley, M. P. Hobson and S. J. Bence, Cambridge University Press, 3rd edition 2006.

$$\psi(x) = \frac{1}{\sqrt{b\sqrt{2\pi}}} e^{\frac{-x^2}{2b^2}} \tag{15.53}$$

is:

$$
\begin{aligned}
\phi(k) &= \frac{1}{\sqrt{2\pi}} \frac{1}{\sqrt{b\sqrt{\pi}}} \int_{-\infty}^{\infty} dx\, e^{-\frac{x^2}{2b^2}} e^{-ikx} \\
&= \frac{1}{\sqrt{2\pi}} \frac{1}{\sqrt{b\sqrt{\pi}}} \int_{-\infty}^{\infty} dx\, e^{\left(\frac{-(x+ikb)^2 - k^2 b^4}{2b^2}\right)} \\
&= \frac{1}{\sqrt{2\pi}} \frac{1}{\sqrt{b\sqrt{\pi}}} e^{-kb^2/2} \int_{-\infty}^{\infty} d(x+ib) e^{\left(\frac{-(x+ib)^2}{2b^2}\right)} \\
&= \frac{1}{\sqrt{2\pi}} \frac{1}{\sqrt{b\sqrt{\pi}}} e^{-kb^2/2} b\sqrt{2\pi} \\
&= \sqrt{\frac{b}{\sqrt{\pi}}} e^{-k^2 b^2/2} \tag{15.54}
\end{aligned}
$$

Note that $\phi(k)$ is exactly the same as $\psi(x)$ providing one substitutes k for x and $1/b$ for b everywhere. It is therefore also normalized. This is not a coincidence as shown by Parseval's theorem, discussed in Sect. 15.3.5.

15.3.3   The Mathematical Uncertainty Principle

The Fourier transform, like the Fourier series, works via constructive and destructive interference of pure waves to enhance or suppress the resultant function where required. This suggests that if $f(x)$ is very narrow, a lot of destructive interference is required to make the function zero on most of the real axis. $F(k)$ will be a wide function of k. On the other hand if $f(x)$ is wide, you will need fewer Fourier modes to achieve the required destructive or constructive interference, and $F(k)$ will be a narrow function of k. Mathematics supports this intuitive argument. There exists a theorem that says the product of the widths of $f(x)$ and $F(k)$ is optimized when $f(x)$ and consequently its Fourier transform $F(k)$ are both Gaussians. The width of a distribution is defined by the standard deviation. For the normalized Gaussian wave function we have already found in Eq. (15.31) that:

$$\Delta x_{sd} = \frac{b}{\sqrt{2}} \tag{15.55}$$

We now calculate the corresponding width associated with its Fourier transform:

$$\Delta k_{sd} := \sqrt{\langle k^2 \rangle - \langle k \rangle^2}$$

$$= \frac{1}{b\sqrt{2}} \qquad (15.56)$$

This follows because the integrals are identical, again with the substitution $x \to k$ and $b \to 1/b$. For a Gaussian, then:

$$\Delta x_{sd} \Delta k_{sd} = \frac{b}{\sqrt{2}} \frac{1}{b\sqrt{2}} = \frac{1}{2} \qquad (15.57)$$

Due to the existence of a theorem proving that the Gaussian optimizes (minimizes) the product, we have that for any other probability distribution:

$$\Delta x_{sd} \, \Delta k_{sd} \geq \frac{1}{2} \qquad (15.58)$$

We emphasize that this is pure mathematics so far. The physical input is the expression for de Broglie wavelength $p = \hbar k$. Multiplying both sides of Eq. (15.58) by $\hbar$ and using the expression for p in terms of $\hbar$ yields the Heisenberg Uncertainty Principle.

15.3.4 Dirac Delta Function Revisited

There exists another very useful representation of the Dirac delta function, namely:

$$\delta(x - x_0) = \frac{1}{2\pi} \int_{-\infty}^{\infty} dk \, e^{-ik(x-x_0)} \qquad (15.59)$$

Proof. Using the above representation

$$\int dx \, \delta(x - x_o) f(x) = \frac{1}{2\pi} \int dx \int dk \, f(x) e^{-ikx} e^{kx_0}$$

$$= \frac{1}{\sqrt{2\pi}} \int dk \, e^{-ikx_0} \frac{1}{\sqrt{2\pi}} \int dx \, f(x) e^{-ikx}$$

$$= \frac{1}{\sqrt{2\pi}} \int dk \, e^{-ikx_0} F(k)$$

$$= f(x_0) \qquad (15.60)$$

since in the above k and x are arbitrary variables, it is also true that:

$$\delta(k - \tilde{k}) = \frac{1}{2\pi} \int_{-\infty}^{\infty} dx \, e^{i(k-\tilde{k})x} \qquad (15.61)$$

Thus, we see that the Fourier transform of the Dirac delta function is the constant function and the Fourier transform of the constant function is the Dirac delta function.

15.3.5   Parseval's Theorem

Parseval's theorem states that if a probability distribution $\mathcal{P}(x) = \psi^*(x)\psi(x)$ of a wave function $\psi(x)$ is normalized, then the Fourier transform $\phi(k)$ of $\psi(x)$ is also a normalized function of k. The proof goes as follows:

$$
\begin{aligned}
\int dx\, \psi^*(x)\psi(x) &= \int dx \left[\frac{1}{\sqrt{2\pi}} \int dk \phi^*(k) e^{ikx} \right] \left[\frac{1}{\sqrt{2\pi}} \int d\tilde{k} \phi^*(\tilde{k}) e^{-i\tilde{k}x} \right] \\
&= \int dk \int d\tilde{k} \phi^*(k)\phi(k) \frac{1}{2\pi} \int dx\, e^{i(k-\tilde{k})x} \\
&= \int dk \phi^*(k) \int d\tilde{k} \phi(\tilde{k}) \delta(k - \tilde{k}) \\
&= \int dk\, \phi^*(k)\phi(k)
\end{aligned}
\tag{15.62}
$$

Changing the integration variable on the right hand side of the last line to $p = \hbar k$:

$$
\int dx\, |\psi(x)|^2 = \int dp\, \Phi(p)
\tag{15.63}
$$

where $\Phi(p) := \frac{1}{\hbar}|\phi(p/\hbar)|^2$. This theorem plays an important role in quantum mechanics. It guarantees that if a given wave-function is normalized (its complex magnitude integrates to unity) then $\Phi(p)$ is also normalized. Thus, Parseval's theorem provides the Fourier transform of the normalized wave function with the physical interpretation as the normalized probability amplitude for momentum. Specifically, the probability amplitude in terms of momentum is:

$$
\tilde{\phi}(p) := \frac{1}{\sqrt{\hbar}} \phi\left(\frac{p}{\hbar}\right)
\tag{15.64}
$$

This tells us that the wave function and its Fourier transform provide complementary descriptions of the same quantum state.

15.4 Waves

15.4.1 Moving Pure Waves

Consider the time dependent pure wave:

$$y(x, t) = A \cos(kx - \omega t) \tag{15.65}$$

We will assume that k and ω are both positive.

- k is called the **wave number**. The wavelength at fixed t is related to the wave number by:

$$\lambda = \frac{2\pi}{k} \tag{15.66}$$

- ω is called the **angular frequency**. It gives the frequency of oscillation of the wave at a fixed point x in radians per second. The frequency of vibrations at fixed x in cycles per second is

$$f = \frac{\omega}{2\pi} \tag{15.67}$$

- **Phase Velocity**: The pure wave $y(x, t)$ in Eq. (15.65) has its maximum value whenever $kx - \omega t = n\pi$, $n = 0, \pm 1, \pm 2, \dots$. As t increases, if ω is positive, this peak value moves in the positive x direction, with speed

$$v = \frac{\omega}{k} \tag{15.68}$$

v is called the phase velocity.
- Similarly, a wave moving in the negative x direction with the same speed takes the form:

$$y(x, t) = A \cos(kx + \omega t) \tag{15.69}$$

- In general one can describe a wave of any shape moving at fixed speed by a function $f(x \mp vt)$. This function does not have to be periodic. However, since it is a function only of the combination $x \mp vt$, as t increases to $\tilde{t} := t + \Delta t$, its shape will not change as it moves a distance $\tilde{x} := x + \Delta x$, where

$$\Delta x = \pm v \Delta t \tag{15.70}$$

Equation (15.70) ensures that the argument of the function remains the same at
$(\tilde{x}, \tilde{t})$ as it was at (x, t).

$$f\left(\tilde{x} \mp v\tilde{t}\right) = f\left((x \pm v\Delta t) \mp v(t + \Delta t)\right)$$
$$= f(x \mp vt) \qquad (15.71)$$

15.4.2 Complex Waves

Quantum mechanics requires us to consider waves that are complex valued. It is
therefore useful to consider complex exponentials such as:

$$\psi(x, t) = Ae^{i(kx-\omega t)} = A\left(\cos(kx - \omega t) + i \sin(kx - \omega t)\right). \qquad (15.72)$$

where A is an arbitrary complex number. Equation (15.72) describes a complex
valued wave with fixed wave number k, angular frequency ω moving at fixed speed
$v = \omega/k$ in the positive x direction.

15.4.3 Group Velocity and Phase Velocity

We have seen in Sect. 15.3.2 that it is possible to generate a normalizable function
of x by taking an infinite linear superposition (i.e. integral) of a large number of
(non-normalizable) pure waves of different wavelengths and amplitudes. This works
because the pure waves interfere constructively and destructively to produce the
desired shape. The same can of course be done for moving waves, which are functions
of $x \pm vt$, to produce moving "packets" of waves, or *wave packets*.

The simplest way to see this is to consider just the sum of two waves of equal
amplitude:

$$y(x, t) := y_1(x, t) + y_2(x, t)$$
$$y_1 = A\cos(k_1 x \mp \omega_1 t)$$
$$y_2 = A\cos(k_2 x \mp \omega_2 t) \qquad (15.73)$$

where k_1, k_2, ω_1 and ω_2 are taken to be positive. The direction of the resultant wave
is determined by the relative sign between the first and second term in the argument
of the cosine function. If the sign is negative, the wave moves to the right, if it is
positive, it moves to the left.

The resultant wave can be put in a useful form using the following trigonometric
identity:

$$\cos(a) + \cos(b) = 2\cos\left(\frac{a-b}{2}\right)\cos\left(\frac{a+b}{2}\right) \qquad (15.74)$$

Applying Eq. (15.74) to the sum of the two waves given in (15.73):

$$
\begin{aligned}
y_{tot} &= y_1 + y_2 \\
&= A\cos(k_1 x \mp \omega_1 t) + A\cos(k_2 x \mp \omega_2 t) \\
&= 2A\cos\left(\frac{(k_1 - k_2)}{2}x - \frac{(\pm\omega_1 \mp \omega_2)}{2}t\right)\cos\left(\frac{(k_1 + k_2)}{2}x - \frac{(\pm\omega_1 \pm \omega_2)}{2}t\right) \\
&= 2A\cos(\Delta k x \mp \Delta\omega t)\cos\left(\overline{k} \mp \overline{\omega}t\right)
\end{aligned}
\tag{15.75}
$$

where

$$
\Delta k := \frac{|k_1 - k_2|}{2} \tag{15.76}
$$

$$
\Delta\omega := \frac{|\omega_1 - \omega_2|}{2} \tag{15.77}
$$

are the magnitudes of the difference in wave number and frequencies, respectively of the two waves, while

$$
\overline{k} := \frac{k_1 + k_2}{2} \tag{15.78}
$$

$$
\overline{\omega} := \frac{\omega_2 + \omega_2}{2} \tag{15.79}
$$

are the magnitude of the average wave number and angular frequency, respectively.

Since we are assuming that both k's are positive, the wavelength λ_g of the first factor

$$
\lambda_g := \frac{2\pi}{\Delta k} \tag{15.80}
$$

is larger than that of the second factor

$$
\lambda_p := \frac{2\pi}{\overline{k}} \tag{15.81}
$$

The first prefactor, with its longer wavelength, can be thought of as changing or modulating the amplitude of the shorter wavelength wave as it moves. It describes the shape and motion of the *envelope*.

The speed of the longer wavelength component, i.e. the speed of the envelope, is called the *group velocity*

$$
v_g = \Delta\omega/\Delta k \tag{15.82}
$$

The velocity of the shorter wavelength component is called the *phase velocity*

$$
v_p = \frac{\overline{\omega}}{\overline{k}} = \frac{\omega_1 + \omega_2}{k_1 + k_2}\ . \tag{15.83}
$$

Example 15.5 Consider two waves with amplitude $A = 1$. Both move to the right (along the positive x-axis). The first wave has wave number and angular frequency:

$$k_1 = \frac{\pi}{2} \text{ m}^{-1} \qquad \omega_1 = \frac{\pi}{6} \text{ rads/s} \tag{15.84}$$

while the second has:

$$k_2 = \frac{\pi}{3} \text{ m}^{-1} \qquad \omega_2 = \frac{\pi}{2} \text{ rads/s} \tag{15.85}$$

1. What are the group and phase velocities of the sum of the two waves?
2. Sketch both waves on the same plot for $t = 0$, $t = 2$ s and $t = 4$ s.
3. Sketch the sum of the waves at $t = 0$, $t = 2$ s and $t = 4$ s.

Solution:

1. The group and phase velocities are, respectively

$$
\begin{aligned}
v_g &= \frac{\omega_1 - \omega_2}{k_1 - k_2} = \frac{\frac{\pi}{6} - \frac{\pi}{2}}{\frac{\pi}{2} - \frac{\pi}{3}} \text{ m/s} \\
&= -2 \text{ m/s}
\end{aligned}
\tag{15.86}
$$

$$
\begin{aligned}
v_p &= \frac{\omega_1 + \omega_2}{k_1 + k_2} = \frac{\frac{\pi}{6} + \frac{\pi}{2}}{\frac{\pi}{2} + \frac{\pi}{3}} \text{ m/s} \\
&= \frac{4}{5} \text{ m/s}
\end{aligned}
\tag{15.87}
$$

2. See Fig. 15.5
3. See Fig. 15.6

15.4.4   Wave Packets

Fourier transform theory tells us that a moving wave packet is made up of many modes, each with its own phase velocity. If the Fourier transform of the wave packet is not too spread out in k-space, then the wave packet consists of a central envelope moving at a specific group velocity. Consider now a wave function that describes a particle with some relatively small spread in wave number (i.e. momentum)

$$\psi(x, t) = \frac{1}{\sqrt{2\pi}} \int_{-\infty}^{\infty} dk \phi(k) e^{i(kx - \omega(k)t)} \tag{15.88}$$

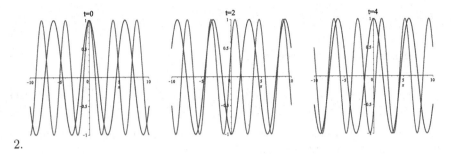

2.

Fig. 15.5 Two waves with same amplitudes, different wave numbers and frequencies moving in same direction at different speeds. *Credit* Figures generated by Maple

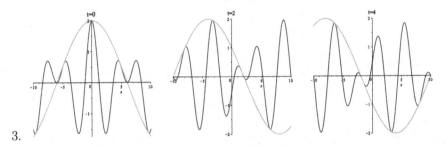

3.

Fig. 15.6 Linear superposition of two waves with different wave numbers and frequencies. The envelope, in green, moves at the group velocity, while the wave itself, in red, moves at the phase velocity. In this case the group velocity is greater than the phase velocity. This can be seen by tracking the green peak and red peak that start together at $x = 0$ in the top figure. *Credit* Figures generated by Maple. (Color figure online)

Note that the angular frequency depends on the wave number. Each component of this wave is a pure wave with fixed wave number, angular frequency and velocity. By allowing the wave number to be negative (and assuming that the angular frequency is positive), Eq. (15.88) sums over waves moving in the positive and negative x directions.

We assume that $\phi(k)$ is sharply peaked, so that one wave number dominates the wave packet. The most probable wave number is determined by:

$$\frac{d(\phi^*(k)\phi(k))}{dk}\bigg|_{k_c} = 0 \qquad (15.89)$$

In this case, the wave is well approximated by a linear combination of two waves with wave number on "either side" of the most probable wavelength k_c so that we can calculate the phase velocity and group velocity. In particular, the phase velocity of the most probable pure wave component is:

$$v_p = \frac{\omega(k_c)}{k_c} \tag{15.90}$$

The group velocity of the envelope is:

$$v_g = \frac{d\omega}{dk}\bigg|_{k_c} \tag{15.91}$$

Finally we deduce $\omega(k)$ for a free particle. If the wave is a quantum wave function that obeys de Broglie's postulate, then:

$$p = mv_g = \hbar k$$
$$\text{and} \quad E = \frac{1}{2}mv_g^2 = \hbar\omega(k) \tag{15.92}$$

This implies that

$$\omega(k) = \frac{k^2}{2m\hbar} \tag{15.93}$$

so that:

$$\begin{aligned} v_g &= \frac{d\omega}{dk}\bigg|_{k_c} \\ &= \frac{k_c}{m} \end{aligned} \tag{15.94}$$

and

$$\begin{aligned} v_p &= \frac{\omega}{k}\bigg|_{k_c} \\ &= \frac{1}{2}\frac{k_c}{m} \\ &= \frac{1}{2}v_g \end{aligned} \tag{15.95}$$

This corresponds to the relationship between the classical velocity of a particle and the velocity of its de Broglie quantum wave described in Sect. 8.4.2. We now see that in the case of a sharply peaked wave packet, the classical particle velocity corresponds to the group velocity while the velocity of the de Broglie quantum wave is the phase velocity.

We now turn to the definition of quantum momentum in a more general context and calculate its expectation value.

15.4.5 Wave Number and Momentum

The Fourier transform allows us to write any quantum wave function $\psi(x)$ as a superposition of pure waves with wave number $k = 2\pi/\lambda$. Moreover, the Fourier transform $\phi(k)$ is a complementary mathematical description of $\psi(x)$: it provides us with precisely the same mathematical information in different form.

de Broglie told us that a particle with momentum p must be associated with a pure wave with wavelength $\lambda = h/p$. In terms of wave number we have the relationship:

$$p = \frac{h}{\lambda} = \frac{hk}{2\pi} = \hbar k \qquad (15.96)$$

Thus a quantum state with fixed momentum must be represented by a pure wave.

The physical significance of the Fourier transform is that it allows us to write any wave function as a linear superposition of states with fixed momentum:

$$\begin{aligned}
\psi(x) &= \frac{1}{\sqrt{2\pi}} \int_{-\infty}^{\infty} dk\, \phi(k) e^{ikx} \\
&= \frac{1}{\sqrt{2\pi}} \int_{-\infty}^{0} dk\, \phi(k) e^{ikx} + \frac{1}{\sqrt{2\pi}} \int_{0}^{\infty} dk\, \phi(k) e^{ikx} \\
&= \frac{1}{\sqrt{2\pi}} \int_{0}^{\infty} dk \left(\phi(-k) e^{-ikx} + \phi(k) e^{+ikx} \right) \\
&= \frac{1}{\sqrt{2\pi\hbar}} \int_{0}^{\infty} dp \left(\phi(-p) e^{-ipx/\hbar} + \phi(p) e^{ipx/\hbar} \right) \qquad (15.97)
\end{aligned}$$

where we have done a change of variables $k \to p = \hbar k$ in the integral to get the last line.

Parseval's theorem (Sect. 15.3.5) then provides $\phi(p)$ with a physical interpretation as the momentum probability amplitude whose square gives the probability density:

$$\Phi(p)\, dp := \frac{1}{\hbar} \phi^*(p/\hbar) \phi(p/\hbar) dp \qquad (15.98)$$

which gives the probability of measuring the momentum of a particle in state $\psi(x)$ with Fourier transform $\phi(p)$ to be in the range $p \to p + dp$. Parseval's theorem guarantees that this probability distribution is normalized: the probability of measuring some momentum, any momentum, is also one.

The many measurements of the momentum of an ensemble of particles prepared in this state yield an average momentum:

$$\begin{aligned}
\langle p \rangle &= \langle \hbar k \rangle \\
&= \int_{-\infty}^{\infty} dk\, \hbar k\, \phi^*(k)\, \phi(k) \\
&= \hbar \langle k \rangle \qquad (15.99)
\end{aligned}$$

The standard deviation of these measurements will be given by:

$$\Delta p_{sd} = \hbar \Delta k_{sd} \tag{15.100}$$

This leads via Eq. (15.58) to the Heisenberg uncertainty principle:

$$\Delta x_{sd} \Delta p_{sd} \geq \frac{\hbar}{2} \tag{15.101}$$

We now see that Heisenberg's uncertainty principle requires a mathematical result (the Fourier transform) and a physical result (the de Broglie wavelength). In brief, the theory of Fourier transforms implies the existence of two complementary descriptions of the state of a particle: the wave function $\psi(x)$ which assigns probabilities to position, and its Fourier transform, $\phi(k)$, which assigns probabilities to momentum. The mathematical uncertainty principle Eq. (15.58) in conjunction with de Broglie's postulate leads directly to the Heisenberg uncertainty principle.

The expectation value of momentum can be calculated either using $\phi(k)$ or directly from the wave function $\psi(x)$ in the following way:

$$
\begin{aligned}
&\int dx \psi^*(x)(-i\hbar)\frac{\partial \psi(x)}{\partial x} \\
&= \int dx \int \frac{dk}{\sqrt{2\pi}} \exp(-ikx)\phi^*(k)(-i\hbar)\frac{\partial}{\partial x} \int d\tilde{k}\phi(\tilde{k})\exp(i\tilde{k}x) \\
&= \int dx \int \frac{dk}{\sqrt{2\pi}} \exp(-ikx)\phi^*(k)(-i\hbar)(ik) \int \frac{d\tilde{k}}{\sqrt{2\pi}}\phi(\tilde{k})\exp(i\tilde{k}x) \\
&= \int dk \phi^*(k)(\hbar k) \int d\tilde{k}\phi(\tilde{k})\frac{1}{2\pi} \int dx \exp(i(\tilde{k}-k)x) \\
&= \int dk \phi^*(k)(\hbar k) \int d\tilde{k}\phi(\tilde{k})\delta(k-\tilde{k}) \\
&= \int dk \phi^*(k)(\hbar k)\phi(k) \\
&= \langle p \rangle
\end{aligned}
\tag{15.102}
$$

Thus, there is an association between momentum and a *differential operator*, which we call $\hat{p}$, that acts on the wave function by differentiation,

$$\hat{p} := -i\hbar \frac{\partial}{\partial x} \tag{15.103}$$

Operators which are polynomial functions of $\hat{p}$ are built by acting with $\hat{p}$ the required number of times. For example:

$$(\hat{p})^2 \psi(x) = (-i\hbar)^2 \frac{\partial^2 \psi(x)}{\partial x^2} \tag{15.104}$$

etc.

The above discussion leads to a more general feature of quantum mechanics: the role of physical observables that involve momentum is played by suitable differential operators. This leads us to the definition of energy in quantum mechanics and also to the equation that determines the time evolution of wave functions in quantum mechanics, i.e. the Schrödinger equation, which is the quantum analogue of Newton's equations.

Example 15.6 For a particle in a state described by a Gaussian Eq. (15.21) calculate Δp_{sd} by calculating the expectation value of the operator $\hat{p}$ and $(\hat{p})^2$ using Eq. (15.103).

Solution:

$$
\begin{aligned}
\langle \hat{p} \rangle &= \frac{1}{b\sqrt{\pi}} \int dx\, \psi^*(x)(-i\hbar)\frac{\partial \psi(x)}{\partial x} \\
&= \frac{1}{b\sqrt{\pi}} \int dx\, \exp(-(x-x_0)^2/(2b^2))(-i\hbar)\frac{-2(x-x_0)}{2b^2}\exp(-(x-x_0)^2/(2b^2)) \\
&= \frac{1}{b\sqrt{\pi}}\frac{i\hbar}{b^2} \int dx\,(x-x_0)\exp(-(x-x_0)^2/(b^2)) \\
&= 0
\end{aligned}
\tag{15.105}
$$

by symmetry, as expected because there is no time dependence.

$$
\begin{aligned}
\langle \hat{p}^2 \rangle &= \frac{1}{b\sqrt{\pi}} \int dx\, \psi^*(x)(-i\hbar)^2\frac{\partial^2 \psi(x)}{\partial x^2} \\
&= \frac{1}{b\sqrt{\pi}} \int dx\, \exp(-(x-x_0)^2/(2b^2))(-i\hbar)^2\frac{\partial}{\partial x}\left(\frac{-(x-x_0)}{b^2}e^{-(x-x_0)^2/(2b^2)}\right) \\
&= \frac{\hbar^2}{b\sqrt{\pi}} \int dx\,\left[\frac{1}{b^2}+\frac{(x-x_0)^2}{b^4}\right]\exp(-(x-x_0)^2/(b^2)) \\
&= \frac{\hbar^2}{b^4}\left(b^2 - \langle (x-x_0)^2 \rangle\right) \\
&= \frac{\hbar^2}{b^4}\left(b^2 - \frac{b^2}{2}\right) \\
&= \frac{\hbar^2}{2b^2}
\end{aligned}
\tag{15.106}
$$

The standard deviation is:

$$
\Delta p_{sd} = \frac{\hbar}{\sqrt{2}b}
\tag{15.107}
$$

in agreement with Eq. (15.56) given that $\Delta p_{sd} = \hbar \Delta k_{sd}$.

Index

© The Editor(s) (if applicable) and The Author(s), under exclusive license to Springer
Nature Switzerland AG 2022
G. Kunstatter and S. Das, *A First Course on Symmetry, Special Relativity
and Quantum Mechanics*, Undergraduate Lecture Notes in Physics,
https://doi.org/10.1007/978-3-030-92346-4

Printed in the United States
by Baker & Taylor Publisher Services